MW01618108

Larry,

To a valued NRO alumnus and great friend. Thanks for your time & for being here.

Best,

Betty

# Beyond Expectations—Building an American National Reconnaissance Capability: Recollections of the Pioneers and Founders of National Reconnaissance

Robert A. McDonald, Ph.D.
Editor

Published in cooperation with the
Center for the Study of National Reconnaissance
Office of Policy
National Reconnaissance Office

American Society for Photogrammetry and Remote Sensing
Bethesda, Maryland

Disclaimer

The views stated in this book are those of the contributors, editorial staff, and editor. The information and opinions expressed in this book should not be attributed to the National Reconnaissance Office, Department of the Air Force, Central Intelligence Agency, Department of the Navy, other government departments or agencies, or any industry or commercial entity. The military grades, government positions, and corporate titles that are associated with the contributors, editorial staff, and editor of this book are used for identification only. The involvement of any individual with this book does not necessarily imply that individual's agreement with the views of others who are associated with this book or that individual's endorsement of the book in part or as a whole.

Library of Congress Cataloging in Publication Data

Beyond expectations : building an American national reconnaissance capability : recollections of the pioneers and founders of national reconnaissance / edited by Robert A. McDonald.

p. cm.

Includes bibliographical references and index.

ISBN 1-57083-065-7 (alk. paper)

1. Military surveillance--United States--History--20th century. 2. Electronic intelligence--United States--History--20th century. 3. Electronic surveillance--United States--History--20th century. 4. Aerial reconnaissance--United States--History--20th century. 5. Space surveillance--United States--History--20th century. 6. Image analysis--United States--History--20th century. I. McDonald, Robert A.

UG475 .B48 2002
358.4'5'0973--dc21

2002027763



Cover design and layout: Carolyn A. Staab

Cover image credits (from upper left to lower right): see chapters Levison, Roy, Geyer, Leghorn, Murphy, Mayo, Roy, Mayo, Buzard, Leghorn, Reese.

## Editorial Support

Principal Writers & Content Editors
Joseph J. Helman, Ph.D.
Thomas B. Nath

Editorial Assistant & Quality Assurance Editor
Jacqueline A. Gray

Contributing Writers & Publication Support
Cherie Herr Jones
Jack Munson

Design & Layout Editor
Carolyn Staab

Indexing & Technical Support
Allen L. Shumway, Jr.

Proofreading Support
Sarah Kindig
Sharon Moreno
William Naylor
Monica Riedt

Research Interns
Lambros Kapoulas
Matthew J. Sanderl
Kathryn Sieh

Security Consultant
Robert Wheeler

Writing Support
Catherine A. Williams
Paul Burgess
Carl Johnson
L. Parker Temple, Ph.D.

Photography
Sara B. Judy
Candia Campbell

Transcription
Sharon Jane McLaren

Interviewers
Paul Burgess
Sandy Daniels
Matt Doering
Kevin Engel
Greg Gluba
R. Cargill Hall
Joseph J. Helman, Ph.D
J. J. Hogan
Cherie Herr Jones
Matt Jones
Garnett Kiser
Julia Kortum
Barbara Male
Rory A. Maynard
Robert L. (Beau) Moulden
Jack Munson
Gale Napoliello
Chanta Quillen
Morris Rosen, Ph.D
Kathryn Sieh
Sheryl Shreckengost
Jules Sussman
Robin Thomas
Catherine A. Williams
James Wilson

# Foreword

## Honoring Our Past, Looking to the Future: The Pioneers of National Reconnaissance

Shortly after being named Director of the National Reconnaissance Office, I had the opportunity to walk through the Pioneer Hall at NRO Headquarters where I saw the medallions and displays that honor the Pioneers and document their contributions. At that time, I realized that I had worked with many of these impressive individuals, and, as a result, had first-hand knowledge of their important contributions. Then on 24 September 2002 I had the privilege to host my first Pioneer Recognition Day at NRO Headquarters.

This book is a compilation of the recollections of the inaugural class of pioneers and the founders of national reconnaissance who were honored on Pioneer Recognition Day in September 2000. These individuals represent the finest talent in government, military and industry. The scope of their significant and lasting contributions covers the full spectrum of national reconnaissance endeavors. Their contributions formed the very foundation of national reconnaissance, and provided the base upon which the NRO was built. Each contribution is significant in its own right, and together they comprise a body of work that is of lasting importance to our nation. We owe a great debt of gratitude to these people who have made the NRO what it is today.

As this book illustrates, the pioneers and founders represent a diverse group within the field of national reconnaissance. They come from different backgrounds and walks of life, and they have made a variety of conceptual, technical, and programmatic contributions to imagery intelligence (IMINT), signals intelligence (SIGINT), communications, and space launch. However, one common thread among these people is their vision and commitment to technical excellence, innovation, teamwork, and the highest standards of integrity. Their dedication and personal and professional values represent important contributions that continue to serve as a foundation for the NRO. In this spirit, this book serves as a roadmap for today's NRO workforce to learn from the challenges and successes of the past, and to apply those lessons in successfully meeting today's challenges.

For over forty-two years NRO systems have served U.S. national security interests very well. These systems were central to the successful ending of the Cold War, and they continue to play a vital function in meeting current national security requirements. Specifically, NRO systems provide critical worldwide situational awareness that help support the full range of national and military intelligence requirements.

Figure. Mr. Peter B. Teets, Director, National Reconnaissance Office, delivering remarks at Pioneer Recognition Day Ceremony, 24 September 2002. (Photo by Sara Judy, NRO Visual Design Center.)

This book provides an opportunity to celebrate milestones of revelation and discovery that define the discipline of national reconnaissance. Because of the necessary secrecy surrounding this work, we have not previously been able to acknowledge openly many of these successes and contributions. Therefore, the publication of this book is a special opportunity for the NRO to offer well-deserved public recognition to these titans of national reconnaissance. We are sincerely grateful for this opportunity.

*Peter B. Teets*
*Director, National Reconnaissance Office*

# Foreword

## Recognizing the Pioneers and Founders of National Reconnaissance

In 1994, I had the opportunity to participate in the planning and subsequent ceremonies that accompanied the declassification of the Corona photoreconnaissance satellite program, and specifically the recognition of the individuals who made significant contributions to that program. I was moved greatly by the pride the honorees took in the recognition of their previously classified accomplishments and achievements, and by the emotional reaction of the honorees, their families, and their guests. In many cases this recognition represented the first public acknowledgement of their accomplishments and the significance of their accomplishments for the country.

The impact of this experience stayed with me, and when I became the Director of the National Reconnaissance Office (NRO), I made it a goal to recognize as many individuals as possible who played a significant role in the success of NRO programs over the years. We began a series of ceremonies to recognize the senior leadership who managed the NRO since its creation. The first ceremony honored the former Directors of the National Reconnaissance Office (DNRO). During this ceremony, former DNRO Hans Mark noted that it was fine to recognize past senior managers of the NRO, but we also needed to recognize the "pioneers," namely, "the people who did the real work." From this suggestion was born the idea of the National Reconnaissance Pioneer Recognition Program. I established the program to provide a means to acknowledge and honor the pioneers who made significant and lasting contributions to national reconnaissance.

In 1999, shortly after we recognized the Directors of the NRO legacy organizational components (i.e., Programs A, B, C, and D), I asked the NRO Director of Policy, and his office's Center for the Study of National Reconnaissance (CSNR), to develop an approach for a pioneer recognition program.[1] We decided to link the initial pioneer recognition event to the NRO 40th Anniversary celebration during 2000. The CSNR, along with the NRO Offices of History, Corporate Communications, Protocol, Security, Facilities, and the program staffs proceeded in the planning, coordination, and implementation of the pioneer recognition activities with much dedication and innovation.

[1] Program A included the U.S. Air Force (USAF) satellite reconnaissance element in the National Reconnaissance Program (NRP); Program B included the Central Intelligence Agency (CIA) satellite reconnaissance element in the NRP; Program C included the U.S. Navy satellite reconnaissance element in the NRP; and Program D included the acquisition and support of NRP aerial reconnaissance assets that were assigned to the Director, CIA Reconnaissance Programs.

The Honorable Keith Hall dedicating the National Reconnaissance Office Pioneer Hall, 27 September 2000. (Photo by Sara Judy, NRO Visual Design Center.)

I appointed an independent pioneer selection board that was representative of the NRO's military, civilian, and industry heritage. The board's task was to recommend pioneer candidates. We extended eligibility to all who made significant contributions, regardless of organizational affiliation. I asked the board members to name forty candidates, and it was entirely appropriate that they proceeded to overachieve by recommending forty-six candidates. The board also recommended that a group of ten "Founders of National Reconnaissance" be recognized separately. These scientists, engineers, and innovators—through their technical expertise—shaped the emerging discipline of national reconnaissance and provided the confidence for creating the NRO in 1960-61.

On 27 September 2000, as part of the NRO's 40th Anniversary celebration, I dedicated Pioneer Hall at the NRO's headquarters facility. This celebration included the inaugural induction ceremony of the first forty-six pioneers, many of whom attended with their families, friends, and former colleagues. I was very pleased to participate in that public recognition of these individuals. The ceremony commemorated their significant contributions, which formed the foundation for the discipline of national reconnaissance and the scientific, engineering, and organizational achievements upon which the NRO was built. Many attendees—who included not only the honorees, but also NRO employees and representatives from industry and mission partners—remarked to me about what a wonderful day it was. Many also expressed gratification that such an effort had been made to recognize these special individuals from the NRO's past.

Sir Isaac Newton once said of his life's work, "If I have seen farther, it is by standing on the shoulders of giants." The NRO employees of the 21st Century stand on the shoulders of the pioneers and founders of national reconnaissance. I salute these reconnaissance giants, and I thank them for their innovative spirit and their abiding willingness to extend the bounds of technology by consistently trying the impossible.

I am pleased that the recollections of these great people have been collected and now can be shared publicly. These stories make clear the legacy of their achievements and serve as standards of excellence for future endeavors.

*Keith R. Hall*
*Director of the National Reconnaissance Office, 1997–2001*

# Preface

*I have entered upon the task with the sincere desire to avoid doing injustice to anyone...other than the unavoidable injustice of not making mention often where special mention is due. There must be many errors of omission in this work, because the subject is too large...*

Ulysses S. Grant[1]

In this book you will find the recollections of individuals whose collective, professional contributions represent the foundation for the discipline of national reconnaissance. On 18 August 2000 the Director of the National Reconnaissance Office (DNRO), Keith R. Hall, selected 46 pioneers who had made significant and lasting contributions to the discipline over the previous forty years. This was the inaugural class for an annual recognition of national reconnaissance pioneers. At the same time, the DNRO acknowledged 10 founders of national reconnaissance—scientists who had contributed to this discipline.

One month after the DNRO announced the selection of the pioneers and founders of national reconnaissance, representatives from the NRO's Center for the Study of National Reconnaissance met with thirty-one of the pioneers, three of the founders, and fifteen of their spouses to record the recollections of their involvement with national reconnaissance. This book is a compilation of those recollections.

When President Ulysses S. Grant wrote his memoirs, it was his desire to do justice to all people about whom he wrote. Following Grant's example, we have tried to do justice to all those about whom we wrote—to those individuals who contributed their recollections to this book and whose lives are reflected in the book's chapters.

**Nature of the Chapters**

The stories in the chapters represent the events and experiences as recalled and understood by the founders and pioneers—the actual contributors to the content of this book. Not all readers necessarily will agree with the completeness of accounts or the accuracy of all the individual statements in the various chapters. We made limited attempts to compare information in the recollections with the documented historical record. As a result, there may be parts of the stories that are missing, or there may be interpretations of events that are inconsistent with other observers' views.

[1] Personal Memoirs of U. S. Grant (1885).

Rather than attempting to document a record of history, we have tried to document the reality that was recalled by the contributors to this book. It is their memories and their perceptions of their experiences that we tried to capture. This means that there may be discrepancies in how the experiences were recalled. These discrepancies can be the result of the passage of time or the uniqueness of individual perspectives.

The passage of time can modify the recollection of any event. Many of the recollections in this book are about events that occurred over thirty years ago, and the passage of that length of time can mute the clarity of details or cause experiences to become lost in the haze of memory. This is only natural. However, in spite of this passage of time, frequently decades, I am amazed at the vividness of detail that these contributors could recall. I would find it difficult to remember with such clarity the activities that I was involved in only months ago. However, these were emotionally significant experiences, often taking place during the high-energy periods of the Cold War. As such, the experiences often became imprinted with great intensity in their memories.

The uniqueness of individual perspectives also can create discrepancies in how people recall their experiences. Our contributors perceived their experiences within their own unique, perceptual framework. On the one hand, the individual's perceptions may or may not be consistent with those of other observers at the time. On the other hand, more recent experiences in the individual's perceptual framework might give new meaning to the past experiences being recalled.

In spite of the passage of time and influence of unique perspectives that might have resulted in diverse recollections, we wanted to document the recollections as the contributors communicated them to us. It is for this reason that we did not attempt to verify the accuracy or completeness of the recollections. The reader must look to the official documents and records of the time to explore a more objective assessment of the events that the pioneers and founders described in this book.

**Structure of the Chapters**

The chapters are of varying length and depth of content. Some chapters are as long as fifteen pages, while others are as short as three pages. The chapters may offer detailed explanations of an individual's work in the field of national reconnaissance, or provide only a brief summary of an individual's career. The chapters also vary in terms of their style, scope, and focus. The narrative may be in the form of a collection of anecdotal stories, historical commentaries, tutorials on some aspect of the discipline of national reconnaissance, or the sharing of lessons that the individual learned. These variations are a function of the personal preferences of the contributors and the guidance that we provided to them when we asked them for their recollections.

Personal preference influenced the nature and extent of the details that the contributors were willing to share. Some contributors were more closed about their careers in national reconnaissance. They were either inherently more private in character or reluctant to be open because of the tradition of secrecy in the field. After thirty years of adhering to strict secrecy oaths, it was difficult for many to begin talking in any detail about this highly classified business. Also, the kinds of people drawn into national security work inherently may be individuals who are more apt to be private in nature.

The variation in the nature of chapters is not solely a result of the personal preference and temperament of the contributors. It also is related to the guidance that we provided. We asked the contributors to rely only on those details that they could remember during the time they were providing their input to us. We specifically asked them not to assist their recall process by doing any research into official records or their personal diaries. We wanted

Figure 1 The Pioneers and Founders at NRO Headquarters during their briefing session before documenting their recollections, 26 September 2000. (Photo by Sara Judy, NRO Visual Design Center.)

to capture the most enduring and vivid memories of their national reconnaissance careers —memories that could be recalled without assistance.

**Preparation of the Book**

This book is the product of a collaborative effort between the editorial staff and the founders and pioneers. This group of contributors not only included the founders and pioneers, but it also included fifteen of their spouses and three of their children. These contributors provided their input in a variety of modes.

The majority of input was in the form of interviews that generally were open-ended, during which the contributors were free to bring their own focus to the dialogue and discuss as much as they might be able to recall or as much as they felt comfortable sharing. In other cases, the contributors dictated their thoughts or submitted written material. Our writing team drew on these inputs to draft our understanding of the contributors' recollections. The writers worked as a team in doing the research, writing the manuscript, and rewriting the numerous drafts. Typically our writers passed their draft chapters to other writers for additional refinement or rewriting. The recollections in the chapters are stories from the contributors, although admittedly the way the words are organized may reflect some of the collective writing style of the editorial team.

Our editorial staff also searched NRO historical files, government archives, and public sources to locate photographs that could illustrate the stories and be added to the chapters. Frequently we were fortunate enough to have the contributors share their own photographic collections with us. It often was a challenge to identify the original source for some of the photographs, but we made every effort to identify the original source and attribute the photographic material correctly.

Another challenge was to derive an unclassified story from recollections that were based on classified experiences. Usually the initial, rough draft contained at least some classified information, which we had to delete. Our requirement was to tell a story without disclosing sensitive engineering details or any other sources and methods—information that might have an adverse impact on the future collection of reconnaissance data if disclosed. We took great care to focus on the unclassified process of developing a national reconnaissance capability, rather than trying to talk about the classified details of these NRO systems.

Figure 2. Jan Geyer and Lynne Mannen during an interview session with Cherie Jones from the CSNR Staff. (Photo by Sara Judy, NRO Visual Design Center.)

Because all of the chapters are based on potentially broad, classified activities, we also subjected our manuscripts to repetitive security and pre-publication reviews. In some cases we found ourselves removing or rewriting extensive information. This made some of the chapters shorter than they otherwise might have been.

After completing a basic security review, we worked with the contributors to ensure that we were faithful to their style and message. We provided them with a draft manuscript for their review. In response they offered us suggestions to correct or amplify what we had drafted. We then incorporated their comments into our final draft. We focused on accurately reflecting what we believed the contributors were saying, and we attempted to maintain consistency of style and readability across manuscripts.

We arranged for the manuscript to be published by the American Society for Photogrammetry and Remote Sensing (ASPRS), the private-sector, nonprofit imaging and geospatial information society. The ASPRS became our partner because the society is the United States' professional, remote-sensing organization and has a long history of drawing its members not only from the commercial and academic remote-sensing communities, but also from the broader government community—both the civil and national security sectors. In fact, in 1954 Arthur C. Lundahl—the Central Intelligence Agency (CIA) officer who founded the CIA's imagery analysis organization—served as the president of ASPRS's predecessor organization, the American Society of Photogrammetry.[2]

At all stages of the preparation of this book, it was a collaborative effort during which we worked with the publisher and contributors to focus on capturing the essence of the contributors' recollections. Where statements in their recollections are accurate and written with precision, this should be attributed to the contributors. However, if there are any errors or lack of clarity in the chapters—whether they are because of omissions or commissions—they are solely my responsibility as the editor.

## Organization of Book

We organized the chapters into three sections: Section I for all the founders of national reconnaissance, Section II for those pioneers who were most associated with signals intelligence

[2] For more information about Arthur Lundahl, see "Arthur C. Lundahl: Founder of the Image Exploitation Discipline" by Dino A. Brugioni and Frederick J. Doyle. In *Corona Between the Sun and the Earth—The First NRO Reconnaissance Eye in Space* (Edited by R. A. McDonald), Bethesda, MD: American Society for Photogrammetry and Remote Sensing, 1997.

(SIGINT), and Section III for those pioneers who were most associated with imagery intelligence (IMINT). Many of the pioneers made contributions in both the SIGINT and IMINT areas, but we placed pioneers' recollections in only one section based on the area of contribution for which the NRO honored the pioneer. Each section begins with an introduction written by the relevant Intelligence Community historian: the NRO Historian for the founders' section, the National Security Agency (NSA) Historian for the SIGINT section, and the National Imagery and Mapping Agency (NIMA) Historian for the IMINT section.

There are a total of forty-two chapters. Thirty-four of the chapters are the first-hand recollections of individual contributions to the discipline of national reconnaissance. The remaining chapters are the recollections provided by spouses of deceased pioneers, or they are accounts of a deceased pioneer's career written by our editorial staff. For five of these remaining chapters, in which wives provided brief recollections about their husbands, our staff wrote short introductory sections that summarize the pioneers' contributions to the discipline. In three other cases, our staff wrote more complete chapters about the careers of deceased founders and pioneers.

The book does not include chapters for five of the founders and nine of the pioneers because some of these individuals were not available to participate in the recollections project, and several were deceased at the time we prepared the book. We did not have the resources or the opportunity to do the necessary research to prepare manuscripts for all of these individuals. The names of the thirty-seven pioneers and five founders for whom there are chapters in this book are listed in the table of contents. The additional fourteen founders and inaugural pioneers who do not have chapters in the book are listed in the accompanying table.

Table. Additional Founders and Inaugural Pioneers

| | |
|---|---|
| **Pioneers** | |
| James G. Baker, Ph.D. | Lee W. Roberts, Colonel, USAF* |
| Howard O. Lorenzen* | Charles "Charlie" R. Roth* |
| James E. Morgan* | Forrest H. Stieg |
| John Parangosky | Roy H. Worthington, Colonel, USAF |
| Val P. Peline, Ph.D. | |
| **Founders** | |
| William O. Baker, Ph.D. | James R. Killian, Jr., Ph.D.* |
| Amrom H. Katz* | Frank W. Lehan* |
| William J. Perry, Ph.D. | |

* deceased

At the end of the book we included three appendices: the first, which provides reference information on the terminology that we used in the book; the second, which includes brief biographical profiles of the pioneers and founders; and the third, which explains the national reconnaissance pioneer program.

## Conclusion

Our objective in compiling these recollections was to capture the stories of those individuals who created the discipline of national reconnaissance in order to gain insight from their experiences—insight that can be helpful to both current and future practitioners as the discipline faces new challenges. We also saw the project as an opportunity to honor these founders and pioneers and to share their stories with their families. Secrecy demanded that their professional lives and contributions be kept private for years. But now we can share many of their individual stories—and they are fascinating.

On behalf of all those who worked on the book, I am pleased that we have the opportunity to look behind the shroud of secrecy and make these recollections available to our various constituencies: the workforce at the NRO, the alumni of national reconnaissance, and the general public—especially the contributors' families. These silent partners now can have insight into the world of national reconnaissance when they read these recollections—and they can be very proud of their spouse, sibling, parent, or grandparent.

*Robert A. McDonald*
*Chantilly, VA*
*December 2002*

**Reference**

Grant, Ulyssses S. (1885). *Personal Memoirs of U. S. Grant.* New York: Dover Publication, 1995 (originally published: New York: C. L. Webster).

# Acknowledgements

*Thanks for the memory*
*Of things I can't forget . . .*

"Thanks For the Memory"
Writers: Leo Robin & Ralph Rainger
Recording Artist: Bob Hope

This is a book of memories and recollections from the founding of national reconnaissance and early years of the National Reconnaissance Office. The collection of these memories, and the preparation of the book, involved the participation of many individuals.

I am grateful to The Honorable Peter B. Teets, Director of the National Reconnaissance Office (DNRO), who has been a champion of the National Reconnaissance Pioneer Recognition Program and a supporter of the pioneer recollections project. I also am indebted to former DNRO Keith R. Hall, who was the driving force for establishing the Pioneer Recognition Program. The NRO's Director of Policy, Mr. Michael Brennan, and his predecessor, Mr. Gil Klinger, have been enthusiastic supporters and sponsors of this project throughout its various phases.

The preparation of the book would have been impossible without the hard work and dedication of the staff members from the Office of Policy's Center for the Study of National Reconnaissance and others within the NRO. This ad hoc editorial staff included individuals who provided extended support over major portions of the project, as well as those individuals who provided more limited support during briefer portions of the project. It was all of these individuals who became the ad hoc team that contributed to the research, writing, and production of the book. I am indebted to all those who were involved with the production of this book. The consistent themes in their work were commitment to the importance of the project and their admiration for the contributions of the Pioneers and Founders.

I am particularly indebted to the two content editors—Mr. Thomas B. Nath and Dr. Joseph J. Helman—for their dedicated support. During the earlier phases of the project Mr. Nath researched and drafted a significant number of the chapters. He was a key contributor during these earlier phases and not only drafted chapters, but also managed the production schedule and the final layout coordination for many of the chapters. As the project progressed, Dr. Helman became the principal writer and content editor who researched and drafted many original and revised chapters. During the final stages of the project I increasingly came to depend on Dr. Helman for his valuable editorial advice.

Production support was critical to the completion of this book. Ms. Jacqueline A. Gray was the quality assurance editor and my editorial assistant, who worked with me from the project's start to its completion. She reviewed the style, format, grammar, and consistency of the manuscripts—and was the glue that held the editorial staff together.

Another early contributor who was involved in extended support and provided a broad range of support was Ms. Cherie Jones. She prepared initial drafts for several chapters, plus she handled a variety of issues pertaining to the figures and managed the pre-publication review process. Mr. Allen L. Shumway, Jr., drawing on over forty years of experience in reconnaissance, offered his historical perspective and also provided production support.

Research and original data collection were fundamental to the writing that was to come later. Our three research interns—Kathryn Sieh, Matthew Sanderl, and Lambros Kapoulas—searched for and identified the photographs for each chapter and were responsible for initial drafting of the captions and source citations. Mr. Michael Rybinski served as the project's signals intelligence (SIGINT) consultant, whose advice contributed to our understanding of the SIGINT recollections. Twenty-five individuals from the NRO Office of Policy and other NRO staff components served as interviewers on Pioneer Recollections Day. The oral recollections that they acquired from the pioneers and founders provided the historical accounts from which the drafters prepared the manuscripts.

Other staff specialists at the NRO made important contributions to the preparation of this book. Dr. Clayton Laurie (former Deputy NRO Historian) and Mr. Matthew Doering (NRO Staff Historian), under the direction of Mr. R. Cargill Hall, the NRO Historian, provided historical research and resource support, including assistance in locating photographs. Mr. Robert Wheeler, under the direction of Ms. Gina Otto from the NRO Office of Security, was our primary security consultant throughout the project. Mr. Dan McGarvey, the Office of Policy program security officer, provided additional security support during the later stages of the project. Throughout the project we also received administrative support from Ms. Charlotte Collinsworth, Ms. Nancy Hall, and Ms. Bonnie A. Williams. Mr. William Naylor provided production support during the final stages of the project.

Mr. Paul Plummer, Mr. Darrell Gates, and Mr. Alexis Warner, under the direction of Ms. Colleen Graver at the NRO Visual Design Center, and with the participation of Mr. Jack Munson, designed the pioneer and founder watermarks. Ms. Sara Judy and Ms. Candia Campbell, also from the Visual Design Center, were the project photographers. Major Julie Norris, Major Andy Klemas, Lieutenant Colonel Ron Marx, and Captain Antonio Gonzalez provided contract support and helped us understand that labyrinth. Mr. Michael Barrett, Assistant NRO General Counsel, was our legal advisor. He responded to our seemingly endless flow of legal questions.

We were obligated to submit the manuscript to the NRO's pre-publication review authorities. Ms. Barbara Freimann's Information Access and Release Center managed this process through the work of Mr. Dennis Cain, Ms. Linda Hall, Ms. Linda Hathaway, and Mr. David Speeney, with extensive support from Mr. Dwight Groggel at the Information Declassification Review Center. During the final stages of this coordination, Mr. Garnett Kiser assisted in resolving security-related issues. Mr. Keith Sheldon and Ms. Patricia Grimes provided critical and timely information systems security support throughout the publication process.

Several individuals associated with the American Society for Photogrammetry and Remote Sensing (ASPRS), our publisher, helped make this book a reality. I extend a special thanks to Ms. Carolyn Staab for her layout and design of the book. She constantly dealt with our changing deadlines and the many adjustments we made along the way. She also offered us numerous valuable suggestions in fine-tuning the manuscript. Most importantly, she used her artistic and graphic talents to transform our manuscript into a visually appealing book. Ms. Kimberly A. Tilley, the ASPRS Assistant Executive Director and Communications Director, had the patience to manage the publication activities, even though they were filled with uncertainties and had ever-changing schedules. Mr. James R. Plasker, the ASPRS Executive Director, was willing to commit the Society to this project at a very early and undefined stage.

Obviously, without the pioneers and founders sharing their recollections, there would be no book. It is their careers in national reconnaissance, and their recollections of the challenges they faced during those careers, that made it possible to prepare this book. Thank you, pioneers and founders, for your contributions that have been so important to our national security—and thanks, of course, for the memories.

*Robert A. McDonald*
*Chief, Center for the Study of National Reconnaissance*
*December 2002*

# Table of Contents

## Appendices

# Introduction

*"[M]y design is not to write histories but lives. . . . [A]s portrait-painters are more exact in the lines and features of the face, in which the character is seen, than in the other parts of the body, so I must be allowed to give my more particular attention to the marks and indications of the souls of men, and while I endeavor by these to portray their lives, may be free to leave more weighty matters and great battles to be treated by others."*
—*Plutarch,* Life of Alexander the Great[1]

## Models For Success—Recollections of Accomplishments

Plutarch's biographies are stories that describe the character and moral attributes of his subjects—soldiers, statesmen, and orators from the Greek and Roman past. His interests were in the personalities of these people and the ways in which they shaped their actions and brought themselves to great accomplishment, or in some cases, unexpected failure. Plutarch's focus was less on the historical events and politics of the time and more on the individual actions and motives of the subjects.[2]

Likewise, the recollections in this book are stories that look beyond historical events associated with a particular time—the developing days of national reconnaissance. In compiling these personal accounts, we have attempted to demonstrate how the founders and pioneers of national reconnaissance built a dynamic national reconnaissance capability. We have sought to document for you, the reader, first-person accounts of motives, challenges, and accomplishments.

### The Distinction Between the Founders and Pioneers

You can come to know these founders and pioneers through their recollections. Both groups have made a different kind of contribution to the discipline of national reconnaissance. The founders, through their scientific competence, gave confidence to national-level decisionmakers. The pioneers, through their front-line work, built the U.S. national reconnaissance capability.

The founders of national reconnaissance are a small group of national-level advisers who provided the technical expertise that shaped the emerging discipline of national reconnaissance in the late 1950s and early 1960s. They are the scientists, engineers, and innovators

[1] Plutarch (1992, p. 139).
[2] Clough (1992, p. xx).

Figure 1. The Year 2000 National Reconnaissance Pioneers and three of the Founders of National Reconnaissance (seated in center), with DCI George Tenet (far right) and DNRO Keith Hall (far left). (Photo by Sara Judy, NRO Visual Design Center)

who served on national-level panels and have been influential advisers to presidents. In that capacity, they were strong advocates for developing a national reconnaissance capability.

The founders also are well-respected experts who, often independent of national reconnaissance, made wide-ranging and significant scientific contributions in their own fields. Among the founders are individuals who made contributions such as inventing a reconnaissance camera, developing instant photography, contributing research that was instrumental in the development of the American hydrogen bomb, and doing research that led to the discovery of nuclear magnetic resonance imaging. The stature of the founders is evident by the fact that each of them has been recognized for his accomplishments through numerous prestigious awards, which have included two Presidential National Medals of Science, one Presidential Medal of Freedom and a Nobel Prize. These individuals are strong personalities with a high degree of intellect and the ability to ask effective questions and see long-term implications.

The pioneers are practitioners who made the national reconnaissance vision a reality. Some of them also could be considered founders. As a group, the pioneers share the qualities of dedication, commitment, a thirst for knowledge, and a focus on success. They are confident and bold personalities, who took the initiative and pushed the frontiers throughout their professional careers.

### The Nature of These People

Both the founders and pioneers, in their different ways, were instrumental in building the U.S. national reconnaissance capability. They are dedicated and hardworking patriots who were committed to their missions. They are people with candor, spontaneity, and readiness to make decisions, even if the decisions might result in mistakes.

All of the founders and all members of the inaugural class of pioneers are men. The fields of engineering and science—so much a part of national reconnaissance—belonged almost exclusively to men during those early days of national reconnaissance. This is reflective of the technical workforce in American society at that time.

Even though this group of founders and pioneers is without women, many women contributed to their successes. These women were the wives. They performed an indispensable role for their husbands by creating a personal support structure that gave their husbands the freedom to make contributions to national reconnaissance. Like their husbands, these wives had to make sacrifices. For some of them the sacrifice was to give up the possibility of a career outside the family. And inside the family, they often had to dedicate themselves to carrying the full burden of family responsibilities so that their husbands could dedicate themselves to national reconnaissance. The wives who contributed their recollections to the chapters give the reader a deeper insight not only into their contributions, but also insight into the lives and characters of their husbands.

While each founder and pioneer is a different person, these individuals have much in common. This will become apparent when you read their recollections. They consistently questioned assumptions, challenged ideas, and used their ideas to build a national reconnaissance capability. They have had a life-long interest in the discipline and related subjects, were uniquely motivated, focused on the goal, and routinely demonstrated their creativity.

*Life-Long Interest.* The founders and pioneers—as well as their spouses—consistently reported about their life-long interest in engineering and space. Their recollections make it clear that these people have a fascination with their work. Louise Davies described her husband, Merton, as someone who, ". . . took to his work like a duck to water and always was fascinated with things related to outer space." This is typical of how these contributors and their family members described their relationships with their careers. Frederic Oder opened his recollections by saying, "I became interested in space from a very early age, long before there was a national space program." Fred Kaufman said, " . . . [I] admired the Romans because their designs and engineering lasted for two thousand years. . . . I always wanted to be an engineer in order to build things that endured." Rob Roy explained that he was interested in missile programs from the time he graduated from the Naval Academy and received his commission in the Air Force. His wife, Patricia Roy, reinforces what Rob told us by explaining, "He would talk incessantly about space and tell my friends that man would be going to the moon in the not too distant future."

*Motivated.* These individuals had strong motivations that were varied in their nature. There is evidence that for some it was the challenge to solve a complex problem; for others it was the pure quest for knowledge; for still others it was patriotism; and still for others it merely was a desire to contribute something useful. In addition, there are other possible motivations: the experience of a powerful rush of adrenalin that can arise from being involved with making national security decisions and launching spacecraft, a desire to avoid another Pearl Harbor, or the drive to do something about Cold War threats of nuclear devastation.

*Goal-oriented.* Attaining the goal was foremost for the pioneers. In almost all the chapters, you will read about how pioneers consistently focused on the goal in a resolute way. They are people who could see beyond the immediate challenge into the solution. For the pioneers no challenge was too difficult, no task was impossible, and all problems had solutions. The stories in their recollections make this point clear. Lee Battle reminds us, "There is no such thing as a random failure;" John Bennett recalls, "Failure was never a consideration;" and Frank Buzard recounts, "I wanted to make sure the damn thing worked on the first launch, and that is what I did."

*Creative.* Creativity and innovation are other qualities that are apparent in the character of the pioneers. Creativity was one of the reasons they could attain impossible goals beyond expectations. It was a core quality that inspired them to find innovative solutions. Their ideas came from anywhere. You will find this in many of their recollections. Pioneers like Fred Kaufman, Marvin Stone, Reid Mayo, John Bennett, and Robert Powell gave us examples of how creativity moved them to find innovative solutions.

For Fred Kaufman innovative ideas would emerge at anytime and in the oddest ways. While commuting in traffic one day, Fred suddenly had a mental picture of himself standing on the hanger deck of an aircraft carrier. He was looking into an open compartment where he unexpectedly saw some kind of secret signals intercept work. When he asked himself why he was thinking about this, it occurred to him that the real mission of a study he was working on was to look for the unexpected. This realization caused him to expand the definition of his project's mission so that its system concept could consider unexpected things that might be detected.

At other times ideas would come as a result of a change in the environment. Marvin Stone did not fully appreciate the significance of some of the things he was doing as a payload systems engineer until he visited an operational ground site. After having an opportunity to see the system in operation, he found that he had developed a new appreciation for his work and understood how key payload specifications affected mission performance.

By visualizing the exaggerated notion of extending a submarine's periscope to orbital heights, Reid Mayo conceived of how the crystal video technologies he had developed for submarines could be applied to an approach for signals intelligence (SIGINT) space reconnaissance. He saw how the submarine periscope system could be modified so that a solid-state version of the intercept system could be mounted in a twenty-inch solar-powered Vanguard satellite. He concluded that the extreme height would give the system coverage that was far greater than that of not only the submarine system, but also of airborne systems. Consequently, he found a way to detect advanced Soviet early-warning and height-finding radar at orbital altitudes.

John Bennett's colleagues were not averse to using nonconventional devices in their experiments. They came up with the idea of using, as experimental testing tanks, very non-professional-looking plastic wading pools with their characteristic colorful, wild patterns. John's team had to do testing that required zero-g balancing of a particular component. The team members came up with this idea of buying inexpensive plastic wading pools and using them as testing tanks. They balanced the component on blocks of foam that they floated in the pools. This very inexpensive test worked superbly and was a key factor in demonstrating the feasibility of using the component in space.

The pioneers also used unusual tools to find solutions to operational and engineering problems. For example, when one spacecraft on the launch pad was suspected of having a malfunction with a spinning component, Robert Powell and a colleague used a stethoscope to make the diagnosis by listening for signs of friction as the system operated. Once they determined that there were no rubbing sounds, they were able to proceed with the launch.

*The Real Person.* Early in this book project, I had the privilege of meeting with this group of public servants to outline what we were asking from them as contributors to the content of the book. While explaining our process, I saw in this audience the human power behind the development of the discipline of national reconnaissance. I saw the reflections of their energy, motivation, commitment to the goal, and the creativity that made possible what should have been impossible accomplishments. When I talked with them, they told me how honored they were to have been a part of the development of a U.S. national reconnaissance capability. They pointed out that they were just people doing their job—quietly, privately, and without any expectation of public recognition. They are complex people who came of age during twentieth-century wartime challenges and uncertainties.

**The Challenges of Their Times**

The founders and pioneers were products of the military challenges of World War II and the uncertain days of the Cold War. These often were experiences that occurred during their formative years and influenced their thinking, career directions, and commitment to public service. In fact, World War II and its aftermath may have been their greatest teacher. We can see this repeatedly in what the pioneers describe in their recollections. The subsequent challenges of the Cold War compelled them to make their contributions to national reconnaissance.

Figure 2. The *USS California* sinking after the Japanese attack on Pearl Harbor, 7 December 1941. Visible in the background are Ford Island (right), *USS Shaw* (left), and *USS Nevada* (left center). (Photo courtesy of U. S. Naval Historical Center.)

*World War II Experience.* In many of the chapters, you will read about the direct influence of World War II in the lives of the founders and pioneers. The recollections of Richard Leghorn, Frederick Kaufman, and Frances Madden provide three different examples of this influence.

Dick Leghorn flew combat reconnaissance missions during World War II and participated in post-war aerial data collection of atomic testing. These experiences influenced the formation of his concept of peacetime strategic reconnaissance. In reflecting after the war, he was impressed with the critical importance of aerial photography during combat. That assessment remained with him when he later observed two test atomic explosions in July 1946. He told us, "To see an actual atomic bomb, and the immediate and complete damage it did, was a fearsome experience." The alarming possibility of a surprise nuclear attack against the United States made clear to Leghorn the urgent need for a peacetime strategic reconnaissance capability— a capability separate and distinct from the tactical wartime reconnaissance of World War II. Leghorn's experiences demonstrated to him that the United States needed more accurate information, especially before the early phases of a war. It was this context that helped form his concept of national reconnaissance and his subsequent contribution to the discipline.

Fred Kaufman was at Pearl Harbor on 7 December 1941 and witnessed the surprise attack with its devastating fires. He wrote, "That intelligence disaster burned an image into my memory that will last forever. As I looked out at the devastation, I remember thinking that if I had the opportunity to do anything to keep that from happening again, I would do my very best."

Frank Madden, along with other pioneers, learned about the value of teamwork during their World War II experience. The importance of working cooperatively became a valuable concept and tool for these pioneers later in their careers. For Frank Madden, his involvement with the Corona project confirmed the value of team members' working cooperatively to get the job done. The individuals on his team knew they were going to do their job, and they had confidence that the others who were working with them were going to do their jobs. In Frank's words, "We developed a sense of comradeship, which is not always easy to do." When Frank reflected on his success with a team approach, he explained, "I believe part of our ability to do that was because we learned about teamwork during our military service days. During wartime we learned to rely on our buddies."

*Cold War Influence.* The Cold War presented the founders and pioneers with encounters that were very different from what they might have imagined from their World War II perspective. With the Cold War, the founders and pioneers were confronting threats and uncertainties that their generation was finding frightful and initially baffling. The Cold War presented the threat of nuclear destruction for the U. S. and the world at large. More critically

Figure 3. The U-2 Reconnaissance Aircraft, unknown date. (Photo courtesy of NRO History Office, likely U.S. Air Force photo.)

for the nation's policymakers, it presented a closed Communist society and an Iron Curtain that denied policymakers the information they would need for sound and confident policy decisions.

The reality of the Cold War propelled this generation of founders and pioneers to the center of what would become an unsettled challenge for information about the Soviet strategic threat. It was the national security threat of the Cold War that moved this generation into action, but it was their knowledge of science, technology, and engineering that provided the means to find answers. The events of the Cold War presented the founders and pioneers with these challenges, which they made into opportunities for greatness that became the foundation for their accomplishments.

**Their Accomplishments**

The accomplishments of the pioneers demonstrate an ability to achieve success beyond an observer's expectations. They made profound contributions directly to the field of national reconnaissance. At the same time, their work contributed ongoing benefits to civilian technology.

*Contributions to National Reconnaissance.* From the earliest days of the Cold War, the pioneers were able to create remarkable national reconnaissance capabilities, whether they were with airborne or space-borne collection platforms, and whether they were for signals intelligence (SIGINT) or imagery intelligence (IMINT) collection. During the earliest period of their accomplishments they developed a remarkable reconnaissance aircraft, the U-2; they developed the first SIGINT satellite, Galactic Radiation and Background (Grab); and they developed the first IMINT satellite, Corona—all examples of meeting the challenge beyond expectations. They continued making these kinds of remarkable contributions beyond the Cold War.

*The U-2 Reconnaissance Aircraft.* The U-2 aircraft was a cross between a glider and a jet—a jet-powered sailplane that was light in weight with a large wingspan and tremendous lift, but a plane that could fly at speeds approaching 400 knots at an altitude of over 70,000 feet with a range of 2,950 miles. These performance characteristics permitted the U-2 to overfly continents and operate beyond the capability of the interceptor aircraft and antiaircraft weapons of the time. However, it was a fragile aircraft. If the original models exceeded 394 knots (Mach 0.8) at their operational altitude, the bolt-on wings or tail section could fail and separate. Remarkably, the pioneers built the U-2 in record time. President Eisenhower approved the U-2 project in November 1954, and within eight months, in July 1955, an initial aircraft was ready for delivery. By 1956 the U-2 was operational.[3]

Not only was the U-2 a technological marvel beyond expectations, so too was its national security impact. By the mid 1960s, the U-2 project already had acquired and processed more

[3] Pedlow & Welzenbach (1998. pp. 39, 53, 76, 79, 93, 100, 316); Baker (2002).

than 1,285,000 feet of reconnaissance film. And the quality of the imagery was amazing. The U-2 flew different cameras, including its "B camera," which was a high-resolution, seven-stepped framing camera that had a 36-inch focal length lens. This B camera could provide horizon-to-horizon photo coverage on two side-by-side, 9X18 inch frames. This was accomplished by taking photographs in seven steps or positions, from vertical to high oblique. The camera could acquire images with a resolution of 60 to 80 lines per millimeter, which could produce from 65,000 feet altitude, a best ground resolution of 2.5 feet at nadir. An 18-inch by 18-inch vertical image could cover over 29 square nautical miles at 65,000 feet. One of the U-2's first major contributions of this imaging capability was to expose the myth of the 1950s' strategic "bomber gap" between the U.S. and USSR.[4]

*Grab SIGINT Satellite.* The Grab satellite collected SIGINT data and traveled in a circular orbit 500 miles above the earth. It collected signals that emanated from ground-based radar antennas, an unbelievable capability at the time. The satellite's tiny antennas intercepted these radar signals at particular bandwidths and transponded a corresponding signal to ground sites within Grab's field of view. As with the U-2, Grab was built in a relatively short period. President Eisenhower approved full development of Grab on 24 August 1959, and by June 1960, within ten months, the Grab pioneers had launched the first successful Grab mission. Grab, like the U-2, collected remarkable intelligence. It was able to collect information about Soviet air defense radars that otherwise could not be acquired by Air Force and Navy ferret aircraft flying the more conventional airborne electronic intelligence (ELINT) missions along Soviet border regions. The technical intelligence derived from these data supported the development of critical Pentagon war plans.[5]

*Corona IMINT Satellite.* Corona was another technological space marvel that was developed beyond expectations and made unprecedented contributions to national security. Like the U-2 and Grab, the pioneers developed it in record time. From President Eisenhower's approval of the concept, it took only two years to design and build an operational spacecraft. The Corona project saw several firsts: the first recovery of an object sent into space when the U.S. Navy retrieved a recovery capsule that splashed into the Pacific Ocean, the first mid-air recovery of an object sent into space when a U.S. Air Force C-119 snatched a recovery capsule that was parachuting back to earth, and the first intelligence photograph of a target imaged from space when its camera photographed the Soviet bomber base, Mys Shmidta, on 18 August 1960.[6]

More significant than acquiring that single, first in-

Figure 4. Grab 2 (top) mated with Injun (middle) and Transit IIIB (bottom) atop the Able-Star portion of the Thor/Able-Star booster. Injun was a scientific research satellite for James Van Allen of the State University of Iowa, and Transit was a U.S. Navy navigation satellite. (Photo courtesy of Naval Research Laboratory.)

[4] Pedlow & Welzenbach (1998. pp. 39, 53, 76, 79, 93, 100, 316); Baker (2002).
[5] U.S. Naval Research Laboratory (2000).
[6] McDonald (1997a); McDonald (1997c).

telligence photograph from space, Corona operated for over a decade and acquired photographic coverage of 600 to 750 million square nautical miles of the Earth's surface. By the end of the program's life cycle, Corona was acquiring photographs with a quality of six-feet resolution from 100 miles in space during extended missions that could last up to 19 days. These missions were supplying the U.S. Intelligence Community with continuous photographic coverage of the USSR—synoptic views of relatively large geographic areas. The intelligence analysts had the opportunity to see and count the Soviet strategic weapons that were threatening the U.S., and they could monitor the status and deployment of these weapons. This new national reconnaissance capability exposed additional mythology about Soviet capabilities, like the myth that there was a strategic missile gap in favor of the Soviet Union. At the same time, these reconnaissance capabilities found new applications, like detecting nuclear proliferation and monitoring arms control arrangements.[7]

*Beyond the Cold War.* These early Cold War successes of the U-2, Grab, and Corona were just a preview of what these pioneers would do for national security during the balance of the 20th century. During NRO's first 40 years it flew some 300 satellites that collected millions of pictures and an untold number of signals. Most of these data and the associated descriptions of the collection capabilities continue to be classified, and I cannot go into detail about these matters because of security reasons. But, in unclassified terms, I can say that at the start of the 21st Century the NRO could acquire near-real-time information from space, and that this information resulted in a stream of high-quality intelligence for national security applications. This intelligence, which originated at a point in time from a platform somewhere in space, often can be transmitted in a matter of minutes to the NRO's customers, whether these customers happen to be in the White House, in a cockpit, or at a civilian disaster-relief operations center. All of this represents capabilities that go far beyond the expectations of the NRO's original customers and the bold decisionmakers who authorized these programs during the earliest days of the Cold War.[8]

*Legacy Contributions to Civilian Technology.* Not only did the pioneers make contributions that directly impacted on reconnaissance and national security, but their activities, which almost exclusively were in the classified world, also led to technological applications in the unclassified, civilian sector. Three specific examples are in meteorology, lunar imaging, and video tape recording. There also were other spinoffs that generally enhanced civilian technology.

Figure 5. A Corona, KH-4B image of the Kremlin in Moscow, USSR, 28 May 1970. On the left portion of the image there is visible a line of people in front of Lenin's Tomb. (Corona Mission 1110)

*Meteorology.* The Defense Meteorological Satellite Program (DMSP), which was developed to support national reconnaissance, became the foundation for a future constellation of civilian weather satellites. In the early days of satellite photoreconnaissance, cloud cover was a major problem that often obscured a clear view of the earth's surface and prevented the acquisition of high-quality imagery. To anticipate this cloud

[7] McDonald, (1997a); McDonald (1997b).
[8] K. Hall (2001).

cover and better manage imaging operations, Thomas Haig and others developed the DMSP satellites. The technology and experience derived from these efforts have evolved into civilian weather satellites, which have increased the accuracy of making long-range forecasts greatly. Following his career in national reconnaissance, Haig went on to use his experience in civilian applications and became the lead person in developing an interactive data system that led to the world's first global weather system. This changed weather forecasting systems throughout the world.

Figure 6. First Picture of the Earth taken from the Moon, by Lunar Orbiter I, 23 August 1966. (NASA Photograph, Courtesy of NRO History Office.)

*Lunar Imaging.* The early Samos photoreconnaissance program resulted in the National Aeronautics and Space Administrations (NASA) being given a photographic capability that it used to image the Lunar surface. Chuck Spoelhof explains how Eastman Kodak developed the cameras for the Samos program. However, when Corona became the first operational satellite photoreconnaissance program, the Samos program was discontinued. Subsequently, in early 1962 the NRO consigned the Samos imaging system to NASA for its use in deep-space exploration. When NASA moved forward with development of its lunar-imaging program, Kodak supported the Lunar Orbiter effort and modified the Samos E-1 camera for use in that NASA program. The camera had a 80mm focal length Schneider-Xenotar lens and a 24-inch Pacific Optical telephoto lens. NASA launched five of these cameras. Three missions obtained detailed lunar photographs that NASA used to select landing sites for Project Apollo's manned lunar landings, and the remaining two missions obtained comprehensive lunar photographs that NASA used to produce photo-maps of the moon's surface.[9]

*Video Tape Recording.* One of the SIGINT national reconnaissance programs included the technology that later became prominent in the consumer magnetic video recording market. One program that Jon Bryson had worked on required a wide-band recorder similar to what one company had been manufacturing for television stations. Unfortunately, that particular recorder weighed about 2,000 pounds, and Jon and his colleagues asked that the company develop a recorder with the same technology but in the size of a briefcase. The company produced the recorder, and the program team used it several times. Unfortunately, the team was unable to master using magnetic tape in a vacuum, and they abandoned that recorder. The company also could not see a market for it and sold its rights. By the 1970s that technology appeared in the marketplace in the form of commercial and consumer videocassettes.

*Technological Enhancements.* There is a range of other national reconnaissance technology that has found applications in the civilian sector. Some of these capabilities continue to be classified, and I may not discuss them here. Four others that are unclassified are examples of this technological enhancement: electro-optical technology, computer-assisted imagery analysis, sophisticated communications technology, and integrated circuits. Let me briefly enumerate. First, the mirror-polishing techniques used to produce electro-optical imaging reconnaissance systems have been used in the development of civilian space projects such as

[9] C. Hall (2001).

the Hubble space telescope. Second, the computer-based change-detection and pattern-recognition techniques used to analyze reconnaissance imagery have been used in medical radiological applications such as for the analysis of mammograms. Third, the technology for wide-bandwidth communications, high-resolution pixel arrays, and high-speed switching used in many classified reconnaissance programs also have been used in the technology for high definition television. Fourth, reconnaissance-related research in gallium arsenide analog integrated circuits has reduced significantly the noise in microwave receivers and produced higher-efficiency microwave power amplifiers. These kinds of gallium arsenide integrated circuits have been used in cellular phones, GPS receivers, and television pre-amplifiers.

*New Technologies for the Civilian World.* These kinds of civilian spinoffs from the pioneers' work are beyond the expectations of the early builders of national reconnaissance systems. These are examples of how these pioneers made lasting and significant contributions, not only to the field of national reconnaissance, but also to the civilian engineering and technology base. The lesson for us is not that they accomplished these things, but how they accomplished them. In reading about how they accomplished these things, you can gain useful insights. I would suggest that there will be more of these kinds of cross-disciplinary benefits from the work of future pioneers in national reconnaissance.

**Lessons for the Discipline—How Did They Do It?**

The pioneers of national reconnaissance, by the example of their accomplishments, offer students of national reconnaissance and future pioneers insights that should inspire further breakthroughs. The pioneers' formula for success was simple: take risks, use mishaps as learning opportunities, employ system engineering concepts, avoid bureaucratic overhead, work as a team, and make personal sacrifices.

*Take Risks.* The pioneers, as part of their character, consistently took risks. The pioneers, who were inspired by their creativity and occasional daring schemes, were prepared to move into the unknown. They made program decisions that at times were fiscally or operationally risky, and they often put their professional reputations in jeopardy through this kind of behavior. The risks might have been high, but they were calculated. If they failed, a program could be lost, and the government might be denied information. If they were successful, the corporate managers and government decisionmakers would be convinced that what initially seemed like impossible tasks could be accomplished.

In some cases, the risks were more professional, such as when pioneers would accept an assignment that would put their careers at risk. One example is the case of Thomas Haig when he accepted an assignment to develop a new NRO meteorological satellite program. Even though he was expected to deliver the first launch within ten months, which was a relatively short period of time, he was presented with resource and technical constraints. In other cases, pioneers were ready to put their programs at significant risk when they made operational decisions. This can be seen in Robert Kohler's recollections when, in describing a problem, he recounted how he selected a risky alternative solution that could have resulted in a catastrophic failure on the launch pad. He chose to repair a failed component on the pad, something that had not been done before. But because he chose this alternative, he was able to meet the schedule in spite of the failed component and was able to do it at one percent of what it otherwise would have cost.

The recollections that you read will give you insight into how the pioneers dealt with risk. In some chapters, like Cornelius Chambers', they offer a tutorial on the pioneers' methodology. Chambers explains that his approach, "required prodigious risk assessments" and outlines specific steps for risk analysis. He reminds the reader that his colleagues were risk takers who, " . . . did their homework and consequently understood the magnitude of the risks." The pioneers had the courage to challenge traditional wisdom and to take these kinds of risks—and they were not undeterred by the prospect of a mistake or a mishap.

*Use Mistakes and Mishaps as Learning Opportunities.* There were mistakes and mishaps, but failure never was an option for the pioneers. Rather, any mistake or mishap was an opportunity to learn more about a problem and move closer to the solution. They knew that what they were doing was complex, but they searched for simplicity in their answers. They instinctively seemed to understand that Clausewitz's observation about war applied to their reconnaissance challenges.[10] Everything is very simple in science and engineering, but the simplest thing is difficult. Their challenge was to find simple solutions to what were difficult problems. When there was a setback, the pioneers had the innovative intellect to analyze the problem and the emotional strength to persevere until they identified and implemented a solution.

The history of national reconnaissance has vindicated the pioneers for the mishaps they encountered and the risks they took to find solutions. They persevered in spite of often unusual or frustrating mishaps. There was no tolerance for unresolved problems. Two anecdotes—one for the SIGINT satellite, Grab, and one for the IMINT satellite, Corona, make this clear.

*Grab's Cuban Cow.* A classic example of an unanticipated mishap was the second Grab mission on 30 November 1960. This Grab satellite, like the first, was launched on a Thor/Able-Star booster from Cape Canaveral, Florida. However, during its launch the Thor booster burned out 12 seconds early, and the range safety officer had to destroy what had become an unguided missile. Fragments landed in the countryside on the north coast of Cuba's Oriente Province about ten miles from the third largest city, Holquin. A Cuban newspaper reported one fragment weighing as much as 40 pounds (18 kilograms), but there were no fatalities—except for a cow. There were subsequent reports that this murdered Cuban cow was given a funeral that included a procession through the local streets.[11]

The Cuban authorities found several of the fragments, including two spheres, two cone-shaped objects, and various cylinders. The English markings on the recovered debris made it difficult to deny their U.S. origin. The Cuban government criticized the incident as "a new Yankee provocation" and a deliberate act of "Yankee aggression." There also were political demonstrations in Havana. Some 250 university students marched cows in front of the U.S. Embassy. One of the cows had a placard with the words:

> *"Eisenhower, Has asesinado a una de mis hermanas. Firma John Kennedy"*
> ("Eisenhower, You have assassinated one of my sisters. Signature John Kennedy")

The sign, obviously, was designed as a double insult to the U.S.[12]

In spite of this launch mishap and its international incident, the pioneers accommodated to the mishap and persevered with their development of SIGINT satellites. One accommodation was to interject a dogleg into future launch trajectories so as to avoid having the launch vehicle pass over the Cuban landmass.

*Corona's Twelve Attempts.* The Corona program, which operated under the cover of the Discoverer program, experienced twelve unsuccessful attempts before a successful mission. From the program's first three Discoverer missions, there was a lack of success. This record continued through the eight initial Corona-numbered missions and one additional Discoverer diagnostic flight. These missions had faults associated with launches, orbital operations, the camera systems, and recovery operations. (See Table.) On one mission the engines burned too long and caused the spacecraft to go into a higher than desired orbit; on another mission

[10] In Chapter 7 of *On War*, Carl von Clausewitz stated, "Everything is very simple in War, but the simplest thing is difficult." (From von Clausewitz, C. (1993). *On 'War*, New York: Knopf, originally published in 1832.)
[11] Phillips (1960); *The Washington Post* (1960).
[12] Phillips (1960); *Time* (1960).

Table. Initial Unsuccessful Corona Attempts 1959–1960

| Date | Discoverer | Corona | Faults | | | | |
|---|---|---|---|---|---|---|---|
| | Mission No. | Mission No. | Launch | Orbit | Camera | Recovery | Other* |
| 24 Jan 1959 | I | N/A | X | | | | |
| 13 Apr 1959 | II | N/A | | | | X** | |
| 3 June 1959 | III | N/A | | | X | | |
| 25 Jun 1959 | IV | 9001 | | | X | | |
| 13 Aug 1959 | V | 9002 | | X | | X | |
| 19 Aug 1959 | VI | 9003 | | X | | X | X |
| 7 Nov 1959 | VII | 9004 | | | X | | |
| 20 Nov 1959 | VIII | 9005 | | X | | X | |
| 4 Feb 1960 | IX | 9006 | | | X | | |
| 19 Feb 1960 | X | 9007 | | | X | | |
| 15 Apr 1960 | XI | 9008 | | | | X | X |
| 29 Jun 1960 | XII | N/A | | | X | | |

*Other Faults: August 19, 1959-Retrorocket malfunctioned; April 15, 1960-Spin rocket failure.
** Capsule ejected over Norway.

the three-axis stabilization system malfunctioned and caused the satellite to tumble, and on another mission the recovery vehicle parachute tore.[13]

Throughout these mishaps, the Corona pioneers used each attempt as a learning experience until Discoverer XIII became the first successful diagnostic flight, and Corona mission 9009, flown under the cover of Discoverer XIV, proved to be the first successful Corona mission. Their persistence paid off when this mission supplied imagery analysts with some 3,000 feet of film that provided more than 1,650,000 square miles of coverage of the Soviet Union.[14]

*Employ System Engineering.* System engineering was a key approach that the pioneers used, not only to anticipate and prevent mishaps, but also to gain insight and to manage their programs more effectively. System engineering is an engineering methodology that focuses on dealing with the integrated whole, as opposed to dealing with isolated, individual parts. It can be considered a way of thinking that attempts to make the elements of a structure fit together and to look for solutions to problems that can be caused by component relationships.[15] The system engineering theme emerges in several chapters. Paul Mayhew explains how having end-to-end responsibility for a program forced him to think early about how to define the system and its subsystems in ways that helped avoid subsequent interface problems. Bob Powell points out how, as the chief systems engineer, he had to sign off on all engineering drawings and interface documents, which put him in a position to see the entire program in an integrated way. Connie Chambers, Frank Buzard, Rob Roy, and others point out in their chapters that an integrated, systems approach made them more effective and was a key to much of their success. You will find in these and other chapters that the systems engineering approach became a means to much of the success that pioneers experienced.

*Avoid Bureaucratic Overhead.* Not only did the pioneers approach their tasks from a systems engineering perspective, but they also maintained the practice of avoiding bureaucratic diversions and distractions. The systems engineering approach, itself, resulted in managers being given end-to-end responsibility for their programs. These end-to-end responsibilities typically seemed to streamline operations and avoid the more traditional bureaucratic over-

[13] McDonald (1997a).
[14] McDonald (1997a,b).
[15] Rechtin (1991; pp. 2,16, 28.).

head. Fritz Oder explains that his approach for success was informality and simplicity: draft a simple statement of work with a clear understanding of what the government wants, and then make an agreement with a handshake. Frank Buzard describes how he avoided bureaucratic diversions by minimizing paper work and keeping his program staff small. "Battle's Laws," which are outlined in Lee Battle's chapter, are a summary of how to avoid bureaucratic entanglement. This theme of avoiding bureaucratic entanglement kept the pioneers focused on national reconnaissance and helped them avoid getting bogged down in administrative paperwork and bureaucratic diversions.

*Work Cooperatively in Teams.* By minimizing bureaucracy, the pioneers were able to create an environment where they could work cooperatively in relatively small, independent teams. They worked not as individuals, but as passionate collaborators. This theme of working cooperatively and in support of common objectives is embedded in most of the stories in the chapters.

Paul Mayhew explains how success at Kodak depended on the commitment, support, and creativity of numerous individuals who were working interactively to find solutions. He acknowledges that individual ideas and contributions helped overcome obstacles, but he points out that it was these individuals' collective and cooperative talent that resulted in the successes. Many of the pioneers explain that the team concept was implemented, not only within organizational components, but also across entities. This meant that cooperation existed between companies, as well as between industry-at-large and the government. Frank Madden explains how when Kodak was trying to solve a problem with Corona film, it was another company, Dupont, that developed the Mylar film base that was critical to Kodak's solution. Don Tang makes the point that participants, whether they were from industry components or government agencies, acted as if they were part of a single team, not players in an adversarial contractual relationship.

The intensity of the team approach is evident in the bonds that developed among team members. These bonds became very strong during the pioneers' work on their programs. Charles Spoelhof and others make references to parties and other social activities that regularly took place among team members and their families. In many cases, the bonding lasted long after individual team members moved on to other activities. Roy Burks describes how in the year 2000, he and his wife attended a reunion with alumni from the Corona program, which had flown its final mission 28 years earlier.

The relative lack of personal competition between pioneers allowed these individuals to free themselves from the prospect of internal rivalries and to focus cooperatively on a different kind of competition with technology. It was this conquest of technological challenges that was more meaningful to the team members than any personal recognition they might have received for their accomplishments. By encountering the challenge and focusing on the objective, they became the leaders who integrated novel ideas, inspired colleagues, and moved teams to do great things.

*Be Willing to Make Personal and Professional Sacrifices.* Ultimately the great things that the pioneers accomplished were possible only as a result of their dedication to the problems at hand and the personal sacrifices they made in order to deal with these problems. Throughout the chapters there are accounts of 60- to 70-hour workweeks and "around-the-clock" workdays. There also are reports of frequent and extended travel away from home. Two hundred days of travel during a year would not be unusual.

At the same time, many had to give up career opportunities. Military officers who accepted assignments with a narrow engineering focus rather than broader military experience often missed opportunities for promotion. The signals they were sending to the military career managers were, "making the grade of general is not important to me."

It was because of these sacrifices that the pioneers were able to meet the challenges and solve what otherwise could have been disruptive problems. Far fewer of their accomplishments would have been possible without exclusive devotion of their energies to their programs. Because of their sacrifices, these engineers and managers were able to do great things that had lasting impacts on the discipline of national reconnaissance and space technology.

**Conclusion**

In this book, you will be reading the recollections of individuals whose sacrifices made amazing things possible. The recollections are stories of creativity, risk, and perseverance from the perspective of these founders and the pioneers. Their stories are accounts of how intellect and ingenuity in engineering, along with a commitment to the nation, built a national reconnaissance capability. Without the collective contribution of the founders, pioneers, and their team members, the U.S. would not have a sophisticated remote-sensing capability that permits it to see and hear the world in spite of our adversaries' attempts to deny us vision and hearing.

In *Look Homeward, Angel,* Thomas Wolfe pointed out, " . . . we are the sum of all the moments of our lives."[16] In the same way, the discipline of national reconnaissance is the sum of all the lives of its founders and pioneers. By reading the recollections of their life experiences, we can gain insight into the discipline of national reconnaissance.

After I read the recollections, I was impressed with what I learned about these individuals' accomplishments. Their efforts exploded new capabilities into space operations, information technology, geospatial information, optics, and communications. I marveled at the power and performance of their reconnaissance systems. I admired them for how they created these systems, and how they made technology work. I saw, in their work, the building of a national reconnaissance capability that was beyond expectations.

These individuals were courageous in their uncompromising dedication to the challenges of science and aerospace engineering. Often they were told that what they were trying to do was impossible. At other times they were criticized for taking risks to reach their goals. In the end, they were successful. Their recollections describe the development of national reconnaissance systems from various stages of conceptualization and operational deployment.

This book not only documents these individuals' contributions, but it also provides the current NRO workforce with lessons for the future and an opportunity for the general public to increase its understanding of space reconnaissance and the mission of the National Reconnaissance Office.

Plutarch said, ". . . It was for the sake of others that I first commenced writing biographies, but I find myself proceeding and attaching myself to it for my own; the virtues of these great men serving me as a sort of looking-glass, in which I may see how to adjust and adorn my own life."[17] These recollections should help us not merely understand our past, but also help us understand the present and find direction for the future. I hope that they will serve as a window in which 21st Century engineers and managers could adjust and adorn their own lives to make pioneering contributions in the 21st century.

When you read these accounts and reflect on their content, you will find that they are insightful stories about the theory, doctrine, and practice of national reconnaissance. You will find information about the lives of these individuals who were committed to a mission and brought the nation extraordinary engineering solutions to the complex challenges of

[16] Wolfe (1929, p. xv).

[17] Clough, Arthur Hugh (1992). "Introduction," *Plutarch's Lives of the Noble Grecians and Romans,*. The John Dryden translation, edited and revised by Arthur Hugh Clough. New York: The Modern Library, Book jacket.

national reconnaissance. Their accounts, typically, do not make explicit the lessons for success; rather, they offer personal recollections about experiences that can serve as models for future success. I challenge you, as you read about these extraordinary accomplishments, to discover for yourself these lessons.

*Robert A. McDonald*
*Chantilly, VA*
*December 2002*

—

## References

Baker, James G. (2002). Professor Emeritus at Harvard-Smithsonian Center for Astrophysics and a National Reconnaissance Pioneer. Telephone conversation with Cargill Hall, NRO Historian, 2 May.

Clough, Arthur Hugh (1992). "Introduction," *Plutarch's Lives of the Noble Grecians and Romans.* The John Dryden translation, edited and revised by Arthur Hugh Clough. New York: The Modern Library.

Hall, R. Cargill (2001). *Samos to the Moon—The Clandestine Transfer of Reconnaissance Technology Between Federal Agencies.* (NRO Monograph) Washington, DC: NRO Office of the Historian.

Hall, Keith (2001). "Reflections on the NRO of Yesterday and Today," *CSNR Bulletin*, No. 2001-1, Winter-Spring 2001.

Matthews, Christopher (1996). *Kennedy & Nixon: The Rivalry that Shaped Postwar America,* New York: Simon & Schuster.

McDonald, Robert A. (1997a). "Corona, Argon, and Lanyard: A Revolution for U.S. Overhead Reconnaissance." In R. A. McDonald (ed.) *Corona Between the Sun & the Earth—The First NRO Reconnaissance Eye in Space*, Bethesda, MD: American Society for Photogrammetry and Remote Sensing, pp. 61-74.

McDonald, Robert A. (1997b). "Corona's Imagery: A Revolution in Intelligence and Buckets of Gold for National Security." In R. A. McDonald (ed.) *Corona Between the Sun & the Earth—The First NRO Reconnaissance Eye in Space*, Bethesda, MD: American Society for Photogrammetry and Remote Sensing, pp. 211-220.

McDonald, Robert A. (1997c). "Lessons and Benefits from Corona's Development." In R. A. McDonald (ed.) *Corona Between the Sun & the Earth—The First NRO Reconnaissance Eye in Space*, Bethesda, MD: American Society for Photogrammetry and Remote Sensing, pp. 255-267.

Pedlow, Gregory W. and Welzenbach, Donald E. (1998). *The CIA and the U-2 Program, 1954-1974* (declassified version). Langley, VA: Central Intelligence Agency.

Phillips, R. Hart (1960). "Pieces of Rocket Landed in Cuba. Havana Sees 'Aggression'—U S. Booster Blown Up After Malfunction," *New York Times*, p. 10, December 2.

Plutarch (1992). "Alexander," *Plutarch's Lives of the Noble Grecians and Romans.* The John Dryden translation, edited and revised by Arthur Hugh Clough. New York: The Modern Library.

Rechtin, Eberhardt (1991). *System Architecting—Creating and Building Complex Systems.* Englewood Cliffs, NJ: Prentice Hall.

*Time* (1960). "Cuba—Spontaneous Combustion," *Time*, p. 34, December 19.

U.S. Navy Naval Research Laboratory, NRL, (2000). *Grab, Galactic Radiation and Background—First Reconnaissance Satellite.* Washington, DC: U.S. Naval Research Laboratory.

*Washington Post* (1960). "Rocket Parts Fall on Cuba Seen Likely," *The Washington Post*, p. A8, December 2.

Wolfe, Thomas (1929). *Look Homeward, Angel—A Story of the Buried Life.* New York: Charles Scribner's Sons. (Renewal copyright 1957, Edward C. Aswell as Administrator of the Estate of Thomas Wolfe.)

# Section 1 **Founders**

## Peacetime Strategic Reconnaissance as a Foundation for National Reconnaissance

R. Cargill Hall, NRO Historian

The concept of peacetime strategic reconnaissance that emerged in the first few years after World War II differed sharply from wartime convention. Army Air Forces (AAF) doctrine associated strategic reconnaissance with the important functions of identifying vital targets for aerial bombardment, pinpointing anti-air threats that might impede the attacking force, and performing bomb damage assessment. But the advent of powerful nuclear weapons prompted a few young AAF officers, civilian scientists, and engineers to consider strategic reconnaissance in quite different terms—as an intelligence tool that could be used to provide advance warning of a surprise nuclear attack on the United States. Strategic reconnaissance initially was termed, "pre-D-Day photography." Then it became known as "pre-hostilities reconnaissance," which also embraced the collection of electronic signals. Strategic reconnaissance then was redefined to mean "acquiring reliable intelligence about the economic and military activities and resources of a potential foreign adversary through periodic, high-altitude overflights in peacetime."

### National Leaders Embrace the Concept of Peacetime Strategic Reconnaissance

In the remarkably short span of fifteen years (from 1945 to 1960), during which the Cold War intensified and largely defined international politics, American leaders embraced this concept of peacetime strategic reconnaissance and made it national policy. Secretly, and at the direction of two presidents, they allocated resources and developed technology to put that policy into practice. First, they used military aircraft for peacetime overflights of potential adversaries. Next, they developed and employed specially-designed, high-flying reconnaissance balloons and aircraft. Ultimately, by 1960 instrumented satellites were gathering signals and imagery of activities around the Earth while operating unobtrusively in space.

The plans for clandestine strategic reconnaissance, the related national policy, and the systems created during the early days of the Cold War provided far more diverse and reliable intelligence than that of traditional indications and warning alone. Collectively, peacetime strategic reconnaissance also made possible the sizing of the United States military establishment to meet actual rather than imagined threats, and the concluding of verifiable arms control treaties with significant cost savings to the nation. Perhaps the greatest single contribution of peacetime strategic reconnaissance was the pivotal role it played in maintaining a delicate peace between nuclear-armed rivals during the forty years of the Cold War.

**Contribution to the Founding of National Reconnaissance**

People make history, but those engaged in the clandestine evolution of post-WWII strategic reconnaissance were not, and could not be, publicly recognized for their contributions until many years later. There are many individuals who made significant contributions to the founding of national reconnaissance. First among them is President Dwight David Eisenhower, who built on the precedent set by President Harry S Truman and made peacetime strategic reconnaissance national policy.

In late 1953 Eisenhower approved an intelligence program that conducted secret overflights of "denied territory." He authorized the funds, identified the teams, and supported unstintingly the development of what was to become the National Reconnaissance Program: the high-flying U-2s, the SR-71s, and space-based reconnaissance satellites. He restructured the Intelligence Community to employ these new systems. Eisenhower also directed the highest level of security for these programs, and, except for the U-2 that became public knowledge in 1960, never publicly divulged the existence of these programs or their secrets.

**The Founders of National Reconnaissance**

This tight security meant that those involved with the founding of this new discipline were hidden from public view. In the past decade, the contributions of the founders and those who supported them have become more visible. During its 40th Anniversary celebrations, the National Reconnaissance Office publicly recognized ten individuals who contributed to the founding of national reconnaissance.

These founders were a small number of advisors and technicians who counseled President Eisenhower and his immediate successors in this intelligence revolution, and furnished key technical ideas that made the revolution possible. They knew the success of peacetime strategic reconnaissance and saw its potential as a national reconnaissance program. These people helped form the foundation of strategic reconnaissance, and set the stage for what was to become the discipline of national reconnaissance

These ten founders interacted with and advised presidents at the critical time when national reconnaissance was becoming a reality. The presidents depended on their guidance in making decisions. It was the insightful and innovative thinking of the founders that moved Presidents Eisenhower, Kennedy, Johnson, and Nixon to support the emerging discipline of national reconnaissance. The recollections of some of these founders appear in the pages that follow.

# Merton E. Davies

Merton Davies was a designer of lenses for field astronomy and an expert on rocket systems. As an early member of the RAND Project, later to become the RAND Corporation, Davies entered the field of electro-optical reconnaissance while it was still in its infancy. Davies was influential in convincing the Air Force to include satellite reconnaissance in its long-term plans. He subsequently persuaded the Air Force to overcome the resistance to the idea of film-recovery reconnaissance systems such as the system used on Corona.

## Developing the Technology for National Reconnaissance[1]

In 1946 the Navy and Army were beginning to look at satellites, and the Air Force decided it wanted to be in that game as well. The Air Force asked Douglas Aircraft Company to write a report on the viability of satellites. The resulting report, *The Feasibility of an Earth Circling Spaceship*, was quite a famous analysis at the time. It was written by a number of Cal Tech professors and scientists, who subsequently started a follow-on effort known as Project RAND.[2]

It was 1948 when I first heard about Project RAND. I was in El Segundo, California at the time, working on Navy aircraft. I applied for a job with Jimmy Lipp, who headed the missiles division at RAND. I told him that I wanted to work on satellites, and he said, "Fine, come on over."

At RAND, we were going through a number of rocket design studies that looked extensively into different chemicals for propulsions and how to compute specific impulses. We also studied what structure should be used, and what types of high temperature materials like stainless steels should be used.

In 1951 RAND presented an updated view of what a satellite should look like. We recognized that a satellite program would be expensive, and we concluded that we would have to justify a project of this magnitude. RAND sent our report to the Air Force, and

[1] This section is based on an interview with Merton Davies at the National Reconnaissance Office Headquarters, 26 September 2000. Davies died on 17 April 2001 before he had an opportunity to review this chapter.

[2] RAND, which stands for "**R**esearch **AN**d **D**evelopment", was a project initiated in 1946 by the U.S. Army Air Force (precursor to U.S. Air Force) under contract with Douglas Aircraft Company. RAND became an independent, non-profit organization in 1948.

Figure 1-1. RAND's first headquarters on Fourth Street and Broadway in Santa Monica, CA. (Photo courtesy of the RAND Corporation.)

recommended that the Air Force develop a reconnaissance satellite. The Air Force came back to us and said, "Go back, do more work, and get more details."

**Project WS-117L Emerges**

We took on the challenge and prepared a more detailed report, which we delivered to the Air Force in 1954. Based on our recommendations, the Air Force decided to initiate a competition. Three companies competed: Lockheed, Martin, and RCA. The contract went to Lockheed after about a year of study. Money was scarce, and it took a year or so to get a contract to Lockheed. This was the beginning of Lockheed's involvement in satellite programs that eventually led to Project Corona.

In 1954 Amrom Katz came to RAND from the Wright Field Reconnaissance Laboratory in Dayton, OH.[3] He gave a presentation about reconnaissance satellites in which he concluded that satellites would not be able to produce usable imagery. To test his conclusion, he used a 35mm Leica camera with a 7.5mm focal length mounted on an airplane to take pictures of Wright Field and Dayton from 30,000 feet. He revised his earlier conclusion when he saw how much we actually could see in these low-resolution pictures. He ended up joining RAND, and we worked together for the next four years on advanced camera and image processing systems for balloons, aircraft, and satellite reconnaissance systems. One of our studies evaluated alternative designs and phasing of overhead systems for peacetime ("pre-hostilities") reconnaissance. We recommended a panoramic camera for high-altitude balloon reconnaissance.

**Balloon Overflights of the Union of Soviet Socialist Republics**

In 1957 we went to visit the Boston University Physical Research Laboratory, where Walt Levison worked.[4] He was working on a panoramic camera that he developed as the high-acuity Hyac-1 camera for a balloon reconnaissance program that flew operationally in 1958. These balloons overflew the Soviet Union as part of Project Genetrix. The camera's resolution was 100 lines per millimeter, and it produced beautiful photography from a very high altitude. After seeing those pictures, Amrom and I spent a lot of time in Washington trying to convince the Air Force (and others in the government) to fly more balloons.

We eventually received permission for more missions. However, instead of having the balloons travel from west to east—as was the norm—we decided to launch them from the Pacific and recover them in Germany. Everything was ready to go, and the timers were set. The Air Force tried to get approval from President Eisenhower to launch, but they could not get to him to act on that day. It took several days to get his approval, and when they finally

[3] Amrom Katz is a Founder of National Reconnaissance.
[4] Walter Levison is a Pioneer of National Reconnaissance. See chapter 31.

got the "Go," they forgot to change the timer that controlled when the balloon would come down. Consequently, the balloon drifted across Siberia, and landed in Poland instead of Germany. The Soviets recovered the balloon and displayed it in Russia. Eisenhower was very annoyed with the embarrassment.

**On To Satellite Reconnaissance**

In March of 1957 I was talking to Fred Wilcox, CEO of Fairchild Camera and Instrument Corporation, at a meeting of the American Society of Photogrammetry (ASP).[5] He wanted to show me a new camera Fairchild was going to mount on an airplane wing tip. The camera lens rotated across the flight path taking panoramic pictures.

While I was on the airplane returning to Santa Monica, I recalled that most of the spacecraft designs at RAND were spin-stabilized. If we employed the camera design that Wilcox had described at the ASP meeting, we could use the spin to perform the panoramic camera sweeps across the 90 degrees from flight direction. This was the origin of the spin-pan camera.[6]

I was still working with Amrom Katz at RAND in the summer of 1957. We put together a study with the help of some other RAND engineers in which we recommended the accelerated development of a film-recoverable reconnaissance satellite with a camera of longer focal lengths (12-inch, then 36-inch focal lengths) for improved ground resolution. We also recommended that the Air Force use a Thor booster for launch with a separate recovery vehicle because, by then, General Electric was producing recovery capsules that were tested and ready to go.

In November Fritz Oder agreed to start the project and formed the WS-117L program office.[7] By December the program office had provided our report to Lockheed, which looked into the concept. Lockheed proposed using an Agena as the spacecraft. This was a surprising proposal because the Agena was not spin-stabilized, and we recommended using a spin-stabilized platform. Instead, Lockheed proposed to stabilize the Agena and then "spin-up" the camera.

Lockheed signed a contract with Fairchild to develop the camera. I thought that was fair because Wilcox gave me the idea to use this rotating camera. Fairchild then designed a camera with a 12-inch focal length, and the Air Force ordered ten cameras from Fairchild.

We submitted the next report on 1 March 1958. The Air Force had trouble getting the funding, and the Central Intelligence Agency (CIA) became involved. Fritz Oder persuaded General Schriever to go to the CIA because the Agency could get funding without as much red tape as the Air Force.[8] At that time, the Department of Defense (DoD) was in the process of being reorganized, and the Secretary of Defense created the Advanced Research Projects Agency (ARPA) in February 1958 to centralize all DoD space research and development activities. This added another layer of bureaucracy.

The relative simplicity of our designs encouraged support for an accelerated film-recoverable reconnaissance satellite, and this support came from the Land Panel of the President's Foreign Intelligence Advisory Board. At the urging of Edwin Land and James Killian, in February 1958 President Eisenhower directed the CIA to take responsibility for developing this satellite (now called Project Corona).[9] When Corona returned its first images, I looked at them and said, "There's Walt Levison's camera."

---

[5] The American Society of Photogrammetry later changed its name to the American Society for Photogrammetry & Remote Sensing (ASPRS).

[6] Davies conceived and patented a spin-stabilized panoramic ("spin-pan") camera for synoptic earth imaging.

[7] Fredric "Fritz" Oder is a Pioneer of National Reconnaissance (inducted into the NRO Hall of Pioneers, 27 September 2000). See chapter 36 for his recollections.

[8] Bernard Schriever (retired as General, USAF) was responsible for the Intercontinental Ballistic Missile development program and the WS-117L satellite program.

[9] Edwin Land (see chapter 4) and James Killian are Founders of National Reconnaissance.

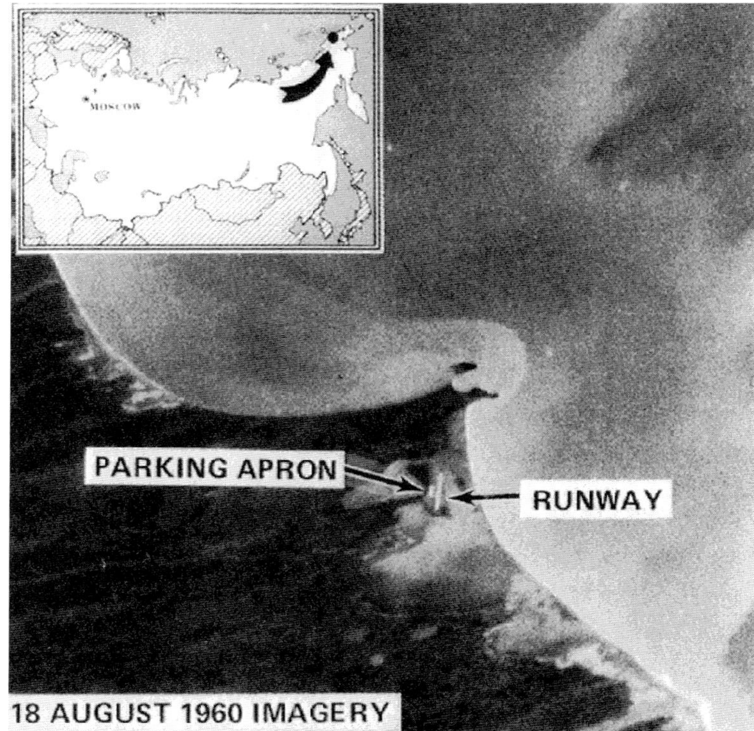


Figure 1-2. First image acquired from the Corona reconnaissance satellite of Mys Shmidta Soviet bomber base, 18 August 1960. (Photo courtesy of NRO History Office, likely CIA-National Photographic Interpretation Center photo).

The year 1958 was very difficult for me because I was not cleared for Corona, and Air Force Intelligence knew nothing about satellites. Air Force Intelligence came to RAND and asked, "Could you send us someone who knows something about satellites? We need technical help." RAND asked me if I wanted to go, and I said, "Yes!" I moved my family to Washington and worked for Air Force Intelligence at the Pentagon from 1959 to 1962.

This assignment brought me into contact with the National Reconnaissance Office (NRO) Program A, and its director, Brigadier General Robert Greer.[10] He sought my assistance in setting priorities, and asked for my recommendations about competing reconnaissance systems for NRO intelligence missions.

**Reviewing Proposed Camera Systems**

When I returned to California in 1962, General Greer and Amrom Katz were running the WS-117L program on the west coast, and they also were following the Corona program very closely. General Greer knew me from the Pentagon, and when he heard that I was back in California he asked me to examine a particular camera to determine how it compared with Corona's. I spent quite a bit of time going over it.

The system sent an image through six mirrors before it got to the film. I told Greer that I thought the system was a dog. The primary reason was that when there are six mirrors, about ten percent of the overall light is lost because some is lost in every mirror. The next day Greer canceled the project. That alarmed me because there were a dozen people working on that effort. I thought, "Oh my, I'm not used to that kind of influence!"

[10] Robert E. Greer (retired as Major General, USAF) served as the first director of the NRO's Program A.

On another occasion, Greer asked me to look at Lanyard photography and compare it to the Corona imagery.[11] I thought the Lanyard imagery was very poor, and it had various problems. From that time on, Greer's office kept a line of communication open between his advance planning people and RAND. Brockway McMillan subsequently sent RAND a letter saying that he wanted me to spend half my time in Washington working for him on satellites.[12] I agreed to this request, and I commuted back and forth between the coasts.

**Involvement With NASA Space Activities**

In the mid-1960s, I became less involved with the national security reconnaissance activities of the NRO and more involved with the civilian space activities of the National Aeronautics and Space Administration (NASA). In the late 1960s, I found myself involved with NASA's unmanned missions to Mars (specifically the Mariner VI and VII programs), because I knew some of the people involved from Cal Tech. They asked me to join the imaging team and I began to increase my involvement with NASA's exploration of the solar system. Nevertheless, I still co-authored a number of papers on the Soviet space programs and published them in *Science*.

By 1970 I retired from the national reconnaissance field, although I remained involved with developments in space through my participation with the Voyager group. That was a beautiful mission and a grand tour. Voyager went to the outer planets and collected a large amount of data, which proved to be both fascinating and fun. Even though I tried to back off, I found myself continuing to participate on the Galileo imaging team.

**Conclusion**

I have been very fortunate to be involved with work that was fun, interesting and exciting. I always enjoyed going to work. Over the years I have visited quite a few places with astronomical observatories, and one time I even took a granddaughter with me. I always thought that is what life should be: doing those things that I most enjoyed.

—

## When Your Husband Has a Secret Career[13]

Merton loved his work. By education and training he was a mathematician, and always was exacting in his work. He took to his work like a duck to water, and always was fascinated with things related to outer space. If there were anything about his work that he did not like, he did not tell me. Of course, he did not tell me much of anything!

I never knew what Merton was working on because of the classified nature of his work. I do not believe being unable to talk about work bothered Merton. Actually, I always felt that he enjoyed being secretive. Often we would joke that I took care of things on earth, and he took care of things outside of the earth in space.

Merton did bring work home once. He was developing a camera, and he had the family out in the backyard flashing things so that he could test something about the camera. We did not know what we were doing. We were just out there holding flashcards while he focused on things. That was life with Merton.

---

[11] Lanyard was a 1960s satellite imagery reconnaissance system that was designed to collect high-resolution photographs. The system was expected to provide two-foot resolution imagery, but only collected a best resolution of six feet. It flew one mission during the summer of 1963.

[12] Brockway McMillan was the Director of the NRO from 1963-1965.

[13] This section is based on an interview with Louise Davies, wife of Merton Davies, at the National Reconnaissance Office Headquarters, 26 September 2000.

Figure 1-3. Founder Merton Davies and his wife, Louise Davies, in front of the NRO 40th Anniversary Commemorative Plaque at Tower 4 Entrance, NRO Headquarters, 26 September 2000. (Photo by Candi Campbell, NRO Visual Design Center.)

**Living With Secrecy**

Everyone who lived through the Second World War understood the importance of secrecy in matters of national defense. I always was aware that Merton worked with classified information, because investigators kept interviewing us about Merton's colleagues. I guess they were doing background investigations. The investigators would ask questions like, "Does this person or that person have a sailboat?" I can't imagine what owning a sailboat had to do with classified work, but apparently it was important because that is the kind of questions we were asked.

I found that when it came to questions and answers, I had to be careful about what I said to people. For three and a half years we lived in Washington, DC where we entertained quite a bit. We had a party one night with some generals, senators, and co-workers. We were having a good time, and I said something about U-2s and Cuba. All at once the whole room became quiet. My husband pulled me aside into another room and asked, "Why did you say that?" I said, "I read about it in *The Washington Post* today!" Nobody told me anything so I did not know what to be quiet about. That turned out to be an advantage.

*Everyone who lived through the Second World War understood the importance of secrecy in matters of national defense.*

If friends or family members asked me what Merton did for a living, I honestly could say, "I haven't the vaguest idea." Aside from the secrecy, his work was so complicated that if he had been speaking in Chinese, it would not have made any difference. Whenever he spoke of work-related activities, I could not understand what he was talking about. If Merton's colleagues would visit to chat about work interests, I would soon go to bed. I never could understand a word they were saying to each other.

I probably was more annoyed that other people knew about the secretiveness of my husband's work. I always thought that people would share when they were married, but I learned that was not the case with a husband who worked with classified information. It does test a marriage when there is so much secrecy. At the time I kept wondering, "Why does Merton have to be so secretive; why is secrecy such a big deal?" It seems to me that all this secrecy strained people's lives. Even now that I have learned that his work involved national security, I still wonder about the details.

### Secrecy and The Family

The children never knew what their Dad did, or where he was going on business trips. I would take Merton to the airport, and the children would ask, "Where did Dad go?" I would say, "To the airport." I did not know where Merton was going myself, so I could not tell the children. As far as we knew, it was the airport. I dropped him off there, and I picked him up there.

In spite of all the secrecy, Merton was supportive of the children, and he was good with them. When they were young, he would play with them. The children had rockets that they could shoot off in the desert, and Merton and the children had fun with these. At the time, we all found it very interesting to look into the sky and watch Sputnik pass over.

### Conclusion

Secrecy does cause problems. I raised the children and lived my life. Merton did his secret work and lived his life. In a sense, this was probably a blessing. It forced me to become a very independent person. I developed my own career after a while. I went into real estate, and I became very busy. I was successful, I might add. I guess it worked out all right; we were married for 54 years. My advice for spouses who are in the same situation is to acquire your own career, develop your own interests, take care of your children, and pay attention to your own activities.

—

## Founder Award Presentation and Citation

Figure 1-4. Founder Merton Davies (second from the left) being recognized at the 2000 Pioneer Recognition Ceremony, 27 September 2000. The founder award plaque was presented by DNRO Keith Hall (left) and DCI George Tenet (third from the left). (Photo by Sara Judy, NRO Visual Design Center.)

**Merton E. Davies**

An engineer, reconnaissance system designer, imagery interpreter, and space cartographer, Mr. Merton Davies participated in all early USAF reconnaissance studies and planning, including the Leghorn-directed Intelligence Development Planning Objectives. He invented the Spin-Pan camera, and collaborated in the film-recovery satellite proposal adopted by the USAF that became Corona. Employed throughout his career at RAND, he continued to serve on NRO and other advisory panels that established reconnaissance requirements and advised on competing reconnaissance systems.

*Career in National Reconnaissance: 1948-1975*

# Sidney D. Drell

Sidney Drell, a celebrated theoretical physicist at Stanford University, has been a scientific advisor to senior United States national security policymakers in presidential administrations from Dwight D. Eisenhower through George W. Bush. Drell's reconnaissance achievements have ranged from making contributions in the physics of detecting ballistic missile launches to the technology of photographing the earth from space. He has encountered these and other technical intelligence challenges and worked with scientists and decisionmakers who brought resolution to the problems.

## Reminiscences of Work on National Reconnaissance[1]

I was introduced to the world of classified national defense problems in 1960, the year JASON was created.[2] The purpose of forming the JASON group was to enlist fresh, then-young and promising scientific talent to work on problems of importance for United States national security. We were in the new age of nuclear weapons, space, and intercontinental missiles, and the challenges they presented to formulating national security policy were significant. The great physicists and scientists who contributed to winning World War II with the developments that led to radar, the atomic bomb and anti-submarine warfare capability had grown older and moved on to other responsibilities. There was a need to train a new generation of scientists who would be willing to commit time and effort to areas in national security that were presenting new and formidable challenges.

I was one of several dozen scientists, mainly physicists, who were invited to join JASON at its inception. I accepted. I then continued to work as a member of JASON and on a variety of government advisory panels where we addressed problems of national security—many of great technical interest—and some of critical importance to our national security.

[1] This section is based on written input that Sidney Drell submitted to the Center for the Study of National Reconnaissance.

[2] JASON is an independent, senior-level scientific and technical "think tank" composed primarily of scientists and academics representing the most prestigious universities and research institutions in the U.S. The term JASON is not an acronym, nor does it have any particular meaning. The principal function of the JASON group is to provide senior-level government managers, primarily in the Departments of Defense and Energy and the Intelligence Community, with scientific and technical expertise.

**Challenge of Detecting Ballistic Missile Launches**

The first problem I worked on during the first JASON summer study in 1960 concerned exploring the possibility of deploying sensors on earth-orbiting satellites to obtain early warning of ballistic missile launches. Together with my friend and JASON colleague, Malvin Ruderman, I studied the infrared (IR) radiation generated in the atmosphere, at high altitudes of roughly 90 kilometers, by an intense burst of x-rays. This in itself was an interesting science problem, but it was motivated by a more immediate purpose. When a megaton-level nuclear bomb explodes at high altitude, it suddenly dumps a lot of x-ray energy into the atmosphere.

That energy initiates a series of interactions that transform a large number of the $N_2$ and $O_2$ molecules that are prevalent at around 90km altitude into nitrogen oxide (NO) molecules in vibrationally excited states. Because they are non-homopolar, the excited NO molecules radiate extensively in the IR spectral region when they decay back down to their ground states. The questions to be answered were how much NO was formed and how extensive, dense, and long-lasting was the IR blanket ("red-out") that was created. This would solve the problem of how effective a high-altitude precursor nuclear burst would be in blinding the IR sensors of a satellite. This system was known as the Missile Defense Alarm System (Midas) in its original incarnation, designed to detect the launch of Soviet Intercontinental Ballistic Missiles (ICBM).

Figure 2-1. Midas 1 lifting off from Launch Complex 14, Cape Canaveral, 1960. (Photo courtesy NRO History Office, likely U.S. Air Force photo.)

Ruderman and I assembled everything known in 1960 about the relevant chemical reactions, high-altitude atmospheric parameters, and the like. We calculated energetically and concluded that it would take a lot of megatons to make an IR cloud that would last long enough and extend far enough to be of strategic value by preventing Midas from detecting, and hence denying us early warning of ICBM launches.[3] For good reason, the Midas program went ahead and has been of great value to the U.S.

As a consequence of this work, I was invited to join the Strategic Military Panel of the President's Scientific Advisory Committee (PSAC), then chaired by the President's Science Advisor,

[3] The work of Drell and Ruderman in this area was published in the scientific journal, *Infrared Physics* (vol. 2, no. 189, Sept-Dec, 1962).

Figure 2-2. Early photo of CIA Headquarters with a portion of the Bureau of Public Roads complex in the upper right corner. (Photo courtesy of NRO History Office, CIA photo.)

Jerome Wiesner. Call it entrapment, commitment, or whatever, but I have remained actively involved in technical national security work for the United States.

### The World of Satellite Reconnaissance

The extent of my government involvement increased by a quantum leap in the fall of 1963 when I was introduced to the technical possibilities of doing photoreconnaissance from space-based satellite systems. Photoreconnaissance satellites, together with signals intelligence (SIGINT) and electronic intelligence (ELINT) satellites that constituted the so-called "national technical means," could pierce the Iron Curtain that had been erected by an obsessively secretive Soviet government. These systems represented a big step toward achieving the Open Skies that President Eisenhower had first called for in 1955. Among their other values, these satellites opened the door to arms control negotiations based on effectively verifiable treaties.

The first generation of photoreconnaissance satellites, known as Corona, was declassified in 1995.[4] I was introduced to this amazing achievement by Bud Wheelon in the fall of 1963. He was then the Central Intelligence Agency (CIA) Deputy Director for Science and Technology (DDS&T) and he telephoned me at Stanford University and asked me to come to Washington.[5] He said he wanted to show me something that was important, and he needed my help. I dutifully obeyed and found my way to the CIA Headquarters Building, then identified by the words Bureau of Public Roads, or something similar on a sign on the George Washington Parkway.

[4] For a discussion of the decision to declassify the Corona program, see "The Declassification Decision," In *Corona Between the Sun and the Earth—The First NRO Reconnaissance Eye in Space* (R.A. McDonald, Editor), Bethesda, MD: American Society for Photogrammetry and Remote Sensing, 1997. *Corona Between the Sun and the Earth* also includes a comprehensive collection of other Corona-related articles, many of which were written by participants in the program. An associated Corona reference is "Corona: Success for Space Reconnaissance, A Look Into the Cold War, and A Revolution for Intelligence," in *Photogrammetric Engineering and Remote Sensing (PE&RS)* (vol. 61, no. 6, June 1995).

[5] Albert D. (Bud) Wheelon served as CIA's DDS&T from 5 August 1963 to 26 September 1966. He is a Pioneer of National Reconnaissance (inducted into the NRO Hall of Pioneers, 27 September 2000). See Chapter 42 for his recollections. Also see Wheelon's article in *Physics Today* (vol. 50, page 24, 1997) for his overview of the Corona project.

I learned, to my amazement, that we could photograph the earth with fairly good resolution from satellites. I also learned that these Project Corona satellites were producing a corona-like electrical discharge that was streaking the film, thereby diminishing the quality of the images and reducing the intelligence value of the photography. Concerned that this problem was not getting the priority attention it needed, Bud had assembled a team of industrial scientists from each of the corporations that had contributed major components of the Corona system, including Itek, Eastman Kodak, and Lockheed. He asked me, as an independent academic, to lead the team in a detailed technical investigation into what was causing this problem and how to fix it. Thus began a most extraordinary experience.

The industrial team that Bud assembled was a group of scientists and engineers, a group that was as outstanding as any I ever have known. I added to my panel two superb physicists, whom I knew as personal friends and respected highly—Luis Alvarez and Malvin Ruderman. The investigation was to be in a scientific realm that was then new to me, and I wanted my two colleagues to add to my confidence that I would stay on track and not go off on useless tangents. Our effort was intense, and lasted about four months, during which time I essentially lived in Washington, except for weekends at home and a weekly Friday physics lecture at Stanford.

Our work was successful in resolving the Corona project's corona problem. In order to avoid the buildup of electrostatic charges, we recognized that extreme care was required to maintain a clean vacuum and balanced electrical and thermal conditions. Particular attention had to be given to the condition and type of material of the rollers across which the film was spooled. All these factors were important to avoid the electrical discharges on the ultra-thin film speeding rapidly across the rollers. We also gained further understanding of the limits in optical resolution that could be achieved with a system based on Corona technology, before moving ahead with more advanced ones. At the conclusion of our study, I briefed our findings to the Director of Central Intelligence, John McCone (frequently referred to in code those days as Earthquake McGoon), and members of the Overhead Reconnaissance Panel chaired by Edwin Land of Polaroid.

**The Land Panel**

I was then made a member of the so-called Land Panel that advised the White House on overhead photoreconnaissance. We reviewed and stimulated the development of new technologies and systems to provide timely, high-quality photoreconnaissance from space. This proved to be a fascinating activity throughout the panel's existence for a decade. Someday, the subsequent advances beyond Corona will have their day in the sun, and people will be astounded to learn of the enormous achievements and incredible value of scientists and engineers in the Intelligence Community.[6]

A critical activity of the Land Panel during the Nixon Administration had to do with the possibility of gaining real-time intelligence from space instead of waiting to recover and develop exposed film, and the choice of a technology to achieve this goal. On one occasion, Dick Garwin and I, with extensive information on this subject as members of both PSAC and the Land Panel, successfully intervened with National Security Advisor Henry Kissinger to argue the case for selecting electro-optical technology and reversing what we believed to be a flawed decision that favored an alternative technical choice.[7] The result proved to be of great value in the years since it was implemented.

[6] Subsequently, Drell also was appointed as a member of the PSAC and served under Presidents Lyndon Johnson and Richard Nixon from 1966-1971.

[7] Richard L. Garwin is a Founder of National Reconnaissance (See chapter 3 for his recollections).

**More of the People**

Throughout my experience on the Land Panel and the PSAC, I had the privilege of working with Don Steininger, a retired Lieutenant Colonel and physics Ph.D. who was a staff man to the President's Science Advisor and a person of extraordinary wisdom and effectiveness. He knew when and how to push the "system" without closing down cooperation. He worked with great dedication, and he was respected for his integrity and objectivity. He was one of my cherished friends, and a colleague in these endeavors for more than a decade. He died tragically at an early age because of cancer, but only after being awarded a National Intelligence Medal for his invaluable service. In addition to Bud Wheelon and Don Steininger, I also want to mention one of Bud's successors as DDS&T, Les Dirks, who was a brilliant government servant in intelligence.

> *Someday, the subsequent advances beyond Corona will have their day in the sun, and people will be astounded to learn of the enormous achievements and incredible value of scientists and engineers in the Intelligence Community.*

I recall, from my early years of involvement, the very strained and occasionally harmful rivalry between the CIA and NRO (then very young and still very secret) for control and turf in the photoreconnaissance program. At times, it was hard to tell who was the enemy. I am glad that this rivalry is a memory of the distant past.

**Challenges for Today and The Future**

Improving intelligence through the better application of science and technology has been a major theme of my government work. In 1988, I had the good fortune to accompany a bipartisan delegation of five Senate leaders, including then-Senator Bill Cohen, to Moscow to meet then-President Gorbachev, Foreign Minister Shevardnadze, and Marshal Akhromeyev. I was one of several advisors to the delegation, and as an expert on technical national security issues, my role was to keep discussions on defense, nuclear policy, arms control, and the like on the straight track with facts. During that week in Moscow, I had a number of opportunities to talk with Cohen about my concerns at the loss of aggressive leadership in our government driving the reconnaissance and intelligence programs to the frontiers of what was technically possible. The Land Panel no longer existed, and the science advisory apparatus in the White House, diminished in importance, had almost no involvement in national security issues.

Upon our return, Bill Cohen, then the ranking minority member of the Senate Select Committee on Intelligence (SSCI), called me to Washington to meet with him and Senator Dave Boren—then its chairman—to discuss my proposal for the SSCI to create a Technology Review Panel to fill this vacuum. In my view, it was needed and could be of great value both in photoreconnaissance as a successor to the Land Panel, and in signals intelligence as a successor to the panel that had been chaired by Bill Baker of Bell Telephone Labs.[8] Senators Boren and Cohen agreed with my concerns and asked me to form such a technology review panel and serve as its first chairman. I did that for four years during the administration of President George H.W. Bush, and we made some genuine contributions. That panel still exists.

In early 1993, I received a telephone call from Admiral Bill Crowe, whom President Clinton had asked to chair the President's Foreign Intelligence Advisory Board (PFIAB). Bill called to ask me if I would be willing to serve with him on the PFIAB as his nuclear expert. I felt greatly honored and said yes. Bill and I had gotten to know each other very well during the 1990-93 period during our collaboration with McGeorge Bundy in writing a

---

[8] William O. Baker is a Founder of National Reconnaissance.

book entitled *Reducing Nuclear Danger: The Road Away From the Brink* published by the Council on Foreign Relations.

My eight years on the PFIAB were interesting and rewarding. A major theme of my work on PFIAB was emphasizing how important the role of science and technology for U.S. intelligence capabilities has been in the past, and most certainly will be in the future.

—

## Founder Award Presentation and Citation

Figure 2-3. Sidney Drell (second from left) being recognized as a founder of national reconnaissance, 27 September 2000. The founder award plaque was presented by DNRO Keith Hall (left) and DCI George Tenet (third from the left). (Photo by Sara Judy, NRO Visual Design Center.)

NATIONAL RECONNAISSANCE

FOUNDER

**Sidney D. Drell, Ph.D.**

A theoretical physicist at Stanford University, Dr. Sidney Drell served on the President's Foreign Intelligence Advisory Board and the President's Science Advisory Committee in intelligence advisory roles. He served as a key scientific consultant to Program B, and served on the Technology Review Panel of the Senate Select Committee on Intelligence where he was instrumental in securing approval and support for several NRO special projects.

*Service to National Reconnaissance: 1960-2000*

# Richard L. Garwin

Richard Garwin served the National Reconnaissance Office and the nation in a wide range of distinguished scientific, technical, advisory, and leadership positions. Dr. Garwin dedicated himself to national service and to helping the United States develop advanced technical intelligence collection systems, weapon systems, analytical techniques, and innovative solutions to complex problems. Several United States Presidents enlisted his scientific knowledge and uncompromising integrity in the cause of national security.

## Dedication to National Service[1]

I was one of a small group of physicists who were trained immediately after the Second World War. We were too young to serve in the war, but our careers were profoundly affected by it. We were trained by great men of science who made remarkable contributions during the war, and who were our professional and personal mentors. As their graduate students and protégés, we could see for ourselves their commitment to this country. Leading by example, they shared with us their convictions about the importance of government service.

We learned two important lessons from these relationships. The first was how to do physics. The second was a conviction that we owe a solemn debt to the country to make our time and our ability available as needed. When the time came for me to pursue a career in industry, I did so with the condition that I would be allowed to dedicate one-third of my time to national service. Our mentors also helped us develop a sense of confidence that almost anything could be done if American technology were properly directed and supported. These men led by example with their deep commitment to, and involvement with, supporting the national security of the United States.

### Academic Beginnings

I earned a bachelor's degree in Physics in 1947 from what was then called Case School of Applied Science (Case Western Reserve University) in Cleveland, Ohio. I became an experimental physicist after earning a Ph.D. from the University of Chicago in 1949. For my Ph.D. thesis I worked on some aspects of radioactive decay. Enrico Fermi was my mentor, and he was a great theoretical and experimental physicist.

[1] This section is based on written input that Richard Garwin submitted to the Center for the Study of National Reconnaissance.

In late 1949, I joined the Physics faculty of the University of Chicago in large part to work with their particle accelerators. The university paid me a salary for nine months of the year, but my family and I had to eat for twelve months per year, so I sought challenging summer employment. In May 1951, during the Korean War, I spent a month in Korea and Japan with Professor Joseph Mayer (a chemist at the University of Chicago). We conducted a study that examined how technology could be applied to support the newly formed Tactical Air Command.

For many summers I worked as a consultant to the Los Alamos National Laboratory, where my projects focused on nuclear and thermonuclear weapons. One of my first discoveries was long-range and long-enduring fratricidal effect of a nuclear explosion on another nuclear weapon. I also contributed to the design of the first thermonuclear weapons, especially "Mike", which was detonated on 1 November 1952, with a yield of ten megatons.[2]

*Eisenhower knew the devastation of war, and he felt that some causes of war could be avoided by better mutual understanding based on peacetime strategic reconnaissance.*

**Joining IBM and Pursuing Opportunities in National Security**

While I was on the physics faculty at the University of Chicago, I was unhappy with the idea of scheduling my experiments on the particle accelerator six weeks in advance, in competition with my colleagues. As a result, I decided to transfer to another field of physics where I could proceed at my own (and, I hoped, a quicker) pace. I turned to work in low temperature physics—superconductors, liquid and solid helium. I decided to pursue an opportunity to work at a new laboratory of the IBM Corporation in New York. Since 1945, this new IBM laboratory had been the "IBM Watson Scientific Computing Laboratory," but it was now moving on to solid-state physics.

As a condition of employment, IBM agreed to permit me to devote one-third of my time to working for the United States Government on projects related to national security. I found national security matters to be interesting, and I seemed to have a flair for this work. For more than forty years, IBM honored the company's part of the bargain, which meant that I committed about half of my time to such activities (counting vacations and weekends spent working on these matters). Furthermore, the agreement stipulated that IBM would not be told what I was doing—at least not by me—although they would receive my consulting fees and travel reimbursement, if any.

Soon after I arrived at IBM, the MIT crowd (Jerome Wiesner and Jerrold Zacharias) put pressure on Tom Watson, Jr. (the head of IBM), to assign me to full-time work on air defense issues. Specifically, the technical challenge was to extend the semi-automated system to the sea lines of approach to the United States and Canada. I spent about half of my time working on this project in the Boston area for more than a year, as I recall. It was in this context that I realized the crucial nature and the value of strategic intelligence and national reconnaissance.

**Advising the White House**

The late 1950s and 1960s saw a very active and influential eighteen-member President's Science Advisory Committee (PSAC), with numerous panels of outstanding experts on subjects ranging from antisubmarine warfare to insecticides and pesticides. I was a member of that committee for eight years, and I served for about fifteen years as a member (and often chair) of various military panels of the PSAC.

President Eisenhower elevated the Science Advisory Committee to the presidential advisory level because he needed the highest quality impartial technical advice available. He often

[2] Mike ("m" for megaton) was the code name for the first true thermonuclear test conducted as part of Operation Ivy.

turned to the PSAC for help, particularly in military matters. One of the most notable of the subcommittees was the "Land Panel" headed by Edwin H. (Din) Land, on which I served.[3] Land chaired the intelligence capabilities section of the Technological Capabilities Panel organized in July 1954. This panel was created in response to President Eisenhower's challenge to the Science Advisory Committee in March 1954 to use new technologies to counter the potential of Soviet nuclear attack on the United States by intercontinental bomber.

When I worked with various technical panels during the development or problem phase of some system, I found that people in the national reconnaissance field would go out of their way to help us understand their problems and prospects. They were focused on achieving a capability at a schedule and cost, and reached out for our help in meeting these goals. Those of us on the panels such as Edwin Land, Bill Baker, Ed Purcell, James G. Baker, and the next generation of such "outside insiders" as Sidney Drell, were not idle givers of advice[4] They turned their great talents and energy to addressing the problems facing our nation and the world. The intervening years have shown that their proposed solutions were solutions that worked.

**National Reconnaissance and "Open Skies"**

As the leader of the Allied forces in Europe following the invasion in 1943, Eisenhower knew the devastation of war, and he felt that some causes of war could be avoided by better mutual understanding based on peacetime strategic reconnaissance. In 1955, President Eisenhower publicly proposed the concept of "Open Skies" to the Soviet Union. The program would have allowed unarmed Soviet aircraft to fly all over the U.S., photographing everything below. Similarly, our aircraft would be permitted to overfly Soviet territory. The objective was to build mutual confidence that neither side was preparing to launch a surprise nuclear first strike, which in turn would contribute to greater strategic stability.

Anticipating the likely Soviet rejection of the Open Skies proposal, President Eisenhower already had initiated a highly classified project to develop the U-2 high-altitude reconnaissance aircraft. Leadership responsibility for development and operation was given to the Central Intelligence Agency (CIA) under Richard Bissell. The first U-2 overflight of the Soviet Union took place in 1956. Security was so tight that even as late as 1 May 1960, when U-2 pilot Gary Powers was shot down near Sverdlovsk, it is likely that fewer than 400 people in the U.S. knew of the U-2 program. In fact, probably a lot more people in the Soviet Union knew of it by then!

In 1956 we could see some of the military capabilities of the Soviet Union, but we did not know their intentions either with regard to a potential attack on the United States, or toward the possible conquest of Western Europe. So, in that year the PSAC recruited me to join William O. Baker, head of Bell Telephone Laboratories, on a new panel he was to chair. The Baker committee was concerned primarily with the domain of the National Security Agency. In the course of our activities, we supposedly received a "no holds barred" introduction to the intelligence collection activities and capabilities of the U.S. vis-à-vis the Soviet Union. Oddly however, there was no mention of the U-2!

**Satellite Reconnaissance and Unrestricted Overflight**

We recognized that the U-2 was not going to provide a long-term solution to U.S. strategic reconnaissance requirements, and this contributed to the research and development of earth-orbiting satellites. I was involved with the CIA's development of the Corona photoreconnaissance system. With this satellite, the CIA was able to obtain repeated coverage of the rocket

---

[3] Edwin "Din" Land is a Founder of National Reconnaissance. See chapter 4.

[4] Sidney Drell is a Founder of National Reconnaissance. See chapter 2. William O "Bill" Baker and James G. Baker are Founders of National Reconnaissance. Edward Purcell is a Founder of National Reconnaissance. See chapter 5.

test facilities from which the Soviets launched Intercontinental Ballistic Missile (ICBM) flights. We received a great deal of information from these flights, and we could accurately identify each missile type and measure its performance. What we could not do was associate a particular flight with one of the many launch pads. We needed to do this because we wanted to associate the correct missile type with the operational sites that were then being built at a rapid pace throughout the Soviet Union. Working with John Sorrels and Bill Perry, we solved this problem by using another data source. Consequently, we were able to identify the missile type for each ICBM launch pad built in the Soviet Union, and did so with high confidence.

Corona was an invaluable resource, but there is more to satellite reconnaissance than pretty pictures. Both the U.S. and Soviet Union employed signal intercepts by satellites. There were significant tactical and strategic benefits derived from locating Soviet radar transmitters, in addition to identifying their signal characteristics. I helped show the national reconnaissance community how to overcome some of the technical difficulties, and the solution was both elegant and practical.

In space-based intelligence collection a form of the Golden Rule should apply, "Do not do unto others what you would not have done unto you." So President Eisenhower was probably delighted when, at the 1960 Paris Four Power Summit, French President DeGaulle questioned Khrushchev about why he was so concerned about the U-2 overflights. DeGaulle asked, "A Russian Sputnik passed over France last week. How do I know it did not have cameras?" Khrushchev replied, "Airplanes, nyet; Sputniks, OK." Eisenhower would have preferred Krushchev to accept his Open Skies proposal, but the Soviet leader's public acceptance of space-based reconnaissance was very valuable in terms of establishing the precedent of freedom of space.

**Value of Intelligence and Interlude in Geneva**

By the start of the 1960s the value of intelligence, and particularly strategic intelligence, had become clear to me. My friend Bud Wheelon observed that, "Intelligence is the most highly leveraged component of national security."[5] I first met Bud Wheelon in the second half of 1958. We both were sent by the State Department as the two most junior members of the American delegation to the Surprise Attack Conference in Geneva, Switzerland.[6] Bud and I became lifelong friends during this conference. The conference represented the first attempt to sit down with the Russians to try to slow the pace of deployment of strategic nuclear weapons. The conference itself produced very few results, but it set in motion an incremental process that eventually led to the SALT treaty and the START agreements for limiting and reducing strategic weapons.

While in Geneva, Bud and I collaborated on a hypothetical research question that asked how many (cooperative) aircraft overflights would be required in order to detect a specified percentage of an assumed number of airfields (with an assumed fraction of cloud cover). We borrowed the services of the Ferranti Mercury computer at the European Laboratory for Particle Physics in Geneva in order to do this calculation, and I remember programming the computer by punching holes in 5-track teletype tape.

**Lessons Learned**

I have had opportunities to participate in a range of issues and technical challenges from a high level, and across several scientific disciplines. When we place highly trained minds in the right places to see the "big picture," we afford ourselves the opportunity to solve similar problems in diverse areas. I think because I had these opportunities I made a contribution

[5] Albert "Bud" Wheelon is a Pioneer of National Reconnaissance (inducted into the NRO Hall of Pioneers 27 September 2000). See chapter 42 for his recollections.

[6] The full name was the Conference of Experts for the Study of Possible Measures Which Might Be Helpful in Preventing Surprise Attack for the Preparation of a Report Thereon to Governments, in November 1958.

by helping to bring people together to find solutions to their problems. As a result of my experiences, I have learned several lessons that I offer for consideration to current and future practitioners of national reconnaissance.

In general, I tried to listen to the people who were doing the work in order to distill what their objective was, and to see whether their methodology was well suited to their goal. At times I would suggest a way to overcome obstacles or perhaps even to skip entire steps, components or stages using the rationale that if you do not have to do something you don't have to worry about the problems associated with getting it done.

> *DeGaulle questioned Khrushchev about why he was so concerned about the U-2 overflights. DeGaulle asked, "A Russian Sputnik passed over France last week. How do I know it did not have cameras?" Khrushchev replied, "Airplanes, nyet; Sputniks, OK."*

Some of the lessons I have learned can be illustrated in these six concepts: learn before deployment, exercise caution when evidence is absent, exercise caution when evaluating hypotheses, resolve the problem of ambiguous data, balance military and national intelligence requirements, and view intelligence as a tool.

*Learn Before Deployment.* I tried to look ahead to see how some element, component, or computer program would function when it met the real world. I developed a reputation as being a contrarian because I frequently challenged assumptions and design choices. I felt it was better to address the problems in a lab or conference room than to encounter a problem for the first time while on orbit. When I thought I had an insight to offer, I would attempt a different analysis to see whether the outcome was the same. My purpose was not to average two approaches but to be alert to any discrepancy in the results that could help identify a flawed approach. The same concept of comparing different analytical techniques also applies to the acquisition, processing, and analysis of intelligence.

*Exercise Caution when Evidence is Absent.* About fifteen years ago I was one of a small group convened in the basement of the Pentagon to study the intelligence obtained on Soviet directed energy weapons. Some useful intelligence had been collected, but we were lacking information in several areas that were deemed important by the group. I was incensed by the suggestion that we view the "absence of information" as an indication that there was no constraint on our estimate of what the Soviets might have accomplished. While I agreed that the "absence of evidence" should not be viewed as "evidence of absence," at the same time neither is the "absence of evidence" evidence of the existence of a major, concerted, and secret program. Additionally, there should be healthy skepticism when a participant making the assertion of the worst-case scenario also happens to be interested in funding his own counterpart program.

*Exercise Caution When Evaluating Hypotheses.* Quite properly, leaders and managers have long sought a process that would evaluate all available information (positive and negative), as well as account for the absence of information, to provide a quantitative estimate (with confidence estimates) of the validity of a hypothesis. I was once asked to attend a presentation by a CIA contractor hired to develop just such a process. It was easy to show that, if applied, the methodology would increase confidence toward any hypothesis with the passage of time and the accumulation of evidence (even of "noise"). Of course, even such a flawed methodology might be prevented from doing harm by supporting in parallel both the hypothesis and its inverse. For example, by applying the methodology over a period of time, two contradictory hypotheses would have been supported with equal confidence: First, that the SS-8 was a large missile; and second, that the SS-8 was a small missile.

I offer an example to illustrate the point. On 22 September 1979, an interesting signal was received from a satellite showing what seemed to be a typical double-humped curve of light output from the Earth's surface over time, similar to what we would expect to see if there had been an atmospheric nuclear test. Satellites had detected this phenomenon some forty times previously. Three days later, I believe, I was invited to CIA Headquarters together with Harold Agnew (Director of the Los Alamos National Laboratory) and Steve Lukasik (former Director of the Advanced Research Projects Agency), as a three-person panel to render some judgment as to whether this actually had been a nuclear test. There were no other data at that time, and there were little negative data because there had not been enough time to determine whether other detection systems had received similar signals or not. My view was that we should wait for more data. Pressed as to my best estimate, I suggested there was a 60% probability that there had been a nuclear test, but with such uncertainty in this estimate it could very well be that there had not been a test.

Some months later the President's Science Advisor, Dr. Frank Press, convened a panel on which I served. The panel was chaired by Jack Ruina of MIT and included, among others, W.K.H. Panofsky, Luis Alvarez, and Rich Muller. I recall a briefing by a sincere and clever analyst who made the error in estimating probability that I indicated. By his analysis, one could just as easily confirm the working hypothesis as the null hypothesis concerning the presence or absence of a nuclear explosion. We ultimately produced an unclassified report on 17 July 1980 in which we concluded that the signal did not indicate a nuclear explosion.

*Resolve the Problem of Ambiguous Data.* How do we resolve the dilemma of ambiguous data? We cannot simply "apply the test of reason" because reason, or common sense, or whatever you want to call it, can be right almost all the time, but it will occasionally overlook important factors. It is quipped that the analysts in British intelligence (charged with saying yea or nay about the outbreak of war the next day) erred only twice in seventy years.

Experiment is the principal approach to making discoveries and changing our theories. In physics, we can advance the field in one of two ways. First, we can advance it by theory, preferably supported by experiment. Alternatively, we can advance it by an experiment that is not compatible with existing theory. Scientific investigation can produce results so compelling by virtue of being reproducible and strengthened by improved experimental conditions that the new phenomenon must be accepted even if it cannot be explained or is inconsistent with current theory. This approach can be applied to intelligence as well as to physics.

In general, we do not make progress in science or in intelligence simply by accumulating all possible data and then thinking about it. There is just too much information and too many possible hypotheses. More often, we have an idea or an inkling or a hunch, and we look for data to support it or at least we ask how it would be evidenced, if true. But finding that evidence does not in itself validate the hypothesis.

*Balance Military and National Intelligence Requirements.* From my own experience, I have concluded that we must be wary of a tendency to over-militarize the Intelligence Community and its function. Our military forces fight our wars (aided by many of people and organizations that are not "military"), but they are not our only tool for influence in this world. Even in peacetime high quality and timely intelligence is required in order to advise the President, Secretary of State, Secretary of Defense and others, regarding actual and potential situations. This cyclical process contributes to informed judgment as to the desirability and potential effectiveness of military activity in times of crisis.

These few officials, with enormous responsibility and limited budgets, are the most important customers of the Intelligence Community. Reliable intelligence is essential in the practice of diplomacy, international treaties, economic competition, and the like. But the intelligence that is required to support diplomacy, international treaties, or economic competition is distinct from military intelligence, and a military-dominated intelligence community might not provide it. I well remember the enormous influence of the Strategic Air

Command during the 1950s and 1960s on the nature and targeting of the intelligence gathering–collecting information that would be immensely valuable in the event of a nuclear war. But it was other information—information that had nothing to do with targeting—that was important to determine which military forces to build and which policies to pursue for the benefit of our nation and our allies.

*View Intelligence as a Tool.* Why have intelligence? Well, why have a map when you are trying to reach a destination? And if you have a map, wouldn't it be a good idea to know whether the roads are washed out, overcrowded, or beset by alligators? We tend to break intelligence apart as a process with sequential steps. But characterizing intelligence in several phases—acquisition, processing, analysis, and distribution tends to trivialize the activity. By focusing on the parts, we often lose sight of the importance of the whole, even for tactical intelligence (that is, intelligence associated with accomplishing a particular task).

*I have concluded that we must be wary of a tendency to over-militarize the Intelligence Community and its function.*

Intelligence is a problem-solving process for which no single approach works best. When I solved problems in physics or in national security areas, I tried to work through the solution in different ways to be alert for discrepancies or build my confidence in the answer. A country or organization should invest a portion of the overall effort into the improvement of tools and techniques for doing the job of applying intelligence to solving problems. But only a portion should be spent on tools. The temptation is (and it is all too common) to constantly exchange an existing tool for a better one, sometimes at the risk of not getting the job done. Intelligence is the job, and tools only help us to work smarter.

**Conclusion**

More than forty years ago I learned from Jerry Wiesner that you could either achieve something, or get credit for the achievement, but not both.[7] I have been fortunate to exercise my interests and abilities across a wide spectrum of problems to include the properties of materials, the characteristics of electronic devices, the design of optical and mechanical systems, and the features of nuclear weapons. However, I think the foundation to my insights has been my understanding of physics.

Since I first became involved in intelligence matters in 1956, I have had the opportunity to work with some of the most intelligent and effective people anywhere, both in government, industry, and in some "outsider-insider" groups such as the Land Panel, the President's Science Advisory Committee, and the JASON group of consultants to the government.[8]

Those of us who have worked in these groups and panels have been bound together by the faith that better and timelier information would enable our government to make better decisions. We have been dismayed when good decisions have not been made. Bad consequences can follow even good decisions, and it would be unrealistic not to recognize this; but neither bad decisions nor bad luck is an excuse for us to provide an intelligence product or process with less integrity than we can readily achieve.

I am sure that it pleases my colleagues to see that our work is recognized and valued. And I am very happy that together we have been able to do a lot of good. I do believe that intelligence plays, and will continue to play, a critical role. We can and must provide better, more relevant, and more timely products. It may seem easier to be self-reliant rather than to

---

[7] Dr. Jerome Wiesner was the science advisor for President Kennedy.

[8] The JASON group was established in 1960. Its purpose was to analyze scientific national security concerns and to train new generations of scientists to replace those who had served in World War II.

demand that others fill your requirements, but this is a dangerous approach. Each national security organization should be a demanding consumer and a reliable supplier.

Reach out! It is fun to solve puzzles, and it is nice to get recognition. But sometimes getting the job done simply requires bringing in an idea or a technique from outside, or calling in a consultant from another part of the organization or from the outside world. Intelligence is of critical importance to our nation and to our world. Let us go ahead with the hard work and the moments of inspiration, working together in the cause of freedom and human betterment.

—

## Founder Award Presentation and Citation

Figure 3-1. Founder Richard Garwin (second from the left) being recognized as a founder of national reconnaissance. The founder award plaque was presented by DNRO Keith Hall (left) and DCI George Tenet (third from the left). (Photo by Sara Judy, Visual Design Center.)

**Richard L. Garwin, Ph.D.**

A physicist, Dr. Richard Garwin, served on the President's Science Advisory Committee. He chaired the committee's panels on military aircraft, anti-submarine and naval warfare, and advised on intelligence aspects and programs in each field. He served as a key scientific advisor to Program B, and established standards and found solutions for electromechanical design of modern long-life spacecraft. As a champion of electro-optical imaging, he helped Henry Kissinger understand the critical role it would have for our national defense.

*Career in National Reconnaissance: 1957-2000*

# Edwin H. Land

Edwin H. "Din" Land was a scientist and inventor who made significant contributions in both industry and government. He invented the instant photograph, and he founded and led the Polaroid Corporation for over four decades. He also served as a trusted scientific and technical advisor to Presidents Eisenhower, Kennedy, Johnson and Nixon. Land's advice influenced the development of both airborne and satellite imagery reconnaissance systems that collected vital national security information. His knowledge and expertise was influential in presidential decisions to pursue "quantum jumps" in national reconnaissance capabilities: the U-2 and SR-71 aircraft, the Corona film-return photoreconnaissance satellite, and an electro-optical imaging reconnaissance satellite.

## Influencing Presidential Decisions[1]

Edwin H. "Din" Land developed an interest and aptitude for physics early in his education. He enrolled at Harvard, but he took a leave of absence from his degree program to pursue scientific and technical interests on his own by enhancing his education at the New York Public Library. In 1929 he returned to Harvard and, although still a freshman, he was given a personal laboratory and gave a lecture at an invited physics colloquium. Land's wife, Helen, assisted in his experiments at a time when women normally were not permitted in physics laboratories. Land left Harvard again in 1932 to establish the Land-Wheelwright Laboratories in collaboration with his Harvard physics instructor, George Wheelwright, III. Although Land never completed his formal degree program, Harvard awarded him an honorary doctorate in 1957.[2] When a Harvard colleague, Bradford Washburn, was asked why Land did not complete his degree in physics, Washburn replied, "He didn't need to."[3]

Land's science interests led to his invention of instant photography, the development of the Land camera, and a career in industry, where he founded the Polaroid Corporation in 1937 and served as President and Chief Executive Officer (CEO). His professional endeavors

---

[1] Edwin Land died on 1 March 1991, and was posthumously honored as a Founder of National Reconnaissance. In lieu of recollections, this chapter contains information based on publicly-available resources. It also draws on the research contributions of Robert Perry and Clarence Smith.

[2] In addition Land received some 20 honorary doctorates from other universities.

[3] Campbell (1994).

Figure 4-1. Edwin Land. (Photo courtesy of Rowland Institute for Science.)

and intellectual interests were focused in the fields of polarized light, photography, and color vision. Land was a prodigious inventor, and by the end of his career he held 537 patents, second only to Thomas Edison.

While at Polaroid, the government recognized his knowledge and abilities in science and photography, and Land became an influential advisor on defense and national reconnaissance issues to several presidents. He led and participated in a number of government study teams and review panels, and he collaborated in these efforts with other notable individuals, including Edward Purcell, James Killian, and James Baker, to name a few.[4]

**Inventing the Instant Photograph and Founding the Polaroid Corporation**

The Land Camera and Polaroid film were revolutionary developments for amateur photographers. For the first time they had the ability to take instant "see-it-now" pictures. The prospect of an instant photograph was inspired by a question from Land's young daughter Jennifer. During a 1943 Christmas holiday in Santa Fe, Land took a photograph of Jennifer, who asked, "Daddy, why can't I see the picture now?" Land put considerable thought into the technical challenge, and developed the concept of instant photography. It took him three more years to develop his invention, and he presented it on 2 February 1947 at a meeting of the Optical Society of America.[5]

Land's invention was the first one-step development of photographs one minute after taking the picture. On 26 November 1948 the first Land Camera, the Model 95, with black-and-white film, was sold at the Jordan Marsh department store in Boston. In the 1950s and 1960s the Polaroid Corporation expanded to subsidiaries in Canada, Germany and the UK. In 1964, the Model 100 camera with color film was produced.[6] Many improvements and modifications on Polaroid products for home, business, and professional use continued throughout the second half of the 20th Century, and several updated versions of Polaroid cameras were on the market at the end of the 20th Century.

**Advising Presidents**

In government, Land served as a scientific counselor and advisor to Presidents Eisenhower, Kennedy, Johnson, and Nixon. Eisenhower expressed his need for frank advice on technical problems, and he later referred to Land and his colleagues as "one of the few groups I

[4] Edward Purcell is a Founder of National Reconnaissance. See chapter 5. James R. Killian is a Founder of National Reconnaissance. James G. Baker is a Pioneer of National Reconnaissance (inducted into the NRO Hall of Pioneers, 27 September 2000).

[5] McElheny (1998); Inlow (1997).

[6] Wurman (1989).

encountered while in Washington who seemed to be there to help the country and not help themselves." Land later recalled that the President said:

> "I am so grateful to you fellows who are out of town. You can't think in Washington. You go away and then you tell me what you've been thinking. There's no way to think if you live here."[7]

Land was a member of the Beacon Hill Study Group and later chaired the Intelligence Subcommittee of the Technological Capabilities Panel (TCP) (the subcommittee was known as Project Three). In addition, he was chairman of the National Reconnaissance Panel of the President's Science Advisory Committee (PSAC). The National Reconnaissance Panel became known as the "Land Panel." Land also served as a member of the President's Foreign Intelligence Advisory Board (PFIAB).

*Beacon Hill Study Group.* In 1952 Land was part of the Beacon Hill Study Group, which focused on reconnaissance and detection methods and technologies. For this study, Edward Purcell wrote a chapter on radar and other electronic means of detection, James Baker of Harvard wrote a chapter on photographic surveillance, and Land contributed a chapter titled "A New Approach to Photo Reconnaissance." In this chapter, Land urged U.S. policymakers to pursue advanced scientific and technical approaches to intelligence collection. He wrote:

> "In the post-war world, intelligence and reconnaissance are more important to the United States, by several orders of magnitude, than in the pre-1945 world. The age of scientific warfare has already produced the intercontinental striking force with atomic weapons. It is now producing intelligence instruments of comparable efficiency and extreme speed. Since they exist, or will soon be developed, they must be used to the maximum; otherwise an enemy could use them better."[8]

The Beacon Hill group urged the development of a high-altitude reconnaissance aircraft that could penetrate and collect intelligence over denied territory. In March 1953 the Air Force issued a design requirement for such an aircraft.

Figure 4-2. In 1947 Land introduced instant photography to the Optical Society of America. (Photo courtesy of Polaroid Corporation.)

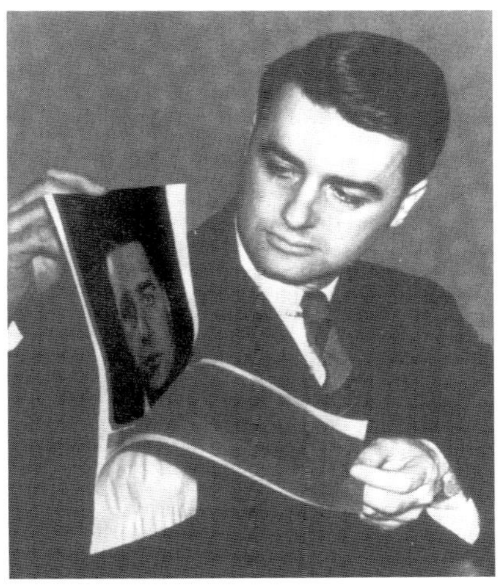

*Technological Capabilities Panel Project Three.* In July 1954 President Eisenhower asked James Killian to recruit and lead a panel of experts to study "the country's technological capabilities to meet some of its current problems." Land was asked by Killian to chair Project Three of the TCP. Project Three was given the task of investigating the nation's intelligence capabilities, and was the smallest of the TCP's three projects, or subcommittees. Land preferred what he called "taxicab committees"—committees small enough to fit into a single taxicab.[9] The members of Project Three included Edward Purcell, James Baker, chemist Joseph Kennedy, mathematician John Tukey, and Allen Latham.

In August 1954 Land and Baker went to Washington to arrange for the various intelligence organizations to brief the Project Three team. As the

7 McElheny (1998, p. 301).
8 McElheny (1998, p. 285).
9 Pedlow and Welzenbach (1998, p. 29).

briefings progressed, the panel members became distressed at the poor state of the nation's intelligence resources. Land later noted, "We would go in and interview generals and admirals in charge of intelligence and came away worried. Here we were five or six young men asking questions that these high-ranking officers couldn't answer."[10]

The TCP submitted its official report on 14 February 1955. Land authored one of the most significant passages of the report, which illustrated the scope of the review and recommendations provided to the President.

> "We must find ways to increase the number of hard facts upon which our intelligence estimates are based, to provide better strategic warning, to minimize surprise in the kind of attack, and to reduce the danger of gross overestimation or gross underestimation of the threat. To this end we recommend the adoption of a vigorous program for the extensive use, in many intelligence procedures, of the most advanced knowledge in science and technology."[11]

*The Land Panel.* The Land Panel, or more formally the National Reconnaissance Panel of the President's Science Advisory Committee, consisted of distinguished scientists and engineers from academia and industry. The panel was responsible for maintaining an overview of the National Reconnaissance Program, paying particular attention to the technical characteristics of intelligence requirements, the status of existing projects, and the adequacy of existing intelligence programs.

Figure 4-3. Sale of the first Land Camera "Model 95" at the Jordan Marsh department store in Boston, 1948, as depicted in the Polaroid commemorative brochure. (Photo courtesy of Polaroid Corporation.)

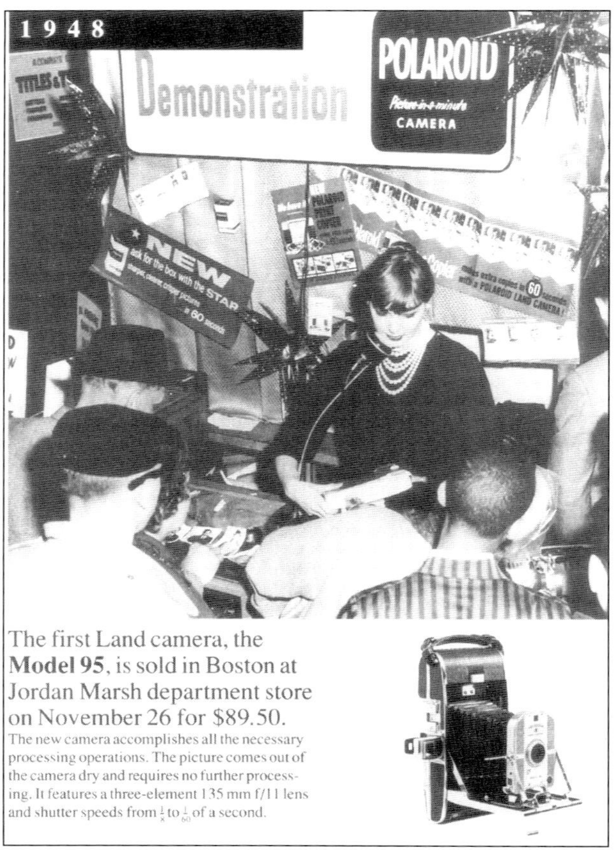


*Land's Impact and Recognition by Presidents.* Land's influence and technical expertise played a major role in the development of both aircraft and satellite imagery reconnaissance systems. Land's groundbreaking work in the area of photographic materials and processes led to his playing a key role in the government's decision to develop an electro-optical imaging reconnaissance satellite. He exemplified the kind of scientist who helped make science advice welcome at the White House. In meetings with presidents his "eloquence and lucid exposition incited their imagination." This rapport prompted them to make decisions of immense consequences in reconnaissance and

[10] Pedlow and Welzenbach (1998), 29.
[11] McElheny (1998), 303-304.

intelligence-gathering technology that saved the nation billions of dollars.[12]

On 6 December 1963 President Lyndon Johnson presented Land with the nation's highest civilian honor, the Presidential Medal of Freedom, at a ceremony in the White House State Dining Room.[13] He received the honor for "bringing his creative gifts to bear in industry, government and education." Fate made this a particularly historic and moving ceremony. Land had been selected for the award by President John F. Kennedy, who was assassinated on 22 November 1963 and Vice President Lyndon Johnson then assumed the presidency. There was concern about having a ceremony for those selected by President Kennedy so soon after his tragic death. The nation was in mourning and the award ceremony provided one of the first opportunities to begin the healing process. Both Jacqueline Kennedy and President Johnson insisted on holding it as scheduled on 6 December. Before the ceremony on 6 December President Johnson had serious misgivings about taking part and almost backed out. However, after others voiced strong support to continue with the ceremony, and pointed out "that it was very important for the presidency as an institution," President Johnson put aside his doubts, went on with the ceremony, and presented the medals.[14]

Figure 4-4. President Lyndon Johnson presenting Land with the Presidential Medal of Freedom at the White House on 6 December 1963. (Photo courtesy of LBJ Library.)

*He exemplified the kind of scientist who helped make science advice welcome at the White House. In meetings with presidents his "eloquence and lucid exposition incited their imagination." This rapport prompted them to make decisions of immense consequence in reconnaissance and intelligence gathering technology.*

### Advocating the U-2 as a New Approach to Photoreconnaissance

In August 1954 Land learned the details of Lockheed's proposed CL-282 aircraft (later renamed the U-2), and also learned that the Air Force had rejected the design. Land presented Lockheed's design proposal to the other Project Three members, and they became enthusiastic supporters of the aircraft's potential. Later that month Land met with Richard Bissell (who at that time was Director of Central Intelligence (DCI) Allen Dulles' Special Assistant for Planning and Coordination) to discuss the U-2 project. By the end of September Land was deeply involved not only in advocating the

[12] Killian (1977).

[13] Subsequently, on 13 February 1968, President Johnson also presented Land with the National Medal of Science at the White House.

[14] Wetterau (1996, p. 18).

project, but also in developing the design into a reconnaissance system. By the end of October, Project Three had reviewed every aspect of the Lockheed design. They viewed the U-2 as an integrated intelligence collection system that could locate and photograph Soviet Bison bombers, and thereby help resolve the "bomber gap" controversy.[15]

Land's team designed the first of the high-altitude optical reconnaissance systems. He worked with James Baker to design a high-quality aerial camera for use on the U-2. However, Land did more than help develop the camera; he also helped design the plane. Land and the plane's designer, Kelly Johnson, collaborated on overcoming the problems associated with high-speed, high-altitude flight.[16] The relationship between the two men became strained at times, and one observer recalled that, "they argued constantly about each other's needs to dominate the relatively small space inside the plane's bays. Land needed space for his bulky cameras, and Johnson needed room for batteries." Johnson was reported to have told Land, "Let me remind you, unless we can fly this thing you've got nothing to take pictures of."[17]

*...they argued constantly about each other's needs to dominate the relatively small space inside the plane's bays. Land needed space for his bulky cameras, and Johnson needed room for batteries." Johnson was reported to have told Land, "Let me remind you, unless we can fly this thing you've got nothing to take pictures of.*

Land's involvement went beyond dealing with the technical issues associated with the U-2 project. It included defining important bureaucratic and organizational arrangements regarding who would conduct the missions and interpret the collected imagery. Land and other Project Three members were opposed to the prospect of the Air Force conducting such missions in peacetime. They believed that military overflights of denied territory could provoke an armed conflict. Consequently, they argued that reconnaissance overflights should be performed by a civilian organization in unarmed, unmarked aircraft. Land's subcommittee recommended the Central Intelligence Agency (CIA) as the appropriate organization to perform this mission.[18]

Despite discussions with Land and other Project Three members, DCI Allen Dulles was reluctant to have the CIA undertake this mission. The DCI believed that the agency's mission could be carried out more effectively through the use of human operatives and other more traditional sources and methods of intelligence collection. Given Dulles' resistance, Land decided to press their case to a higher authority, namely President Eisenhower. Eisenhower had commissioned Project Three through the TCP to study intelligence matters and make recommendations, and the committee was fulfilling its mandate through the TCP chairman, James Killian.[19]

Early in November 1954 Land and Killian met with President Eisenhower to discuss the issue of national reconnaissance. Killian recalled the discussion:

> "Land described the [U-2] system using an unarmed plane and recommended that its development be undertaken. After listening to our proposal and asking many hard questions, Eisenhower approved the development of the system, but he stipulated that it should be handled in

---

[15] Pedlow and Welzenbach (1998), 20, 31. The Soviet Myasishchev-4 bomber, designated Bison by NATO, was first sighted by western intelligence in 1953, and was the first Soviet jet-powered, long-range bomber.

[16] Campbell (1994).

[17] McElheny (1998, p. 299).

[18] Pedlow and Welzenbach (1998, p. 32).

[19] Pedlow and Welzenbach (1998, p. 33).

an unconventional way so that it would not become entangled in the bureaucracy of the Defense Department or troubled by rivalries among the Services."[20]

Eisenhower wanted to be sure the plane would perform as advertised, and he also questioned whether the intelligence collected would be worth the political risk of the overflights, recognizing that sooner or later a plane would be shot down. Land reassured the President, "If we are successful, it can be the greatest intelligence coup in history."[21] Land recalled to a friend that he persuaded President Eisenhower to support the U-2 project by describing his new camera's capabilities in terms familiar to the President, specifically with an illustration from golf, a game that the President enjoyed. Land maintained that the capabilities of this new system would provide the ability to detect a golf ball at 2000 yards distance, in comparison with the 150 yards from which the typical golfer can see a golf ball.[22]

Land and his colleagues continued to try to persuade DCI Dulles that the CIA should run the proposed reconnaissance program. On 5 November 1954 Land wrote to Dulles:

> "I am not sure that we have made it clear that we feel there are many reasons why this activity is appropriate for CIA, always with Air Force assistance. We told you this seems to us the kind of action and technique that is right for the contemporary version of CIA: a modern and scientific way for an Agency that is always supposed to be looking, to do its looking. Quite strongly, we feel that you must always assert your first right to pioneer in scientific techniques for collecting intelligence—and choosing such partners to assist you as may be needed. This present opportunity for aerial photography seems to us a fine place to start."[23]

In a cover letter and an attached memorandum, the Project Three members stressed the value of the U-2 as an intelligence collection method relative to more traditional methods:

> "We believe that these planes can go where we need to have them go efficiently and safely, and that no amount of fragmentary and indirect intelligence can be pieced together to be equivalent to such positive information as thus can be provided...A single mission in clear weather can photograph in revealing detail a strip of Russia two hundred miles wide and twenty-five hundred miles long and produce four thousand sharp pictures."[24]

The strong advocacy of Land, Killian, and other Project Three members, combined with President Eisenhower's support, finally persuaded DCI Dulles to accept the U-2 design and associated organizational recommendations. The final agreement on the U-2, namely to proceed with development as a joint CIA-Air Force program, came at a luncheon on 19 November 1954 with representatives from the CIA, Air Force, and Lockheed, and Edwin Land in attendance. Five days later Eisenhower gave authorization to proceed with the project, and he told Dulles that that project was to be managed by the CIA with Air Force assistance.[25]

*Land persuaded President Eisenhower to support the U-2 project by describing his new camera's capabilities in terms familiar to the President, specifically with an illustration from golf.*

---

[20] Killian (1997), 82.
[21] McElheny (1998), 301.
[22] Campbell (1994).
[23] Project Three letter to DCI Allen Dulles, 5 November 1954, in Pedlow and Welzenbach (1998), 33.
[24] Pedlow and Welzenbach (1998), 34; McElheny (1998), 298.
[25] Pedlow and Welzenbach (1998), 37.

**Reviewing the Design of the A-12 Advanced Aerial Reconnaissance System**

As anticipated by Eisenhower and others, the Soviets developed the capability to intercept and shoot down the U-2. This event occurred on 1 May 1960 when Francis Gary Powers was shot down, captured, and put on trial for espionage.

Although research and development of space-based reconnaissance satellites was underway in the late 1950s, the first successful mission did not occur until August 1960, and only after repeated failures. Given the uncertainty and unavailability of space-based reconnaissance at that time, an improved aerial platform was required that could succeed the vulnerable, subsonic U-2. The design that met the requirement was a supersonic, high-altitude aircraft. In 1957 Land became the head of a small committee that reviewed designs for a 2,000-mile-per-hour aircraft, which was known by the designator A-12 (also referred to as Project Oxcart).[26]

*Levison recommended that Land inform Eisenhower of the failure of the Corona program, and face the prospect of terminating the program. Land refused to concede defeat.*

In addition to speed and altitude, the aircraft also incorporated stealth technology, which would make it less detectable to radar than the U-2. Lockheed and General Dynamics were contracted to design the A-12, and Land was actively involved in the design of the aircraft and its systems.[27] Although operational use of the A-12 began after the successful demonstration of space-based reconnaissance systems, the aircraft made significant contributions to aerial reconnaissance through the end of the 20th Century.

**Influencing Development of Corona and Establishing the NRO**

By the late 1950s the U.S. was dedicating considerable resources to developing a space-based reconnaissance system. The political risks of U-2 missions, coupled with the Soviets' ability to track and eventually intercept these aircraft, illustrated the need for a less-provocative method of collecting critical intelligence. The Soviet launch of Sputnik in 1957 established the precedent of freedom of space, which largely resolved the questions that emerged related to the definition of territorial sovereignty in the era of satellites. Eisenhower exploited the precedent of freedom of space established by Sputnik for the benefit of U.S. reconnaissance efforts. Nevertheless, serious technical challenges remained to be overcome in areas such as launch, film survivability, data transmission, and capsule re-entry, before a U.S. operational photoreconnaissance satellite system could be deployed.

*A Joint Air Force-CIA Project.* Similar to developing the U-2 program, Land urged President Eisenhower to place the Air Force's WS-117L satellite reconnaissance program under the CIA as a joint endeavor. Eisenhower accepted this advice, cancelled WS-117L in February 1958, and restarted the program as Project Corona under the management of Richard Bissell at CIA with guidance provided by Land's PSAC National Reconnaissance Panel. In fact, Bissell recalled that he first learned of Corona and his own management role in the program "in an odd and informal way" from Land. Specifically, Bissell was called into Allen Dulles' office and, in Land's presence, was informed that Corona had been added to his other responsibilities.[28]

*Failure Before Success.* The next twelve months proved difficult both for the Corona program and for Land. At the beginning of 1959 George Kistiakowsky replaced James Killian,

[26] Pedlow and Welzenbach (1998); McElheny (1998), 325. The Air Force developed a military version of the A-12 and gave it the designator, SR-71 (which was often referred to as the "Blackbird").

[27] McElheny (1998), 325.

[28] Hall (1997), 42-50; McElheny (1998), 329.

Land's trusted friend and colleague, as the President's science advisor. The relationship between Land and Kistiakowsky was tense at times, and the style and substance of Land's advice was not received as warmly as it had been by Killian. The repeated failures of Corona's various systems and components complicated matters, and contributed to a mood of frustration and, at some points, despair.[29]

In November 1959 the Corona program was experiencing such difficulties that the Itek project manager, Walter Levison, visited Land at his weekend home in New Hampshire to report failure.[30] Levison recommended that Land inform Eisenhower of the failure of the Corona program, and face the prospect of terminating the program. Land refused to concede defeat. Instead, he recommended redoubling the effort over three months in an effort to rescue the project.[31] Levison returned to Itek and, along with Frank Madden (his chief engineer), rebuilt most of the system. This work included the development of new processes and materials that were required to address the challenges of operating in the harsh environment of space.[32]

By the spring of 1960 technical failures and bureaucratic infighting associated with the Corona and Samos programs had created a negative situation that required presidential intervention.[33] On 26 May Eisenhower intervened and ordered an ad hoc panel to review the issues that were contributing to the technical failures and delays. Eisenhower directed that the panel report directly to Kistiakowsky, rather than Secretary of Defense Thomas Gates. Dr. Joseph Charyk, Under Secretary of the Air Force, was directed to work with Kistiakowsky.[34] The Defense Department resented that this review was to be performed by outside scientists rather than by the military, but Eisenhower was firm in his decision.[35] Land and Killian conducted the review, under Kistiakowsky's supervision. Their findings pointed to problems in management and bureaucratic organization. The report stated:

> "The multiple layers of military hierarchy involved caused delays by sudden changes in specifications and other difficulties. It was the situation that our panel was trying to change; the Air Force, aware of our intentions, was coming up with its own plan to clean up some aspects of the situation. Our efforts coalesced, and we could recommend a clear administrative arrangement, bypassing various military echelons and enabling the individual in charge of development projects to report directly to the new National Reconnaissance Office within the Office of the Secretary of the Air Force."[36]

Land and Killian were arriving at conclusions and recommendations that would be unpopular with many in the Department of Defense and the Services. The underlying notion was to create an authoritative organization of "a national character" responsible for managing

---

[29] McElheny (1998), 330-331.

[30] Itek was the contractor responsible for building the camera system. Walter Levison is a Pioneer of National Reconnaissance (inducted into the NRO Hall of Pioneers 27 September 2000). See chapter 31. See also McElheny (1998), 331.

[31] McElheny (1998), 331.

[32] Frank Madden is a Pioneer of National Reconnaissance (inducted into NRO Hall of Pioneers 27 September 2000). See chapter 32 for his recollections.

[33] Samos was an early satellite imagery reconnaissance program. The program previously was named Sentry.

[34] Joseph Charyk served as Under Secretary of the Air Force and the first Director of the NRO (DNRO). He served as DNRO from 1961-1963.

[35] McElheny (1998), 335.

[36] McElheny (1998), 335.

national reconnaissance. On 22 August 1960 Kistiakowsky met with Land, Charyk, Bissell, and others to review the memorandum that was being prepared for Eisenhower. Kistiakowsky visited Land to work out the remaining details for the establishment of the NRO, with Charyk as its first director.[37] The previous week, on 18 August 1960, the fourteenth Discoverer mission achieved success, which was the first successful Corona mission.[38]

*Successful Mission and a New Organization.* On 25 August 1960 Land joined Kistiakowsky, Killian, Allen Dulles, and others in a brief meeting with the President, which immediately preceded a meeting with the National Security Council. They presented Eisenhower with the pictures from Corona's first successful mission. Land captured the moment with a bit of drama when he unreeled a spool of Corona pictures across the Oval Office carpet and declared, "Here are your pictures, Mr. President." Eisenhower approved Land and Killian's recommendations regarding the establishment of the NRO. The President expressed regret that the recommendations had not been made two years earlier, "so that he would be the guy to see the lovely pictures we would be making. As it is, it will probably be Dick [Nixon] and not him, and he [Eisenhower] won't even have the clearances necessary to see them."[39] The group then proceeded to the National Security Council meeting at which the approval was granted for the establishment of the NRO, the agency which Land and Killian had recommended have primary responsibility for the design, acquisition, and operation of reconnaissance satellites.[40]

*Land's Impact.* Land's influence during the development of Corona was significant in both technical and bureaucratic aspects. He was influential in the decision to pursue national reconnaissance from space with the Corona film-return system, and then he was influential in the creation of the NRO as an Intelligence Community organization charged with the design, development, and operation of Corona and subsequent reconnaissance satellites.

**Influencing the Decision for Near-Real-Time Satellite Imaging**

Following the success of the early film-return systems, the Intelligence Community recognized a need for a reconnaissance system that could provide more timely imagery for intelligence exploitation. There were numerous failures and some successes during the 1960s and 1970s with efforts to develop new satellite systems that could surpass the resolution, broad-area coverage, and time delay of film-return systems. However, at the end of a decade of continued research and development, a near-real-time imaging system still was unavailable for use in satellite reconnaissance.

The mission duration of film-return systems like Corona, which for some missions was almost a month, required the successful recovery of a reentry vehicle or "bucket." If the bucket with exposed film were recovered successfully, it had to be transported to a processing site and then to an intelligence exploitation facility before the intelligence could be "read out" and disseminated. This process could entail seven or more additional days. Early retrieval of the bucket in a time of crisis meant sacrificing the unused film, a potentially costly decision. If the retrieval were delayed, the imagery might provide useful intelligence, but it was not timely. Hence, the intelligence was of limited value to policy makers and war fighters during a fast-moving crisis or conflict.

---

[37] Hall (1999); McElheny (1998), 336.

[38] Discoverer was the cover name for the Corona program. McDonald (1997), 301.

[39] McElheny (1998), 338. Apparently, Eisenhower believed that his Vice President, Richard Nixon, would win the November 1960 presidential election that ultimately was won by John F. Kennedy.

[40] McElheny (1998), 337.

The 1967 Arab-Israeli Six-Day War and the Soviet invasion of Czechoslovakia in 1968 stimulated new concerns about reconnaissance capabilities during a crisis. The Six-Day War ended by the time U.S. policymakers and military planners received satellite reconnaissance imagery of the conflict. In the Czechoslovakia episode, a Corona satellite imaged unmistakable Soviet military preparations for invasion. The problem was that the imagery was not recovered from space, processed, and disseminated to analysts and decision-makers until after the invasion. Both incidents illustrated the critical value of imagery for warning and indications, while at the same time demonstrating the importance of timeliness.[41] As a consequence, the United States Intelligence Board (USIB) asked the NRO to look into the feasibility and cost of a collection system applicable to warning-and-indications needs.

*The 1967 Arab-Israeli Six-Day War stimulated new concerns about reconnaissance capabilities during a crisis. The Six-Day War ended by the time U.S. policymakers and military planners received satellite reconnaissance imagery of the conflict.*

*Committee on Imagery Requirements and Exploitation Defined the Requirement.* The NRO looked to the Intelligence Community and the Director of Central Intelligence's (DCI) Committee on Imagery Requirements and Exploitation (COMIREX) to help define the requirement for a follow-on imagery satellite that could support warning intelligence collection.[42] COMIREX also noted that the requirement for a follow-on imagery satellite should be for a flexible system that could carry out the warning-indications role and at the same time be capable of satisfying routine, current intelligence, and special reconnaissance tasks. Additionally, the imagery warning and indications system should be operated continually and in conjunction with constantly-ready interpretation and analysis facilities. If those requirements were not met, the considerable cost and difficulty of developing such a system could not be justified.

Roland Inlow, who served as the chairman of COMIREX in the late 1960s and 1970s, recalled that Land had reduced the explanation of the requirements for satellite imagery to three simple phrases: *"See it All, See it Well, and See it Now."* [43] These phrases proved to be an accurate prophecy for near-real-time satellite imaging systems. In effect, it was the philosophy for COMIREX system requirements: "See it All" drove development of a system with area coverage; "See it Well" drove development of high-resolution surveillance systems; and "See it Now" drove development of the near-real-time system.

*Identifying the Need for a "Quantum Jump" in Technology.* The candidate state-of-the-art technology that was available in the late 1960s to fulfill the COMIREX requirement for "see it now" was an adaptation of a concept developed in the 1950s for the WS-117L satellite program. That system was designed to take pictures, develop the film on board, then scan the film electronically and transmit the images to ground stations. The WS-117L (later renamed Sentry then Samos) had many technical difficulties and eventually was terminated

---

[41] Inlow (1997, p. 228).

[42] COMIREX was established by DCI Directive 1/13 on 1 July 1967 to advise and assist the USIB on matters involving overhead reconnaissance and imagery exploitation. The chairman was designated by the DCI, and the committee was composed of representatives from departments and agencies on the USIB, including CIA, DIA, NSA, State, Army, Navy, Air Force, and Marines. Other agencies involved in imagery reconnaissance such as NRO, NPIC, and DMA served as consultants. COMIREX ceased to function after formation in 1992 of the Central Imagery Office (CIO) which in October 1996 was incorporated into the National Imagery and Mapping Agency (NIMA).

[43] Inlow (1997), 228.

in 1962 in favor of the Corona film-return system. However, the WS-117L concept was promising in that it could develop film in space and provide images electronically to Earth, thus removing the need for a premature film-capsule recovery during a crisis.[44] In October 1968 the Land Panel completed its initial review of a technical proposal for an imaging system with near-real-time readout. This system would incorporate technologies that were untested in environments outside the laboratory, thereby representing a "quantum-jump" approach.

Two bureaucratic groups emerged: those who favored the ambitious "quantum-jump" approach, and those who favored a more incremental approach. Those favoring the ambitious approach seemed less concerned with the risks of technical, schedule, and funding difficulties, whereas proponents of a strategy of more deliberate, incremental technology advances tended to be more concerned about those risk factors. This competition led to a series of proposals, counterproposals, debates, and infighting between senior officials, organizations, and policymakers (mainly among NRO, CIA, PFIAB, USIB, and the President's Science Advisor). It was during these debates that Land's credibility and influence were of great importance in reaching a final decision.

*Land's Perspective.* Land's personality orientation led him to be extremely focused and dedicated to the projects in which he was involved. For example, in the 1960s and 1970s he committed his fortune and the fate of Polaroid to the development of "absolute one-step photography," embodied in the SX-70 camera. This camera incorporated his inventions in film with a camera that was based on a system of micro-electrons, optics, and an automatic ultrasound range finder. In writing about Land's commitment to the project, McElheny observed:

> "This vast SX-70 project...was Land's greatest technological triumph. It brought together the most varied array of technical talents that Land had ever assembled. In terms of money and jobs that would have been jeopardized by failure, SX-70 involved the biggest risks he had ever run."[45]

This personal dedication to a Polaroid camera that he claimed would "do the impossible" was consistent with his drive for the Intelligence Community's development of a comparably-ambitious and technically-advanced reconnaissance system.

*In a memorandum to President Nixon in May 1969, Land urged the President to direct the NRO to initiate development of an EOI imaging satellite at the highest priority.*

In reporting to the President's Science Advisor, Land argued, "the necessary technology for a 'see it now' system has become available," and its development could start in 1969. Of the technologies reviewed, the Land Panel favored electro-optical imaging (EOI)—a solid-state array system, with no moving parts. An EOI system could record images on charge-coupled devices and send them electronically back to Earth. Land and his panel emphasized the simplicity of the system, which featured solidity and reliability when compared to other systems that had problems such as film transport, mirror vibration, heat stabilization, and corona discharge.

A comprehensive review and overhaul of budgets and priorities for satellite reconnaissance was undertaken following Richard M. Nixon's inauguration as President in 1969. Subsequent to this review, President Nixon endorsed the concept of an operational near-real-time

---

[44] Hall (1997), 42-44; Oder and Belles (1997), 76.

[45] McElheny, (1998, p. 342). See pages 341-408 for a discussion of a history of the SX-70 effort and its repercussions on Land and Polaroid.

readout capability for imaging reconnaissance satellites. Proponents of this system seized on his endorsement as a license for action.

Figure 4-5. President Nixon (l), who endorsed the near-real-time concepts promoted by Land, during a visit to CIA Headquaters. DCI Richard Helms (r) is escorting the President. (Photo courtesy of CIA.)

*The Bureaucratic Dialogue.* In a memorandum to President Nixon in May 1969, Land urged the President to direct the NRO to initiate development of an EOI imaging satellite at the highest priority. Although Land recommended that the near-real-time readout system was an urgent national requirement—and that EOI was the most effective technique for satisfying that requirement—senior officials in the Department of Defense, the NRO and CIA were slow to express their support for his views.

During the winter of 1969-1970 development of EOI technology made progress that, in the judgment of the Land Panel, strongly reinforced earlier recommendations for the start of a system-definition phase. Noting favorable technical developments, Land maintained that it was feasible to schedule a 1974-75 operational date "if we get on with development."

The NRO was less enthusiastic about the prospects of EOI. In 1971 the NRO provided Dr. Henry Kissinger (the President's Special Assistant for National Security Affairs) a report he requested on crisis-response capabilities. This report advised that for the near term, the only promising approaches were from existing systems and technology. This perspective of crisis response was similar to the State Department view that a less expensive or responsive "Model-T" system (2-3 days) using off-the-shelf technology would suffice, as opposed to a near-real-time capability. However, the report of a January 1971 meeting of the Executive Committee (ExCom) was inconclusive, and did not support the Department of State view.[46] In contrast, by the summer of 1971 members of the Land Panel, including Sidney Drell and Richard Garwin, were meeting personally with Kissinger to press the case for pursuing EOI without further delay.[47]

*The President's Decision.* Subsequently, the Director of the Office of Management and Budget (OMB) wrote to the Secretary of Defense to emphasize the President's interest in a near-real-time or crisis-capability system. As OMB interpreted the President's wishes, "It would be desirable if such a system could be operational at an early date and at a reasonable cost, and... appreciable utility during the President's administration...should be a goal of development." Minutes from an April 1971 ExCom meeting stated, "...the President has expressed vigorously the desirability of near-real-time and the President...wished to have this

[46] The Executive Committee was composed of the Deputy Secretary of Defense, DCI, the President's Scientific Advisor, and the Director of the NRO (non-voting member).

[47] Sidney Drell and Richard Garwin are Founders of National Reconnaissance. See chapters 2 and 3 respectively for their recollections.

capability available during his second term in office." At that time, the 1972 presidential election was eighteen months away. However, there was still contention over the best system to pursue.

Land told Nixon that the bureaucrats were unwilling to assume large financial risks without "strong Presidential backing," and added that EOI was "a quantum jump that would give the U.S. an unquestioned technological lead in this field." Nixon was influenced by Land's personal advocacy of this new approach to imagery collection, more than by the more practical considerations of risk, cost, or requirements.

President Nixon announced on 23 September 1971 that he had decided to pursue the quantum-jump approach. On that day, Kissinger advised all concerned that the President had concluded the EOI system should be developed with the goal of 1976 for an operational date "under a realistic funding program." The President also decided that there would be no further development of alternative systems. Development began and the system became operational in 1977. Former NRO Historian Robert Perry, in his assessment of Nixon's motivations for setting 1976 as a target goal for EOI system operation, noted that Nixon may have had a desire to be remembered for having fostered near-real-time reconnaissance capabilities, and therefore wanted the system to be deployed before he was scheduled to leave office.

*EOI Declared Operational.* Although Nixon resigned before the end of his second term, the EOI system became operational on what would have been his last day in office. After its launch on 19 December 1976 and subsequent engineering check-out, the system acquired its first image on 19 January 1977. On the next day, and probably as his last official act, DCI George H. W. Bush signed a cable declaring the EOI system operational. Because 20 January was inauguration day for President Jimmy Carter, acting DCI Henry Knoche waited until the following day to take several examples of imagery from the new system to the White House. He then briefed President Carter on this latest addition to the intelligence collection capabilities of the United States.

**Conclusion**

Edwin Land had the ability to translate complex technical issues into simple, understandable concepts. He was able to use that skill in communicating with senior U.S. leaders. Edward Purcell recalled Land as,

> "very effective…not so much for his technical advice but for his ability to convey in a personal way the relative importance of the things we were talking about. And his boldness, which, of course, was part of his business career: talking about what you really could do if you wanted to. [Land would ask] "Why don't you try that instead of horsing around?" He was very important because, in the first place, he communicated very well with Eisenhower."[48]

Edwin Land's technical knowledge, foresight, and skillful persuasion in policy debates made significant contributions to the decisions by Presidents Eisenhower and Nixon to pursue new airborne and satellite reconnaissance systems. These new capabilities represented major advances in helping the Intelligence Community to collect, process, and disseminate critical information to national decisionmakers. Additionally, EOI systems remain the backbone of satellite imaging reconnaissance capabilities into the 21st Century.

—

[48] McElheny (1998), 328.

## Founder Award Citation

FOUNDER

**Edwin H. Land**

Edwin Land was the Polaroid CEO and chairman of the Intelligence Subcommittee of the Technological Capabilities Panel. He was a scientist, inventor and counselor to Presidents Eisenhower, Kennedy, Johnson and Nixon. As Chairman of the President's Science Advisory Committee's National Reconnaissance Panel, he advised the government on new airborne and satellite reconnaissance systems and improvements to existing ones. He played a vital role in advising President Nixon on the capabilities of electro-optical imaging.

*Career in National Reconnaissance: 1952-1980*

—

## References


Campbell, F.W. (1994). "Dr. Edwin H. Land," *Biographical Memoirs of Fellows of the Royal Society*, 40, 195-219. Reprinted by the Rowland Institute for Science, http://www.rowland.org.land/, accessed 4 March 2002.

Hall, R. Cargill (1999). "The National Reconnaissance Office: A Brief History of Its Creation and Evolution," *Space Times*, March–April.

Hall, R. Cargill (1997). "Post War Strategic Reconnaissance and the Genesis of Project Corona." In Robert A McDonald, ed., *Corona, Between the Sun and the Earth—The First NRO Reconnaissance Eye in Space*, Bethesda, MD: American Society for Photogrammetry and Remote Sensing.

Inlow, Roland S. (1997). "How the Cold War and Its Intelligence Problems Influenced Corona Operations." In Robert A McDonald, ed., *Corona, Between the Sun and the Earth—The First NRO Reconnaissance Eye in Space*, Bethesda, MD: American Society for Photogrammetry and Remote Sensing.

Killian, James R. Jr. (1997). *Sputnik, Scientists and Eisenhower—A Memoir of the First Special Assistant to the President for Science and Technology.* Cambridge, MA: MIT Press.

McDonald, Robert A., ed. (1997). *Corona, Between the Sun and the Earth—The First NRO Reconnaissance Eye in Space.* Bethesda, MD: American Society for Photogrammetry and Remote Sensing.

McElheny, Victor K. (1999). *Edwin Herbert Land 1909-1991—A Biographical Memoir, Biographical Memoirs*, Volume 77, Washington, DC: The National Academy Press.

McElheny, Victor K. (1998). *Insisting on the Impossible—The Life of Edwin Land, Inventor of Instant Photography.* Reading, MA: Perseus Books.

Oder, Frederic C.E. and Martin Belles. (1997). "Corona: A Programmatic Perspective." In Robert A. McDonald, ed., *Corona, Between the Sun and the Earth—The First NRO Reconnaissance Eye in Space*, Bethesda, MD: American Society for Photogrammetry and Remote Sensing.

Pedlow, Gregory W. and Donald E. Welzenbach (1998). *The CIA and the U-2 Program, 1954-1974.* Washington, DC: Central Intelligence Agency.

Wetterau, Bruce (1996). *The Presidential Medal of Freedom: winners and their achievements.* Washington, DC : Congressional Quarterly.

Wurman, Richard Saul (1989). *Polaroid Access: Fifty Years.* Bedford, MA: Polaroid Corporation Massachusetts. Access Press, Ltd.

# Edward M. Purcell

Edward Purcell was a renowned physicist and astronomer, and co-recipient of the 1952 Nobel Prize in physics for the development of the nuclear magnetic resonance method of measuring certain nuclear properties. In addition to his distinguished research and teaching career, Purcell dedicated a significant amount of time as a consultant and advisor to the United States Government. He served Presidents Eisenhower, Kennedy, and Johnson as a member of the President's Science Advisory Committee from 1957-1965. For his many contributions, Purcell was honored posthumously as a Founder of National Reconnaissance.

## An Educator, Advisor, and Catalyst for the National Reconnaissance Program[1]

Edward Purcell's early interest in engineering was inspired by the telephone equipment and technical journals his father, a telephone exchange manager, brought home from work. Purcell's aptitude and interest in science earned him a scholarship to Purdue University in Indiana, where he began his studies in 1929 at the age of seventeen. Although his chosen field of study was electrical engineering, Purcell became interested in physics, and this interest had a profound influence on the rest of his academic and professional life.[2]

**Harvard, Massachusetts Institute of Technology, and the Nobel Prize**

After spending a year in Germany as a foreign exchange student, Purcell entered Harvard University in 1934 as a graduate student in physics. He studied under Professor Kenneth Bainbridge, with whom he worked on the first Harvard cyclotron. Upon completing his Ph.D. in 1938, Purcell accepted a teaching position in Harvard's physics department, where he would remain for the next forty-two years except for various leaves of absence for special projects.

Purcell took a leave of absence from Harvard in 1941 to join Bainbridge at the newly formed Radiation Laboratory of the Massachusetts Institute of Technology (MIT), where he worked on developing advanced microwave radar. The Radiation Laboratory conducted military research during World War II, and Purcell became the Head of the Fundamental Developments Group. This group's wartime research efforts focused on the exploration of new frequency bands and the development of microwave techniques.

[1] This chapter is based on material drawn from public sources.
[2] Robert V. Pound (1997, p. 2).

The outbreak of World War II interrupted Purcell's work on nuclear magnetic resonance (NMR), but the post-war years saw a renewal of progress. In the United States, two groups of physicists independently sought to develop a new method for observing the magnetic resonance in the nuclei of molecules in liquids and solids. At Harvard, Purcell and his colleagues studied the possibility of detecting the absorption of radio frequency energy by atomic nuclei. Their efforts led to the discovery of the nuclear magnetic resonance method of measuring particular nuclear properties. At Stanford University, Felix Bloch led a team that developed a similar method known as nuclear induction. In 1945, within three weeks of each other, both groups managed to create the conditions necessary to observe the phenomenon.

*In 1952, Purcell joined Edwin Land, James Killian, James Baker, and other noted American scientists on the Air Force Beacon Hill study group [that] advised the government on scientific matters, primarily related to the development and improvement of aerial reconnaissance and high-resolution photography.*

In 1952, Purcell and Bloch were co-recipients of the Nobel Prize in Physics for their landmark discoveries. In his Nobel acceptance speech Purcell said, "To see the world for a moment as something rich and strange is the private reward of many a discovery." Since Purcell's discovery, NMR had become an important component in biology and chemistry. In medicine, magnetic resonance imaging gives physicians a non-invasive tool for diagnosis. The NMR is also used in such diverse scientific disciplines as wildlife biology, archaeology, and nutrition.[3]

Purcell returned to Harvard in 1946 as Associate Professor of Physics. He continued to make breakthroughs with his work in the area of nuclear magnetism, and he made significant achievements and contributions in several areas of physics to include biophysics, molecular physics and astrophysics. In the field of astrophysics, Purcell and Harold Ewen (then a graduate student) collaborated in being the first to detect the spectral line of neutral hydrogen in interstellar space as had been predicted earlier by H.G. van de Hulst. Purcell's experiment detected radio emissions from hydrogen clouds in space. Because hydrogen is the dominant material of the universe, such observations became a prime astronomical tool. Purcell made this discovery with the benefit of a five-hundred dollar research grant, which he used to construct a horn antenna on the roof of Harvard's Lyman Laboratory.

### The Beacon Hill Group

In 1952, Purcell joined Edwin Land, James Killian, James Baker, and other noted American scientists on the Air Force Beacon Hill study group.[4] The Beacon Hill Group advised the government on scientific matters, primarily related to the development and improvement of aerial reconnaissance and high-resolution photography. The group published a report that stressed the importance of aerial reconnaissance while, at the same time, recognizing the political risks inherent in the unauthorized overflight of denied territory.[5] The group concluded:

---

[3] Brebis Bleaney (1999, pp. 45, 437-447).

[4] Edwin H. Land was the CEO of Polaroid and an advisor to Presidents Eisenhower, Kennedy, Johnson, and Nixon. James R. Killian was the President of Massachusetts Institute of Technology and an advisor to President Eisenhower. James G. Baker was an astronomer at Harvard University who designed cameras and lenses used in U.S. aerial reconnaissance. Land, Killian, and Baker are all Founders of National Reconnaissance.

[5] In 1951, the Air Force assembled a group of fifteen experts in response to the Strategic Air Command's requirement for information about targets in the Soviet Union. These reconnaissance experts located their headquarters in Boston on Beacon Hill, which was adopted as the group's code name.

> To avoid political involvements, such aerial reconnaissance must be conducted either from vehicles flying in friendly airspace, or—a decision on this point permitting—from vehicles whose performance is such that they can operate in Soviet airspace with greatly reduced chances of detection or interception.[6]

Figure 5-1. Edward M. Purcell (Photo courtesy of Harvard University News Office.)

The Intelligence Community's response to the Beacon Hill group's recommendation was not overwhelming. However, two years later many of the same scientists (including Purcell) would have another opportunity to press their recommendation, this time with the backing of the President.

**The Technological Capabilities Panel**

In the early part of 1954, President Eisenhower was troubled by what he considered to be one of the most daunting national security concerns: the possibility of a surprise nuclear attack by the Soviet Union. Eisenhower contacted James Killian, President of MIT, with whom he had become friends during Eisenhower's tenure as President of Columbia University from 1948 to 1950. Eisenhower asked Killian to gather the nation's top scientists and engineers to work on an approach to reduce the risk of a surprise nuclear attack.[7]

As instructed by the President, Killian gathered the group of experts and formed what became known as the Technological Capabilities Panel (TCP). The panel's members included many of the scientists from the Beacon Hill study group two years earlier, including Purcell. The TCP met for the first time on 13 September 1954, and the panel was divided into three subcommittees. Purcell was part of the Intelligence Subcommittee (headed by Edwin Land), also known as Project Three, which was tasked to study U.S. intelligence capabilities. Land's group concluded that advanced warning of a Soviet surprise attack could not be assumed given current U.S. strategic intelligence capabilities.

During a fact-finding trip to Washington, Land learned the details of a classified Lockheed proposal for a high-altitude reconnaissance aircraft: the CL-282. When he briefed the rest of the subcommittee about the project, they were enthusiastic about the opportunities presented by this potential new intelligence collection capability.

Both the Central Intelligence Agency (CIA) and the Air Force had misgivings about the CL-282, but Project Three members became aggressive advocates of the program. It was this team's intense lobbying and committed support that eventually convinced Director of Central Intelligence Allen Dulles to approve the CL-282. Nearly two and a half years after Purcell and the other Beacon Hill members made their recommendations for an aerial reconnaissance program, the CIA initiated Project Aquatone, which was the code name for the U-2 program.[8]

---

[6] Gregory W. Pedlow and Donald E. Welzenbach (1998, p. 19).

[7] W.K.H. Panofsky (1999).

[8] Pedlow and Welzenbach (1998, pp. 27-32). CL-282 was the original project name. The aircraft was designated the U-2 in 1955.

### Movement to Stealth

On 4 July 1956, the first U-2 overflight of the Soviet Union took place and, despite assurances to President Eisenhower that most U-2 overflights would be undetectable by the Soviets, this first mission was detected. The Soviet government lodged an official complaint with the U.S. Embassy in Moscow on 10 July. Later that month, both Poland and Czechoslovakia formally protested overflights of their countries as well. By mid-July, it appeared that Eisenhower was losing his enthusiasm for the U-2 program in light of the diplomatic problems caused by the overflights.[9]

*The space program was in its infancy, and it was incumbent on Purcell and his colleagues to persuade others of the opportunities and value of pursuing a robust space program.*

President Eisenhower's reluctance to continue the overflights of Soviet Bloc countries could be mitigated if the aircraft were less vulnerable to detection. Given his knowledge and experience with radar, Purcell wanted an opportunity to work on the problem. He requested that Edwin Land arrange for him to begin to work on the problem in Boston. Land agreed, and Purcell and several other experts in the field of radar technology began to study and experiment with radar-absorbing materials and techniques.

Purcell and his group, known as Project Rainbow, devised materials and processes that reduced the radar cross-section of the U-2, thus rendering the aircraft difficult to detect and observe by radar. These novel materials and methods greatly contributed to the development of various stealth aircraft, including the U-2, A-12, SR-71, F-117 and B-2, as well as other special projects.[10]

### President's Science Advisory Committee

Purcell served on the original President's Science Advisory Committee (PSAC) at the request of President Eisenhower. In 1964, as the Chair of the Intelligence Subcommittee, he recommended development of the NRO Program B follow-on film recovery photoreconnaissance satellite.

Purcell also chaired the Subcommittee on Space. He and Edwin Land co-authored a pamphlet often referred to as the "Space Primer" that was used to educate people on the possibilities of space exploration. The space program was in its infancy, and it was incumbent on Purcell and his colleagues to persuade others of the opportunities and value of pursuing a robust space program. Space exploration was an unfamiliar, technically daunting, and expensive proposition, yet Purcell identified concepts and methods that persuaded others to believe that this new frontier was within reach. Purcell was proud of the degree to which their projections proved accurate as the space program developed in the years that followed. This foresight was particularly true of the moon landings, whose possibility they had articulated.[11]

Purcell and his PSAC subcommittee colleagues exercised considerable influence on the organization of the National Aeronautics and Space Administration (NASA), the space exploration program, and later the conduct of the Apollo missions. One contribution involved their persuading NASA to provide the astronauts with specially designed color stereo cameras to make photographs of the undisturbed lunar surface around the landing site.

---

[9] Pedlow and Welzenbach (1998, pp. 96-111).

[10] The A-12 Oxcart is the CIA version of the Air Force SR-71 Blackbird reconnaissance aircraft.

[11] Pound (1997, p. 5).

**Extraterrestrial Communications and Space Exploration**

Drawing on his earlier breakthrough work with detecting radio emissions from hydrogen clouds in space, Purcell hypothesized on the prospect of communicating with alien civilizations in space, if they existed. Purcell argued that given the vast distances involved, communicating with distant civilizations by radio would be "ridiculously easy, utterly benign, and fun," as compared with the prospect of visiting them in spaceships. However, because of the vast distances involved, many decades would elapse between sending a message and receiving an answer.

Purcell recognized that such a project could fail for political and financial reasons. He observed, "If you spend a lot of money [on attempts to communicate], and go around every ten years and say: 'We haven't heard anything yet,' you can imagine how you would make out before a Congressional committee." In 1993 his concerns became reality when a year-old NASA project to search for alien signals was abruptly cancelled by Congress. Purcell was relieved when the project continued with private funding.

Purcell was fascinated by the question of what radio frequency aliens might use to communicate, since ignorance of this on Earth might make it impossible to detect their signals. Purcell observed, "It would be like trying to meet someone in New York when you have been unable to communicate and agree on a meeting place." He believed his discovery of radio emissions from hydrogen clouds in space had largely solved this problem by offering the equivalent of a Grand Central Station in deep space.

He also confronted the related dilemma of how to determine if a detected signal is artificial or natural, a dilemma he referred to as the "problem of the axe-head." Purcell rhetorically asked, "If an archaeologist finds a lump of stone that looks vaguely like an axe-head, how does he know that it is an axe-head and not an oddly-shaped lump of stone?" His answer was that aliens would likely use the language of mathematics to distinguish their signals.

As far as the prospect of sending people into deep space was concerned, Purcell was quite skeptical. He rejected various exotic schemes posited for interstellar travel, such as the use of anti-matter as fuel to drive vehicles across the distances between vast planetary systems. Purcell attracted the scorn of space travel enthusiasts during a lecture at Brookhaven National Laboratory in 1960. He declared that traveling to the stars in manned spaceships was so difficult and required such prodigious amounts of time and energy, that the idea, "of traveling around the universe in spacesuits—except for local exploration—belongs back where it came from, on the cereal box." Almost three decades later two space engineers continued to take exception to Purcell's remark when they referred to his comment as, "the boldest denial of interstellar flight on record, a distinction of dubious honour."[12]

*Although Purcell spent a great deal of time advising presidents and members of the Intelligence Community, he retained his teaching position at Harvard and maintained his class load. Purcell's son Frank recalled, "Dad was often on the night owl train between Boston and Washington."*

**Relishing His Role as an Educator**

Purcell provided a significant contribution to the United States through his service to the government, but his interest was primarily scientific rather than political. His dedication and natural talent in the research and teaching of science brought him the greatest satisfaction. Although Purcell spent a great deal of time advising presidents and members of the Intelligence Community, he retained his teaching position at Harvard and maintained his

[12] "Edward Purcell: Breakthrough in Space Exploration," *Electronic Telegraph*, 15 March 1997.

class load. Purcell's son Frank recalled, "Dad was often on the night owl train overnight between Boston and Washington."[13]

Among the many honors and awards bestowed upon Purcell during his life, the one that perhaps gave him the greatest satisfaction was the Oersted Medal of the American Association of Physics Teachers.[14] A former student recalled Purcell's commitment to ensuring that all of his students learned:

"A characteristic of Professor Purcell's teaching style and dedication to his students was his constant process of trial and error to ensure understanding. His goal was that everyone in his class would grasp and own all the basic principles of the subject. If a student did not "get it" one way, Professor Purcell would try another, until the understanding came. Professor Purcell was a born teacher whose style infected his students with an insistent curiosity about the ways of the universe."[15]

Following his retirement from teaching, Purcell remained intellectually active and engaged. From 1983 to 1988 he authored a pedagogical column in the *American Journal of Physics* under the heading "Back of the Envelope." In this column he posed problems for thought, the premise being that the problem could be solved quantitatively by a few lines of calculations on the back of an envelope. In the following issue he provided the solution to the problem along with an insightful explanation. This forum was another medium for Purcell to play his most coveted role as educator.[16]

**Conclusion**

Edward Purcell was admired and respected by those who knew him. In speaking of Purcell and his contribution to the PSAC, James Killian wrote, "When [President] Eisenhower was later to speak in memorable tribute of 'my scientists' he was surely recalling among others this quiet, modest, lucid man."[17] Another colleague, Robert Kriedler, referring to Purcell's involvement with the PSAC recalled that, "Ed Purcell did not speak often, but when he did there would be enormous silence in the room, because everybody knew that whatever he said was going to be worth listening to with careful attention."[18]

Purcell taught a large number of students and colleagues with his special insight for explaining complex phenomena in simple ways. In addition to his passion and talent for teaching science, he had a desire to see wars avoided. In 1965 he resigned from the PSAC and severed his ties to other military advisory panels because of his disagreement with U.S. involvement in Vietnam. In 1985, Purcell was among those scientists who opposed a space-based strategic defense system (popularly known at the time as "Star Wars"). He believed such a system would not work and would only serve to escalate the U.S.-Soviet arms race.[19]

Edward Purcell died on 7 March 1997 in Cambridge, Massachusetts. On 27 September 2000, Mr. Keith Hall, the Director of the National Reconnaissance Office, and Mr. George Tenet, the Director of Central Intelligence, posthumously recognized Edward Purcell as a Founder of National Reconnaissance.

—

[13] Telephone interview with Mr. Frank Purcell, 24 July 2001.
[14] Pound (1997, p. 7).
[15] Sussman (2001).
[16] Pound (1997, p. 7).
[17] James R. Killian (1977).
[18] Pound (1997, p. 8).
[19] *Harvard University Gazette*, 13 March 1997.

## Founder Award Citation

**Edward M. Purcell, Ph.D.**

A Harvard Nobel Laureate and radar expert, Dr. Edward Purcell worked on all early overhead reconnaissance projects (U-2, A-12/SR-71, and others) that operated at extreme altitudes. His most significant contribution involved methods to make these vehicles, if not invisible to radar, hard to observe with radar using new materials. This feat led to later efforts that produced Have Blue, the F-117, B-2, and other U.S. special projects. He also chaired the Land Panel subcommittee that selected the Program B follow-on film-recovery reconnaissance system.

*Career in National Reconnaissance: 1950-1965*

—

## References


Bleaney, Brebis. *Biographies of the Members and Fellows of the Royal Society*, 45, 437-447. London: The Royal Society, 1999.

"Dr. E.M. Purcell, 84, Shared Nobel for Work on Hydrogen," *The New York Times*, 10 March 1997. http://nobelprizes.com/nobel/physics/obit-purcell.html (15 November 2001).

"Edward Purcell: Breakthrough in Space Exploration," *Electronic Telegraph*, Issue 659, 15 March 1997. http://www.telegraph.co.uk (15 November 2001).

Killian, James R. *Sputnik, Scientists, and Eisenhower*. Cambridge: MIT Press, 1977.

"Nobel Prize Winner Edward Purcell Dies at 84." *Harvard University Gazette*, 13 March 1997. http://www.news.harvard.edu/gazette/1997/03.13/NobelPrizeWinne.html (15 November 2001).

Pedlow, Gregory W. and Donald E. Welzenbach. *The CIA and the U-2 Program: 1954-1974.* Washington, DC: Central Intelligence Agency, 1998.

Panofsky, W.K.H. "Physics in Arms and Arms Control." *American Physics Society* 1999. http://www.aps.org/units/fps/aoct99.html. (13 August 2000).

Pound, Robert V. "Biographical Memoirs: E. M. Purcell." *National Academy of Sciences*. August 1997. http://www.nap.edu/html/biomems/epurcell.html. (15 May 2001).

Purcell, Frank. Telephone interview with Mr. Jules Sussman, 24 July 2001.

Sussman, Jules. "Re: Prof. Purcell," E-mail correspondence with the Center for the Study of National Reconnaissance, 15 August 2001.

# Section 2 **SIGINT Pioneers**

## Bringing the SIGINT Revolution to Reality

David Hatch, NSA Historian

The world of secret intelligence, like the larger government world its organizations serve, has been marked over time by many revolutions in concept and practice. In the U.S., the revolutions in intelligence have been of two types. Because the American people traditionally have disliked secrecy in government, the first type of intelligence revolution has been the grudging recognition that such organizations are necessary at all. The second type of intelligence revolution has been one of technical innovation.

In some cases, technical innovation has been a reaction to scientific or technical developments from outside government, but in many instances, the Intelligence Community has been on the "cutting edge" of progress in scientific or technical fields. This statement is true for signals intelligence (SIGINT) in general, but it is especially true for SIGINT collection from reconnaissance satellites.

### Historical Context for Conventional SIGINT

Message intercept and cryptanalysis, the staples of signals intelligence production, trace back at least to the Renaissance. Communications intelligence (COMINT) was a product of the 20th century and its amazing new invention, the wireless radio. Almost all countries that adopted wireless communications for domestic purposes also initiated eavesdropping on their neighbors—and the United States was by no means the last to do so.

Signals intelligence has undergone several revolutions. Among these revolutions were: the application of modern mathematics to both making and breaking cipher systems; the development of machines to encrypt and decrypt; and the invention of machines to assist cryptanalysis.

The U.S. Government, which traditionally had not had a civilian intelligence organization, became dependent on signals intelligence during the middle of the 20th century. The ability to exploit the communications of America's enemies during World War II saved thousands of lives and shortened the war by many months. American leaders who depended on SIGINT during wartime also heavily relied upon SIGINT during the Cold War.

**Satellites as a Revolution in SIGINT Collection**

Signals intelligence underwent another revolution at the beginning of the 1960s, namely, the deployment of satellites to collect signals. The first SIGINT reconnaissance satellite, which publicly was known by the unclassified project name Galactic Radiation and Background (Grab) experiment, was launched successfully on 22 June 1960. It represented a powerful new tool in the clandestine arsenal of the Free World. Even though they were somewhat limited in range and coverage, Grab and the other early SIGINT satellites greatly extended the capabilities of American intelligence. This space-based revolution manifested itself in several ways.

The first, most important, and lasting benefit of SIGINT satellites was their ability to extend intercept coverage to previously denied areas. Satellites could conduct recurring intercept from areas well beyond the range of existing types of "conventional" intercept, whether fixed or mobile.

In the early 21st century, which has been characterized by information abundance—some might say information overload—it has been difficult to remember how little was known for certain about the Soviet Union during the early and middle periods of the Cold War. The fear of America's powerful Soviet and Communist adversaries, combined with a lack of data or even misinformation about them, could have led to war by accident. Information was scarce, and reliable human intelligence (HUMINT) was difficult to obtain in those closed societies. Even SIGINT, the stellar achiever of intelligence during World War II, could penetrate only so far.

The collection of SIGINT by conventional aircraft or naval vessels had opened up new vistas for intelligence collection, but sometimes with a terrible price attached. American reconnaissance aircraft and ships were vulnerable, and there were some ugly incidents from the start. Adversaries downed several of these aircraft with the loss of many lives.

As a result of the technical achievements that allowed the U.S. to extend SIGINT into orbit, intelligence analysts gained access to unique types of data on subjects of great interest to policy makers. Whether standing alone, or in combination with photographs from overhead IMINT and reports from HUMINT sources, this new SIGINT data filled many of the big gaps in knowledge about the Soviet Union.

Thus, SIGINT data collected from satellite reconnaissance were of great importance in protecting the security of the United States and its allies. Ultimately, by providing definite information about the strategic and conventional military capabilities of our adversaries in place of half-informed analysis or fear-based speculation, satellite data contributed significantly to stability and the avoidance of war.

Advances in space-based satellite collection capabilities also facilitated the restructuring of the SIGINT community in beneficial ways. Satellite collection made possible the downsizing in the number and range of piloted reconnaissance flights. This meant a reduction in the number of aircrews who would be placed in harm's way. Another benefit was the closing of expensive field sites, an important consideration at a time when the outward "gold flow" was becoming a worrisome problem.

**Contributions of Pioneers**

On a higher level, the demonstration of successful SIGINT collection via satellite during the mid-20th century was a promise for the future. With the feasibility of the concept confirmed, there almost was no limit of what could be accomplished by pioneering scientists and their colleagues.

These pioneers in the space reconnaissance revolution have marvelous stories to tell. Viewed one way, what they narrate for us is the story of how the newest technology was converted to the service of our country. From another perspective, they tell how vision became reality, and how that vision opened up new vistas for American intelligence.

The historical record tells some of this story, as a visit to the archives would show. But sterile records often leave out the most interesting aspects of any process, including this one. The recollections of the national reconnaissance pioneers can tell us what the policy and technical debates were about, why decisions were made, and what the difficulties overcome in implementing them were.

The recollections of the SIGINT pioneers of national reconnaissance in this section bring the story of an important revolution, the early use of space for SIGINT reconnaissance, to life.

# John T. Bennett

John Bennett served as a systems engineer at TRW, where he designed signals intelligence spacecraft for the National Reconnaissance Office Program B. He developed innovative system designs based on the Intelligence Community's requirements, and advanced the nation's national reconnaissance capabilities by achieving technical goals previously thought to be unattainable.

## System Design from Requirements to Launch: Conquering Technical Feasibility[1]

I am eighty years old now, but the memory bank still works remarkably well. What was my impact on national reconnaissance? The spacecraft reflectors and my so-called structural brilliance were significant achievements, but they were not the things of which I am most proud. For some reason, the reflector concept seems to stay in most people's minds as my outstanding achievement, perhaps because it was the most visible.

The thing I am most proud of is my success with the process of taking a set of requirements and coming up with a spacecraft and associated launch components as a system, and then trying to achieve maximum efficiency and performance. In reality, the integrated propulsion, control, and solar power stage—as the first such design to fly in space—was the most significant technological advancement. The launch stage was the key to the success of the mission. If the spacecraft cannot get on orbit, the mission equipment is of little value.

While individual technological achievements are commendable and necessary, it is the selection, blending, and coordinating of all elements into a system that is necessary to achieve the desired goal. I learned that this process generally is not well understood. Over the years, I found that most industry people, as well as the customers, did not know or understand how a set of requirements evolved into a system definition for the spacecraft and launch vehicle.[2] As I saw it, I did a very good job choosing the ingredients, developing recipes, mixing the ingredients, and baking cakes. For me, the cakes were spacecraft and launch vehicles. The reflector was only the frosting on the cake.

[1] This section is based on John T. Bennett's written input as amplified by an interview at the National Reconnaissance Office Headquarters, 26 September 2000.

[2] The term "customer" generally refers to the United States Government and its representatives.

My forte for many years was taking performance requirements and conceiving an overall system to do the task, specifically the spacecraft and launch vehicle configuration. Of equal importance was the selection of a team of individuals to advance the concept studies and to resolve the technology problem areas. I was fortunate to have a long and varied experience choreographing the former, and I developed an extremely capable cadre of teammates who were dedicated to the task.

Figure 6-1. Hawk surface-to-air missile system. (Photo courtesy of US Army, Redstone Arsenal Historical Information.)

We did some terrific things through World War II and the years that followed. I am very proud of being a part of the team of seven guys who put together the Hawk surface-to-air missile, which remained in the inventory for 35 years. In my early years I was involved with materials research, specifically fiberglass laminates, metal-to-metal bonded, and honeycombed structures. Eventually, I got into space work, not only commercial and National Aeronautics and Space Administration (NASA) work, but also classified intelligence systems work.

**My Contributions to Two Signals Intelligence Satellite Programs**

I have strong memories of two signals intelligence (SIGINT) satellite reconnaissance programs, and I believe the story of how we developed these programs is an interesting one. My recollections, naturally, reflect my own perspective. Others who were involved might recall things differently. I will discuss my contributions to the development of these programs within the context of my approach to work in this field.

*My Basic Approach.* My outlook and approach to the work that I did for national reconnaissance never changed, and I have simple objectives that never changed. When I tackled a job, I never tackled it from the standpoint of what I would get out of it. The main draw was the challenge. Of course, I still could take a humdrum job and complete it to satisfaction. People who have self-discipline can do that. What I wanted was to do as good a job as I possibly could. I wanted to do a job that my mother and father would be proud of, if they knew. I think that goes back to being a farm kid during the Depression.

I enjoyed the satisfaction of doing a job with a great group of people whom I respected, and who respected me, and that included the customer. Another thing that I am proud of is my fundamental principles, like honesty. Anything I ever said to a customer was honest. I never arm-waved or exaggerated, or (what they used to call) "dry-mixed" or stacked the data.

My team and I gained considerable experience getting things done "in spite of the system." We did an excellent job (in the customer's opinion) of justifying the establishment of a program. We overcame the company's bureaucratic procedures, management dictators, and a horde of self-styled experts in order to produce a spacecraft configuration and technology demonstration.

The successful placement on orbit and fantastic mission success were the keys to dispelling the previously-perceived technological barrier, thus opening people's minds to better mission possibilities, much as the dispelling of the sound and thermal barriers did for aircraft. This progression was a logical step in the evolution of a better national reconnaissance capability, and I am pleased with contributing to its realization.

*The First Satellite Program.* My most significant contributions to this SIGINT satellite program were the conceptual design of the spacecraft and launch vehicle system, and the direction of conceptual and technical efforts that demonstrated the program's feasibility. At the time, the SIGINT field was one of great interest and need, and had progressed to a point where it was constrained by a perceived technical feasibility barrier. Fortunately for me, I was at the right place at the right time, and had the pleasure of being a contributor to establishing the feasibility of this most important system.

Before the program was initiated, Dr. Bill Carlson (TRW's Chief Scientist) briefed Mr. Fred Kaufman and me on a mission of great interest to the government, but which the scientific community overwhelmingly believed was not technically feasible.[3] Bill stated that the government wanted to conduct a small feasibility study to "put it to bed" (one way or another) once and for all. Having waved that gauntlet in our face, he slyly asked if we were interested in doing the study.

After conferring with the customer, Fred took out his circular slide rule and started doing mission calculations while I hunkered over my trusty drafting table and put together a preliminary configuration of a launch vehicle, spacecraft and injection stage. We defined the preliminary systems, and identified five or six problem areas for further feasibility study. We held discussions with Dr. Lloyd Lauderdale and others.[4] These discussions produced a mutual feeling that this idea just might be feasible in spite of the "experts." We assigned additional people to work on the problem areas, and conducted feasibility tests on hardware components and mechanizations for both mission and spacecraft areas. We updated the system, and Fred and I made a formal presentation (including a terrific mechanically-operating model inspired by Don Nolan), to the customer.

**Figure 6-2. Example of circular slide rule that Fred Kaufman used to calculate preliminary mission parameters. (Photo by NRO Visual Design Center, slide rule courtesy of John Bennett.)**

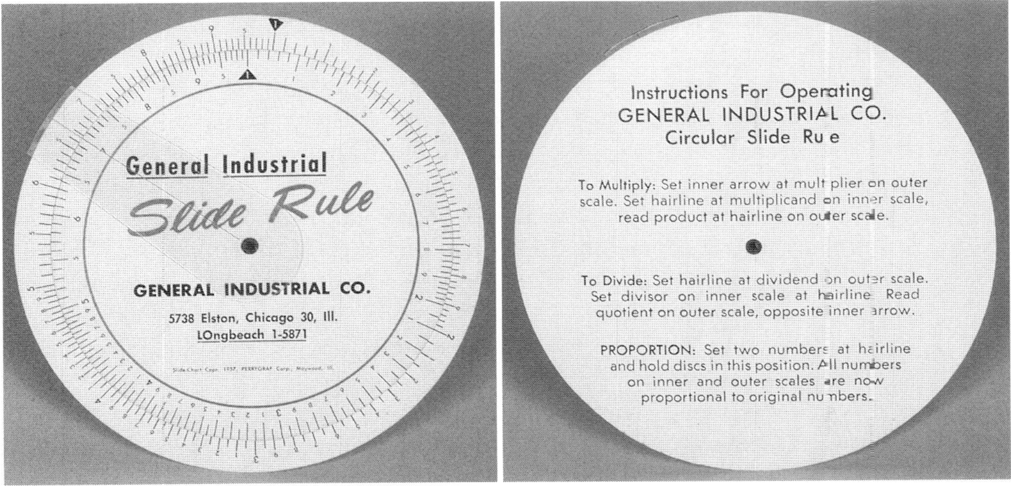


[3] Fred Kaufman is a Pioneer of National Reconnaissance (inducted into the Hall of Pioneers, 27 September 2000). See chapter 12 for his recollections.

[4] Lloyd Lauderdale is a Pioneer of National Reconnaissance (inducted into the Hall of Pioneers, 27 September 2000). See chapter 13.

After much discussion, I was asked if I thought it would work. I answered that I could see no problems that could not be solved by good engineering, and that I would stake my reputation on it. The next day we were happy to hear that the government had decided that the mission was feasible, and that a program would be initiated. Establishment of the program did not quiet the skeptics, and many in the scientific community continued to predict that the system would fail. Even TRW's upper management had difficulty feeling comfortable with the task, but Dr. Carlson recommended to Dr. Reuben Mettler that the program was important to our country, and he decided to take on the task. He had faith in our abilities.

*This may sound a bit cocky, but this cadre never considered anything but success. Failure was not a consideration, and our record was very good.*

The spacecraft and the injection stage were inventions borne of necessity. As I put the first configuration together, I could not make the launch work. I went to a completely different type of reflector structure, which is probably the lesser of the advancements I made. Then I reviewed the injection and the solid rockets, which everyone liked because they are reliable. We had to improve the specific impulse of the solid rocket by 30 or 40 percent, which was a high risk. Then I looked at some liquids, and put together some bi-propellant liquid. This started getting us into the correct orbit, but did not solve the problem of the spacecraft getting too long because of the big engine. I asked myself, "What's the most efficient way I can get into orbit?"

I pondered about a more efficient structure, because everything follows the laws of science. I looked at downsizing the vehicle until it was reduced to a 100-pound thruster. I found that RocketDyne was working on an engine for a control thruster. It was not a regenerator; it was simply radiation-cooled. It was bi-propellant with very small engines. I added two propellant tanks, two oxidizers tank spheres, and put the engine, semi-buried, up in the middle. I put the sphere for the fuel-pull pressurization in the top. Now the vehicle was very short, and we had a stage that used almost everything. The design consisted of a long burn, radiation-cooled, semi-buried, simple pressure-feed system, and I saved more weight and reused the pressurization gas for added spacecraft control gas downstream. Except for the actual tanks, spheres, and bladders, everything was stretching the state-of-the-art. Weight was the real key.

*The Second Satellite Program.* My contribution to the follow-on system was of the same nature as the first program, with the exception of having a much better database and an expanded pool of cleared people to select from. I conducted numerous customer-sponsored and independent studies, which provided insight into the technology required for more advanced missions. This process also helped to identify several outstanding people from the first program's conceptual feasibility phase to add to this team. In the development of this system, the old feasibility issue raised its head again, and once more our team demonstrated the program's technical feasibility.

This follow-on conceptual work was done with most of the team from the first effort. My greatest personal satisfaction on this effort was winning the contract back, and coming through with a successful system after our chief competitor failed. Everyone that contributed to these two successful systems should have great pride and satisfaction in doing the "unfeasible," as well as contributing to the good of our country.

*Fulfilling My Program Goals.* My goals for these programs were always rather simple: To do something challenging and worthwhile, to do the best job possible within my control, to expand my knowledge and experience level, and to share with those involved the satisfaction of a job well done. I believe these goals were fulfilled on both of these programs. We approached each new challenge with an open mind, enthusiasm, dedication, application of time and effort, and a resolve to do the best job possible. That is a hard combination to beat! This may sound a bit cocky, but this cadre never considered anything but success. Failure was not a consideration, and our record was very good.

**A New Outlook and Transition to Retirement**

Right before I retired, my philosophy, and consequently my outlook on my career, changed. There was one aspect of the last program I worked on that impacted my future. It was apparent that the pioneering days of the space business were waning. This trend was particularly evident at TRW, where the predominance of arm-wavers, song-and-dance artists, half-doers, and don't-make-any-waves people had penetrated even the highly classified areas. This development, coupled with the unappreciative attitude of the division Vice-Presidents, who never said "good job," "congratulations," or "thank you" to me or my team for the two largest spacecraft contracts they had ever received, gave rise to some serious thinking on my part that went something like this:

"John, you are 63 years old, in good mental and physical health, and have been in the business for over 42 years in one role or another. You have had a very enjoyable career in a large number of fields with many fascinating challenges. You have known a number of great innovators like Jack Northrop and Theodore Von Karman, and many wonderful colleagues. You have put in roughly 67 years of working time for the 42 years you have been in the business. Is what you just went through worth the self-satisfaction any longer? There are other fascinating things to be learned, such as geology, mineralogy, archeology, paleontology, and the history of our country. There are also fascinating things to be experienced, such as exploring all the old forts, viewing the towns and beauty of our country, enjoying hunting and fishing, and spending time with your wife, friends, and adult children. Why not spend some of the rest of your life doing some new and interesting things? John, you won't get bored!"

I had lengthy discussions with my wife and with my long-time colleagues, whom I found were having similar thoughts. Consequently, I decided to call it quits after I completed one last technology challenge in a relatively new field of space technology. Once this challenge was successfully completed, I retired in July 1983 with no regrets and a healthy outlook for a new phase of my life. We built a comfortable house in the beautiful red rock country in the mountains of central Arizona, in the small town of Sedona, and proceeded to enjoy life fully. At 80 years old, I am still in good health. My wife and I are active in civic affairs, and we pursue the other interesting things we always wanted to know about or do.

*It was apparent that the pioneering days of the space business were waning ... the predominance of arm-wavers, song-and-dance artists, half-doers, and don't-make-any-waves people had penetrated even the highly classified areas.*

**Benefits to TRW and the Country**

Overall, both TRW and the country benefited from our work in national reconnaissance. The company benefited by opening up a new field of business (i.e., covert spacecraft) and by receiving two of the largest contracts for hardware programs in its history. Indirectly, the contracts quadrupled its manufacturing and test facilities, and increased both its skilled labor pool and the transfer of unclassified technology to other uses. The techniques, processes, and basic knowledge learned through these efforts became part of people's craniums (not computer data banks), and naturally influenced their work in the company's other areas. There also was a great potential for developing new product areas, such as ground and space antennas and an all-purpose spacecraft bus and power stage. Similar opportunities were available in the payload section. Unfortunately, the company did not choose to pursue the antennae and space bus, but that is another story.

The country benefited from our national reconnaissance efforts in the areas of military preparedness, military operations, and other aspects of national security. In addition, the intelligence from our programs helped shape United States foreign policy. There also were applications of our work for the private and commercial sector.

**Impact on My Family**

How did my involvement in national reconnaissance impact on my family? My wife and I had been married since 1947, so in effect she had 18 years of pre-conditioning prior to the period where I became heavily involved with the NRO. Much of that earlier period was similar to the NRO period, which included typical 70-hour weeks, but I had 80-hour spurts and an occasional 100-hour week. It also included, of course, periodic out-of-town travel for two to three days at a time.

There were, however, much worse periods. One of these was about a year after our marriage when we both were working 50-hour weeks at Northrop, and I undertook the design and building of a Formula 1 racing airplane. We ate breakfast, drove to work, and occasionally had lunch together. But after work my wife would go home, and I would go to work on the airplane and usually arrive home at about 11:00pm. Saturday was our 8-hour day on the airplane, and Sunday was no more than an occasional 4 hours. After a year, this tapered off to only a Saturday schedule, unless we were attending a race. Now that is what I call getting a marriage off to a good start!

This life was not really as bad as it seems. We were living in a furnished apartment with few household chores. Dorothy made all of her own clothes, and did some modeling. She also could spend some prime time with her grandmother, who had raised her. Being young, full of energy, and enthusiastic also helped us weather this period gracefully.

In 1949, we entered into a business with a partner designing and manufacturing industrial machinery. I was able to do much of the design work at home. I only had to spend Saturdays at the factory for coordination with my partner, who ran the plant. Our wives joined us in catching up on correspondence and distributing a 20,000-30,000-piece advertising mailing. At this time, I was working a 50-hour week at Northrop, but Dorothy had dropped to a 40-hour week, which left her with some spare time. At least now I was at home, even though I was working on the machine and tool design. We generally were able to take vacations and make time for fun.

Figure 6-3. The Bennett children, daughter Nancy at age two and son, John Thomas at age one. (Photo courtesy of John Bennett.)

The real killer came in 1954. My daughter was two years old and my son six months (Dorothy no longer worked—except at raising our children). We had just moved into our first new house, which had dirt for a yard. At this time, six of us from Northrop went to Raytheon in Boston to start the conceptual design of the Hawk surface-to-air missile system. As a result of some misunderstanding at the management level, this two-week trip ended up lasting four months. With daily telephone calls, and Dorothy being the great trooper she is, we survived all the builders' pickup tasks on the house and the responsibilities of having two young children.

A couple of years later, I went to Europe for four months to provide technical direction for sixty companies in five countries in setting up the Hawk missile program for the North Atlantic Treaty Organization (NATO). I was saved this time by returning home three days before Christmas and bearing French perfumes for Dorothy and the most wonderful toys I had ever seen for the kids! What a jewel my wife is. I have been married 54 years now, and I can truly say we never had a fight in the whole 54 years. The reason is mutual respect.

There obviously was a loss of prime time with my wife and children, but never a loss of love, concern, guidance, or security. Dorothy and I provided our children with an appreciation for music, art, and literature. We also gave them the opportunity to experience many things. They not only experienced the usual sports, but they also experienced hunting, fishing, camping, riding, canoeing, shooting the old-time guns (muzzleloaders—flintlocks, etc.), and developed a love and respect for nature. All of this paid off in different ways. For example, we found our children had a better understanding than their classmates (and even many adults) of machines from bicycles to airplanes. They learned how to fix or make things, and about the history of our country. My wife and I shared with them tales of our youth during the Great Depression. This gave them a sense of the real world and an appreciation for what it took to get where we were. We instilled in our children not only a desire for knowledge and learning, but also values, self-discipline, and above all, the ability to think.

*I would offer two lessons for the future of national reconnaissance. Beware of information-age technology and beware of bureaucratic growth.*

Overall, our family appreciated its time together, and we developed confidence, stature, and understanding. My family was never spoiled or selfish, and we were very close. We were happy—what more can I ask?

**Lessons I Would Offer for the Discipline of National Reconnaissance**

I would offer two lessons for the future of national reconnaissance. Beware of information-age technology and beware of bureaucratic growth.

*Information-Age Technology.* I am concerned that information technology and the drive of the computer world is causing people to depend on computers to do their thinking. Computers can be wonderful tools if people do not expect them to do their thinking. Despite the opportunities for a wonderful education that this country offers, we are decreasing the quality of education by making everyone a keypuncher. We need people who can think and not just identify problems, but identify the key problems.

*Bureaucratic Growth.* Over time, bureaucracy infiltrates an organization. In fact, increased staffing sometimes boosts bureaucracy along. If an organization gets too many people, it usually breaks down the jobs so finely that pretty soon it is happy with mediocre people. Then bureaucracy starts to rise. Creeping bureaucracy always changes the outlook of the whole organization. Eventually, the people at the top tend toward self-preservation. The way to succeed is the bureaucratic way and not through technical achievement. That is true for contractors, government civilians, and even military. I was in the Marine Corps during World War II, and I saw it there too. In that way, bureaucracy leads to conservatism and risk aversion.

During the period that I worked for the NRO, from 1965 to 1983, there was relatively little bureaucracy, and I saw people consistently willing to take risks. Given the importance of the mission, people in the organization were expected to take risks, especially technical risks. The front-line personnel had a mission requirement and wanted to know what was feasible, and what were the key problems. I did not see much change in bureaucracy and risk-taking during the years I was associated with national reconnaissance. However, I have no doubt that since then there has been an evolution toward bureaucracy.

**Thoughts on Program and Team Management**[5]

My style of managing a project was rather informal, in that everyone knew our goal and the part he or she played. People were brought up to date on changes, results, and new problems in brief stand-up meetings once or twice a week, or as needed. I communicated with people every day, or every other day, to check their needs and progress, and for coordination. Everyone was always well informed, as was I, so it was never a problem to orchestrate the whole task.

I made it a point to involve team members in customer briefings, test demonstrations, and one-on-one technical discussions, because this contributes to good morale and builds mutual respect. The customer was faced with some difficult decisions and needed to have confidence in the information that was being provided. One constant rule on my projects was to give the facts, good or bad. New ideas and suggestions always were welcomed from everybody, regardless of whether it was in someone's area of expertise. I learned that, "out of the mouths of infants come the most profound statements of the world." I encouraged brainstorming, because I never knew when an answer would pop up or a statement would trigger another thought that might lead to an answer. It was possible to be very productive with creative people.

*Another rule I established was no memos. If someone had something to say, I told that person just to say it. If someone had data or results to transmit, I told that person to provide a copy and explain it in person. For one program, I wrote a total of one internal memo. The team liked that.*

One thing I always spent a lot of effort on was getting people to look diplomatically at alternative ways of doing the job in the interest of simplicity, functional improvement, and easier analysis. We engineers love complicated machines! Sometimes a person works on something until a mental block develops, and then concludes that "there is no better way to do it." Often a gentle nudge here or there will help to get things off center and moved in the right direction.

Another rule I established was no memos. If someone had something to say, I told that person just to say it. If someone had data or results to transmit, I told that person to provide a copy and explain it in person. For one program, I wrote a total of one internal memo. The team liked that. We just dug in our heels and did the best job we could. As far as person-to-person interaction, there were no problems.

The lousy facilities in which we worked under the guise of security contributed to disgruntlement among members of the team. We worked in basements with poor ventilation, water seeping through the walls, often no phones, lack of emergency lights and exit signs. One facility had a disconnected sprinkler system. In some facilities we had to do our own janitorial chores. We never had windows. At times, some of this was a source of humor, and sometimes it was just plain disgusting. I must be honest and say that management support was most obvious by its absence! On one project, I got moved six times to six different locations as few as three or four miles away from our basic plant. Even the unclassified elements I coordinated with became a time-consuming chore. The problem was that we did not have time because we were so focused on achieving our program objectives. It became an irritation to my crew and me. It is a credit to the crew that they persevered and did the job.

The overarching secrecy contributed to some discontent, although working in the classified world never was a problem for me. In the spring of 1941, about a year after I went to

---

[5] See Annex at the end of the chapter for elaboration on the criteria used for selecting a team and team leaders.

Northrop, I was working on the design of one of the little research flying wings because I had some experience in wing design. It was classified Secret, but not need-to-know like we have, because we were working on third-scale experimental airplanes for the B-35 Flying Wing. Of course, my cohorts would want to know what I was doing in preliminary design, and I would say, "Oh, I'm just working on some wood-wing problems." The vague replies bothered a lot of people. Also, there were very good people I would have liked to have on teams, but they could not take the responsibility of the classified information, worrying about a slip of the tongue or other slip-ups.

**My Colleagues and Their Efforts**

For the most part I worked with the same people for seven or eight years, and with a couple of them for up to 23 years. We recognized that some of the younger staff had promising talent, and we accelerated their development by coupling them with old timers. All of them were top-notch people who liked the high-pressure, challenging type of work.

Although I previously worked on ground and airborne reconnaissance programs, Dr. Bill Carlson, TRW's chief scientist, introduced me to space reconnaissance. Bill Carlson was a man I greatly respected and admired for his technical capability and his dedication to the welfare of his company and country. His dedication exacted a very high price, as he literally worked himself to death. In spite of his position and busy schedule, he always took the time to listen to an idea I might have and to bend my ear seeking another opinion on something he was involved with. I hope the company, the NRO, and the country appreciated him.

I have included a few examples of the innovation and humor my colleagues brought to the job. These events and activities show how they got things done, added fun to the workplace, and kept us on our toes.

*Finding a Solution in the Lingerie Department.* Mark Newman entered the Navy just before World War II, and was a quad mount gunner throughout the war. After the war, Mark started out as a draftsman, and with experience and secondary classes, became a very good mechanical designer. Mark had a very droll sense of humor, and was also quite a punster in a reserved manner. When we were building the spacecraft model, we often would go to the shopping center for dinner, and then visit the stores looking for things we might use on the model. Mark was trying to figure what he might use for the downlink antenna surface, so we had wandered around in the fabric department looking at the fabric on several pieces. As we passed the lingerie department, we stopped and looked at some women's slips, then stopped to look at some hose on display. After feeling the material and noting the mesh fineness, Mark said that material might be worth trying, so I said give it a shot. We actually ended up using the hose as an engineering tool! The sales clerk would not sell us just one that we needed, so we had to splurge and buy a pair!

*Using Plastic Wading Pools as Testing Tanks.* Another example of my colleagues' innovation involved the design of the reflectors. One of the tests we performed on the main reflector rib required that we zero-g balance the rib to isolate gravity effects from test effects. Don Corbett came up with a very clever approach that placed the plane of the rib parallel with the floor and incrementally balancing the rib on blocks of foam that float in tanks of water. To conduct the test we added aft guy lines, circumferential lines, and mesh-simulating members. The crew quickly set up the fixed root and support, cut the foam blocks to size, and then went looking for tanks in which to float the blocks, only to find nothing suitable at TRW.

As usual, our guys rose to the challenge. Don Corbett dashed off to a store that sold plastic wading pools and brought back our needs, plus backup spares for a very nominal sum. We placed them on the floor, filled them with water, and set the rib and blocks in place.

Voila! The test worked superbly and was a key factor in showing the feasibility of the main reflector. Later, Don briefed Tony Sablehaus and crew on the test design and test results, and then they were given a tour of the test fixture and witnessed a test in real time. They were really impressed with the simple test set up, and thought the foam blocks floated in wading pools was quite ingenious. They even thought the wild patterns and colors of the pools added an upbeat note! They were impressed with the low cost of the test setup and congratulated us for a very successful component of our feasibility program. An "A" for the good guys!

A couple of weeks later, the test lab supervisor was showing two of our division Vice Presidents around the lab and showed them the rib test set up. When one of the Vice Presidents saw the brightly hued wading pools, he threw a fit and chewed out the lab director about the non-professional appearance and ridiculous equipment. He told him to replace the tanks immediately! An "F" for the good guys! I talked to the Vice President and told him of the customer's previous reaction and that we could change the tanks for $4400 and a 2-week delay in the testing if approved by the customer. Needless to say our "frivolous" tanks were appreciated for quite some time.

*If a conflict is inevitable, set it up so that it will not escalate and no one will lose face. Also, go to the battle with the deck stacked. Who said a successful design is just good engineering?*

*Doing Our Homework and Stacking the Deck.* Not all of the teams' innovations were met with enthusiasm by fellow engineers. NASA had problems with premature failure of tubular fiberglass and graphite reinforced structural members with metallic end fittings. TRW received a contract to investigate this problem, but had not yet developed a solution. Having conducted materiel research for Northrop in the early days of glass reinforced plastic and adhesive bonded metal structure, I believed I understood the problem and the logical solution. I offered my services to the head of the NASA contract, but he was not really interested. Knowing that this could become a feasibility item on our graphite reinforced tubes, we set up our own tests to validate our design. Simple tests on a 2-inch diameter carbon graphite tube with a .040-inch hole in one side resulted in a 20-25 percent reduction in failure load versus a tube with no hole. The addition of doublers around the hole made insignificant changes.

What was the culprit? The laws of physics say that all materiel fails by creep (i.e., exceeding the elastic limit). In most materiel, local yielding to the adjacent materiel will redistribute the load in the materiel interrupted by the hole. However, with carbon graphite having an elastic modulus of 44 million and the fibers themselves being almost classed as a brittle materiel, the yielding for load redistribution around the hole did not take place and the adjacent fibers failed, much like a zipper. The solution was to eliminate the holes or add tube materiel to lower the stress level (more weight). So how did we put fittings on the end of our rib tubes? We made the fittings out of the same materiel as the tubes, so it had the same modulus and coefficient of expansion. We designed a carbon graphite mat and injection molded fitting to be bonded to the tube, which was very light and had no mechanical fasteners.

Now after all this palaver, what is the story? Sooner or later, the materiel and manufacturing people would have to look at the design to provide manufacturing and cost data for a program schedule and cost. Because we also wanted to test a sample, several of us tossed around the best way to obtain one or two test parts. Unknown to us, Luchy Roteliuk made a sketch with some dimensions, and Saturday he called in and said he was sick and would not be in until Monday. Over the weekend, he made a wood fitting, took a female plaster-of-Paris cast, laid up a fitting of automotive repair glass fiber and resin, and cured it in his stove oven.

On Monday morning he came over and laid it on my desk, and it looked very professional. We quickly agreed to band the fitting to a short length of tube and run a quick joint strength test, which it passed. I then called a meeting with the manufacturing and materiel people. We told them that we had decided to use a molded fitting bonded to a tube for the ribs and showed them the design drawings. The materiel man went ballistic! He made declarations like "It won't work—can't make the part—won't carry the load—I won't approve it —I will fight it to the end!" At the first pause, Luchy brought out our specimen and explained that he had made it at home in his garage, and that it had already been tested to design loads. The gentleman picked up the specimen, looked at it, banged it on the table and walked out. So much for a supreme case of NIH (not invented here) ego. Manufacturing found sources to injection-mold the parts, and, as expected, they worked perfectly. This was another "first" to fly in space. What is the moral of this story? If a conflict is inevitable, set it up so that it will not escalate and no one will lose face. Also, go into battle with the deck stacked. Who said a successful design is just good engineering?

*Keeping Humor in Our Work.* Most people in our group, including Don Nolan, were excellent pranksters. Don was born and raised on a dry dirt farm near El Centro, California. After graduation from high school in 1938, he went to the Los Angeles area and studied to be an airplane mechanic. When John Northrop started building airplanes in Hawthorne, California in 1940, Don was hired for sheet metal fabrication. I met him in the spring of 1940 when I was hired for engineering as the new company's 345th employee. Northrop's rapid growth after that left a great shortage of engineers and draftsmen, so the company started their own drafting classes. Don completed the class and came to work in the armament group, where he rapidly advanced to designing armament installations for the B-35 flying wing bomber. Don and I worked on many programs at Northrop and TRW, up to the first NRO project. He became an excellent engineer, and also could build almost anything he designed.

I often leaned over Don's drawing board while discussing a design, and I would pick up a pencil and scratch pad, and forget and walk off with them. Some great person said that intelligence and pranks run in the same mind. One day when I picked up Don's pad it had "Engineer's Pad" neatly printed in red ink across the top. As usual, I took it back to my desk. The next time I reached for Don's pad, he quickly handed me a pad cut in half and a pencil of half-length. Neatly printed on the top of the pad was "Half-Ass Engineer's Pad"—touché! One evening I cut the half pad into quarters and wrote on the top "Engineer's Max Capacity Pad" and left it on Don's desk, along with a one-quarter-length pencil stub. A week later when I came in, there was a quarter-inch wide pad with a note taped to it that said "Important Data at Top of Pad." With the aid of a 4X magnifying glass, I was able to read "A Pad For Engineer That Doesn't Need One." Attached on the end of a thread was a miniature pencil about 1/8 inch diameter and 1/2 inch long with a black point and a little eraser glued on top. At our mid-afternoon coffee time, I had everyone gather around my desk where I put the pads on display, and I presented Don with a handshake and a cake that said "The Winner." After cake, coffee, and a hearty laugh, it was back to work.

**Conclusion**

I thoroughly enjoyed facing the challenges presented to me during my career, and working with people I respected. I got satisfaction from a job well done and the knowledge that we made a significant contribution to the welfare of our country and the advancement of national reconnaissance.

I believe the recognition the NRO gave to the pioneers was very deserving for the selected group of individuals. But there were many other people who were excellent team players.

Many of these people never received one bit of recognition. If they did, it seemed that even senior management was jealous of their success. One of my biggest regrets is that the team we put together never was recognized for its contributions. The team members did not look for attention or glory. Companies had a tendency to pass them by, and I always worked hard to get my teammates recognized. When the customer knows it is not a "song-and-dance job," when contractor and customer can get to know and respect each other, and when the contractor can depend on what the customer says, the situation is beneficial to both. I am convinced that the National Reconnaissance Pioneer Program is a very good idea, and I would have been happy even if I were not selected as a pioneer. In fact, my selection came as a surprise to me.

I wish the great people on my team could have shared this honor. I am truly appreciative that the NRO chose to recognize me for my efforts, and for the type of job I did best. I am deeply honored to be considered a Pioneer of National Reconnaissance, and somewhat awed to be installed in the Hall of Pioneers. I also would like to thank the NRO for the deserved recognition given to the Pioneers' families. This was the first time that I ever encountered such an acknowledgment.

I am quite impressed with what I have seen of the NRO as it moves into the 21st Century. Much has changed since my retirement in 1983. It almost makes me wish I were "tilting windmills" again. The use of NRO systems and technology for other purposes is commendable. Having done considerable intelligence analysis, I realize the difficulty of determining whether disclosure of specific capabilities will or will not contribute to "giving away the store." This can be a tough call to make, but I believe the future looks bright, and the NRO is in capable hands.

Figure 6-4. Dorothy and John Bennett, January 1965. (Photo courtesy of Dorothy Bennett.)

—

### He Worked Long Hours, but No Regrets[6]

During the World War II years, I worked in Flight Test at Douglas Aircraft in El Segundo, California, and later went to Northrop Aircraft, where I worked in the Engineering Department from 1944 through 1952. During this period, I worked in several offices, mainly as project secretary in the B-49 Flying Wing Bomber Office. After that project ended, I was secretary to the Test Administrator and held a Top Secret security clearance.

During the years 1944-1946, we often worked long hours and worked seven days a week for several months. I met John at Northrop when he returned to the Engineering Department after a couple of years

[6] This section is based on written input that Dorothy R. Bennett submitted to the Center for the Study of National Reconnaissance.

in the Marine Corps. We were married in 1947. We had two children. Our daughter was born in 1952, and our son was born in 1953. I stayed home for 18 years and then went back to work for our local unified school district. I worked there for 12 years as the Director of Community Services in the Business Division, and retired in November 1982.

During the years John worked at Northrop, he worked long hours and traveled often. During the twenty-plus years he worked at TRW, he always worked long hours and also traveled often. We never discussed the classified work he was doing and, frankly, this never bothered me. I am sure that for wives who never have worked long hours, especially in classified work, this could be difficult and hard to understand. Having worked in aerospace, I knew John's work was important to the company and our country, and that classified projects required dedication to the job.

Of course, long hours come at some sacrifice, such as less time to spend with family, lost vacations, missing school events and games, and an increased burden on me as the mother to handle things normally taken care of by the father. I became quite good at mowing the lawn and actually came to enjoy it!

Because both of us were Depression children, the work ethic was inherent in our upbringing and actually a source of pride and accomplishment. We passed this on to our children, especially our daughter who is in management at Southern California Gas Company. She is a dedicated workaholic like her dad! Unfortunately, we lost our son who was a Vietnam veteran.

Both John and I had very satisfying careers and I have no regrets. John always has been recognized as a top-notch engineer, and he thoroughly enjoyed the challenges of the jobs. I thank the NRO for its recognition of his efforts and accomplishments.

—

### Memories of Growing Up with a Pioneer as My Father[7]

These are a few memories of growing up with my father. He had a strong work ethic and was immersed in his work, which was a mystery that I knew nothing about. Because my dad worked on classified projects, we rarely had family discussions about his work like most of my friends' families. Another unusual thing was having the neighbors ask me what was going on whenever government men came around asking strange questions about my dad and our family. At school, I was never able to participate in the "share and tell" sessions when students explained what their fathers did for a living. I did not know what my father did at work except that he was an engineer, and I was not supposed to talk about that.

I have other memories about the secrecy surrounding my dad's work. I never had his work address because he seemed to get transferred to a different basement or windowless room every six months or so. On the few occasions I needed to call him at work, I was switched or transferred three to five times to other lines before dad would answer. He always told me not to accept written material from people I did not know, not to talk to strangers, always watch what I say, and not to join any clubs or activities without his approval.

However, years later I realized I had picked up many of dad's work habits and his ethic. The best quality I inherited from my dad is always doing just a little extra in putting out the best work I can. Other habits I picked up from him are working late into the evening and sometimes on weekends, enjoying my work without resenting the time it commands. Clearly, these traits can be both positive and negative.

---

[7] This section is based on written input that Nancy Bennett Walsh submitted to the Center for the Study of National Reconnaissance.

I acquired these qualities from my dad even though his work kept him away much of the time. I sometimes wondered if he actually came home at night or snuck in early in the morning just to have breakfast with me at 6:00 a.m. (which was never my best time back then). I also had to take over a lot of dad's chores at home. I would see him off some mornings on a business trip to an unknown place, for an unknown length of time, for an unknown reason. The bottom line is that although I sometimes did not have dad around for things like school events or Halloween trick-or-treating, he did help turn me into a successful, hardworking, intelligent person.

—

## Pioneer Award Presentation and Citation

Figure 6-5. Pioneer John T. Bennett (second from left) being recognized at the 2000 Pioneer Recognition Ceremony. The pioneer award plaque was presented by DNRO Keith Hall (left) and DCI George Tenet (third from the left). Don Winters (Executive Vice-President and Manager, TRW) joined in the presentation. (Photo by Sara Judy, NRO Visual Design Center.)

**John T. Bennett**

As TRW's chief engineer in support of Program B, Mr. John Bennett conceived the spacecraft design, including the advanced reflectors, used in signals intelligence satellite programs. His advanced designs met severe weight, volume, and dynamic constraints, and revolutionized overhead signals intelligence collection.

*Career in National Reconnaissance: 1965-1982*

—

## Annex

**Criteria for Selecting a Team and Team Leaders**

My criteria for selecting a team is finding people with the following attributes:

Those that have learned the art of thinking and the practice of self-discipline. People cannot do anything without both of those.

Technical capability and well-rounded experience (sheepskin is not a primary criterion—my top two people had only a high school formal education, and a large number were trade school graduates). Performance is the requisite.

Interested in and dedicated to doing a good job.

Capable of determining and defining key problems, rather than just solving a problem given to them. Someone may be brilliant at solving problems, and I could hire those people like they hire secretaries and clerks, but usually I could not find the other ones.

Inquisitive, inventive and ingenuous.

Self-starter who needs little direction.

Open minded to suggestions and other alternatives.

Good team players (no prima donnas, no matter how good they are).

A good sense of humor helps.

The leader of the team should:

Always treat people fairly.

Have respect for the people.

Keep them fully informed.

Give them maximum exposure to the customer's people for good morale and the development of mutual respect.

Solicit their opinion on key items.

Give them credit and appreciation for their effort in the presence of their home base boss, upper management, the customer, and other workers.

Isolate them from company bureaucracy as far as possible.

Set an example by not asking people to do more or make personal sacrifices that you are not willing to do or make yourself.

Have a big coffee pot and keep it full and hot!

# John W. Browning

John Browning led a project that established a new and powerful signals intelligence capability for the United States. He was responsible for the first successful launch and subsequent launches and operations. Browning was honored posthumously as a Pioneer of National Reconnaissance. His wife, Myrnie, recalls some of the qualities she saw in her husband. Her recollections follow a brief description of Browning's contributions to national reconnaissance.

### Trustworthy, Persistent, and a Love for His Work[1]

John Browning had a distinguished military career that spanned four decades. Prior to his work in national reconnaissance, Browning was a highly decorated aviator who saw combat in World War II. He also served as a fighter pilot in the Korean War.

Browning served as the Program Director in the National Reconnaissance Office (NRO) for a critical new signals intelligence (SIGINT) satellite program. After joining the program, Browning organized and led the team that designed, developed, and deployed this SIGINT system. The system provided an original SIGINT gathering capability and dramatically enhanced SIGINT collection. Browning continued as the program director for three years, and he later became the deputy commander of satellite operations at a ground station.

Browning became known as "Colonel Electric" for his many space satellite program achievements. He retired from the Air Force in 1982. He was awarded the Distinguished Service Medal, an honor usually reserved for senior general officers.

Browning died in January 1996. On 27 September 2000, the Director of the National Reconnaissance Office (DNRO) Keith Hall and the Director of Central Intelligence (DCI) George Tenet, posthumously honored Browning as a Pioneer of National Reconnaissance. Browning's wife, Myrnie, accepted the award on his behalf.

—

[1] This section is based on information contained in the material submitted with the nomination of John W. Browning for selection as a National Reconnaissance Pioneer.

Figure 7-1. Colonel John W. Browning, circa 1970s. (Photo courtesy of NRO History Office, likely U.S. Air Force photo.)

## My Husband Knew His Work was Important[2]

John often would have to leave home unexpectedly, and sometimes he would be gone for quite a while. Of course, I never knew where he was going or what he was doing. It was that way from the very beginning of our life in the Air Force, and I never thought anything about it. His absences never bothered me, perhaps because we had a strong marriage.

The traits that made John a success in the Air Force are also the reasons for which I married him. I knew him since I was nine years old. John was a trustworthy person, and we could always count on his word. He was truthful, and he abided by the rules. John lived by his own creed. He had to have good health—that came first. He had to have something interesting to do, he had to have something to look forward to, and he had to have somebody to share it with.

John majored in electrical engineering in college, and one of his nicknames in the Air Force was "Colonel Electric." His colleagues respected John, and he certainly loved what he did in the Air Force. Of course, he could not tell me much about his work, but one thing he told me repeatedly was that he was doing something important. His patriotism was 100 percent!

John never quit. He was persistent, and he persevered. One of John's strongest features was his sense of humor, even toward the end of his life. A couple of years before he died he said, "I not only won World War II single handedly, but Ronald Reagan and I are responsible for the fall of the Soviet Union."

Figure 7-2. Myrnie Browning standing by John Browning's pioneer medallion in the NRO Hall of Pioneers on Pioneer Recollections Day, 26 September 2000. (Photo by Sara Judy, NRO Visual Design Center.)

[2] This section is based on an interview with Myrnie Browning, wife of John W. Browning.

—

## Pioneer Award Presentation and Citation

Figure 7-3. Myrnie Browning (second from left), wife of pioneer John Browning. Mrs. Browning received the pioneer award plaque for her husband who was recognized posthumously at the 2000 Pioneer Recognition Ceremony. The award was presented by DNRO Keith Hall (left) and DCI George Tenet (third from the left). Brigadier General Craig P. Weston (Director, Corporate Operations Office and Chief Information Officer, NRO) joined the presentation. (Photo by Sara Judy, NRO Visual Design Center.)

**John W. Browning, Colonel, USAF**

Colonel John Browning directed a key signals intelligence satellite project for Program A. He managed the first launch of this satellite and oversaw its successful operation. This effort produced a satellite that marked a breakthrough in signals collection.

*Career in National Reconnaissance: 1967-1975*

# Jon H. Bryson

Jon Bryson directed the development, acquisition, and operation of a Program A signals intelligence satellite system that handled rapidly increasing data rates. He helped reduce the time necessary for signals intelligence data processing and dissemination to military and national intelligence users and decision makers through the use of on-orbit satellite processing of complex signals.

## Providing Timely Intelligence[1]

In the early 1960s, I was a Georgia-Tech co-op student working for the National Security Agency (NSA) at Ft. Meade, Maryland. I returned to work at Ft. Meade after I received my degree in electrical engineering. My wife also worked there, and one day she phoned me to tell me about a new NSA job posting. I was a second lieutenant at the time, and the posting was for a major or a GS-13 civilian. I read the job description and said to myself, "I can do all that stuff." I ran down to the office that posted the job. The door had a red seal on it, which meant there was classified work going on inside. They asked me some questions, and the next thing I knew, I was on my way to work with a small detachment of NSA people who supported the National Reconnaissance Office (NRO) signals intelligence (SIGINT) program. From that exposure, I knew I had found my line of work, and in about two years I had formally joined the NRO.

**Soviet Strategic Threat**

It was the mid-1960s at the time, and by today's standards the technological problems were much more challenging and the working environment was still evolving. Basically, everybody was focused on the Soviet strategic threat. Specifically, how many ballistic missiles were they building?

By the time I arrived, the focus had shifted a bit to the Soviet anti-ballistic missile (ABM) program. We were trying to figure out how to apply both imagery and SIGINT to that problem, while continuing to monitor their strategic weapons construction. To accomplish this and other missions, we had to deal with both industry and the Intelligence Community, and then try to figure out what intelligence problems could be addressed by technologies that industry was developing. We tried to keep industry aware of the hot topics of the day,

[1] This section is based on an interview with Jon Bryson at the National Reconnaissance Office Headquarters, 26 September 2000.

Figure 8-1. Soviet SS-13 Intercontinental Ballistic Missile. (Photo courtesy of Defense Intelligence Agency.)

and we encouraged them to keep us informed of any relevant ongoing projects. One of the notable figures that I recall was Bill Perry who, at that time, was the engineering manager for the Electronic Defense Laboratories of Sylvania/GE in Mountain View, California.[2] His work produced several solutions to problems associated with collecting intelligence on foreign communication systems. He later went on to greater glory as the President of the Electronic Systems Laboratory, and then as Secretary of Defense.

Another difficult challenge related to foreign ballistic missile programs. We knew the Soviets were test-firing missiles once in a while, because we would take pictures of these missiles. Frequently, the missiles were on the pad one day and gone the next time we looked.

**Reducing Time Lines**

The minimal oversight that characterized the early days of strategic reconnaissance tended to favor initiative over rank. One of the great things about working in this field in the 1960s and early 1970s was that if someone had the right combination of objective and spirit, the bosses tended to let that person take some fairly risky approaches to things. High morale and enthusiasm proved to be as valuable as credentials in terms of crafting solutions to the many challenges we faced.

When I first came aboard there was an extended period of time between the collection of satellite-derived intelligence and its availability for use by the consumer. There were plenty of people around who said the intelligence could never be delivered faster. Some of us in SIGINT took that on as a challenge. We studied how the data was transmitted from the collector to the processing center, how it was processed, and how it was disseminated. We defied conventional wisdom by reducing the processing time.

We pushed ideas for speeding up the process based on the notion that intelligence is good, but timely intelligence is better. That was our focus. In that same timeframe, leaders in the field of imagery intelligence were confronting similar challenges. Edwin Land was advocating an electro-optical system, and there was a general feeling across the NRO that we were collecting good intelligence, but that it was necessary to get it to the users faster.[3]

Once we improved the time lines, we started to evaluate the range of customers that could benefit from "good and fast" intelligence. Of course, the military was foremost in our minds, since tactical users need rapid intelligence more than just about anyone. As we began to demonstrate these improved capabilities, the military showed increased interest in our efforts. Those of us who were military officers took it upon ourselves to effectively market our "good and fast" approach. We did so with reasonable success, although there still were plenty of skeptics in the military who viewed us as "spooks" from Washington.

---

[2] William J. Perry is a Founder of National Reconnaissance.

[3] Edwin H. Land is a Founder of National Reconnaissance. (See chapter 4 for a discussion of his contributions).

**Importance and Impact of Our Work**

The military and national leadership put a lot of emphasis on our work. We received excellent logistical support even though the Vietnam War was raging. If we needed an airplane to fly a satellite to a launch base, the Air Force would take it out of the Vietnam re-supply pipeline to accommodate our needs. I was impressed that they were willing to disrupt the transport of supplies to the war zone to fly our satellite. That showed us how important they considered our work, and we felt a tremendous obligation to do the best job we could.

Our work gave a big boost to a variety of technologies. We worked hard to develop technology that would mitigate our primary constraints of weight and power—large-scale integration electronic circuits and denser memories. It took us a long time to develop reliable, solid-state memory. In contrast, you now can buy memory and a host of circuitry for your home computer in quantities we never imagined.

There is a particular story that especially demonstrates how our work gave a boost to technology. I was working on a program that nobody remembers anymore. The type of wide-band recorder we needed was being produced by one of the contractors for television stations, but each of the recorders weighed about 2,000 pounds. We asked the contractor if they could make the same technology in a briefcase size. They worked on it, and sure enough they did it. We used it several times, but we could never quite master the magnetic tape in a vacuum, so the program eventually ended. The contractor took a look at this thing, could not see a market for it, and sold the rights. Eventually that technology showed up in what became video cassettes.

**Conclusion**

One of the aspects that made it interesting and rewarding to work for the NRO between 1966 and 1990 was the "walling off" of our organization from the normal government bureaucracy. Not only did that simplify our job, but also it gave us a sense that we were part of something special. There was a personal drive to make things work—to provide more intelligence, to provide it at higher quality, and to provide it faster. I moved from program to program and found that there were different technical challenges, but the objectives tended to be the same. The mission was exciting for a young officer right out of college. Everybody had substantive assignments and responsibilities, and nobody got lost in the bureaucratic swirl. We were really accomplishing something, and we were able to see the fruits of our

Figure 8-2. Pioneer Jon Bryson (left) interviewed by Dr. Moe Rosen on Pioneer Recollections Day, NRO Headquarters, 26 September 2000. (Photo by Candi Campbell, NRO Visual Design Center.)

labor. We also were able to understand how our work was being used to make important decisions and contributions to national security.

In being recognized as an NRO Pioneer, I see myself as a representative of a great number of people who did a tremendous amount of good work. They succeeded at difficult and challenging tasks that resulted in many family hardships. We accomplished many good things, and we achieved impressive results. Collectively, we won the Cold War, and that is a pretty positive accomplishment.

—

## Pioneer Award Presentation and Citation

Figure 8-3. Pioneer Jon Bryson (second from left) being recognized at the 2000 Pioneer Recognition Ceremony. The pioneer award plaque was presented by DNRO Keith Hall (left) and DCI George Tenet (third from the left). Brigadier General Craig P. Weston (Director, Corporate Operations Office and Chief Information Officer, NRO) joined in the presentation. (Photo by Sara Judy, NRO Visual Design Center.)

**Jon H. Bryson, Colonel, USAF**

Colonel Jon Bryson directed the development, acquisition, and operation of a Program A signals intelligence satellite system that handled rapidly increasing data rates. He also helped reduce signals intelligence data processing and dissemination through the use of on-orbit satellite processing of complex signals.

*Career in National Reconnaissance: 1966-1992*

# John O. Copley

John Copley guided the development and refinement of the National Reconnaissance Office Program A signals intelligence satellites from the earliest experiments to the phase-in of constellations that provided broader coverage. His improvements in payload design and ground data processing permitted the Intelligence Community to exploit more effectively Program A's signals intelligence capabilities.

## Developing Early Signals Intelligence Programs[1]

World War II profoundly influenced my life. I often wonder what would have happened if Germany had not invaded Poland in 1939, or the Japanese had not bombed Pearl Harbor in 1941. Two things I am sure of: I never would have joined the military, and I never would have learned to fly an airplane.

I recall that I resisted my father's encouragement that I apply to West Point. I did not apply, but even had I done so, I certainly would have joined the Signal Corps rather than the Air Corps. I was absorbed with things electronic since the seventh grade. How did I find myself in the fall of 1944 flying a B-24 out of Grottaglia, Italy, and then spending the next thirty-one years in the Air Force? The answer is that in 1942 my best thoughts told me to make the trek to Springfield, Massachusetts, in order to apply for Aviation Cadet Training. It seemed to be the surest way to avoid finding myself in a foxhole, under fire, and wishing I were anywhere else.

My thirty-one years with the military are a little harder to explain. By the summer of 1945 I was serving as the base communications officer at Hondo airfield in Texas, and trying to convince the Air Force that I would be much better off as a civilian. One thing led to another, and I soon found myself assigned to Scott Field, Illinois, training radio operators and mechanics (though I had no formal military training in any area of communications or electronics). Then came a year on Johnston Island (a dot in the Pacific 800 miles southwest of Hawaii and much farther from anywhere else), a year in Korea, and a few months in Hawaii thrown in for good measure. I always worked in the areas of communications and electronics.

In 1952 I was fortunate to be assigned to Wright-Patterson Air Force Base in Dayton, Ohio, which was—among other things—the home of the Air Force Institute of Technology. Through dogged persistence, I finally received my B.S.E.E. By September 1955 I was

[1] This section is based on written input that John O. Copley submitted to the Center for the Study of National Reconnaissance.

transferred to the Rome Air Development Center in New York. I was now ready to start my career in the research and development of Air Force electronic systems.

If World War II was the most calamitous event of my life, the launch of Sputnik in September 1957 had to be the second. Following Sputnik's launch, tremendous effort was expended to develop U.S. programs to assure that the Soviets did not achieve superiority in the satellite arena.

**Air Force Initiates Space Reconnaissance**

In 1954 the Air Force Research and Development Command (ARDC) initiated the first satellite reconnaissance program, Weapon System (WS) 117-L. Colonel Bill King was assigned a five-man detachment and charged with the establishment of a project office at Wright Air Development Center (WADC) in Dayton, Ohio. Lieutenant Colonel Vic Genez at ARDC Headquarters was convinced of the feasibility of reconnaissance satellites, and that conviction finally resulted in the initiation of a development program. However, funding was limited, and the members of the program office were generally dismissed as "space cadets."

In the spring of 1957 the Air Force sponsored a competition for the development of a spacecraft capable of orbiting reconnaissance payloads. Lockheed won the competition and was awarded a multi-million dollar study contract with the Secretary of the Air Force's specific direction to "bend no tin." The study resulted in the definition of the Agena spacecraft and an eventual program of over 300 vehicles. However, the atmosphere changed abruptly following Sputnik. The multi-million dollar study contract became a hardware contract worth five times the original amount, and with a launch date as soon as possible. Thus began my association with space reconnaissance systems.

**My Introduction to Space Reconnaissance**

When the Soviet Union launched Sputnik, I was stationed at the Rome Air Development Center and involved in the development of ground-based electronic intelligence systems. By January 1958 I found myself in Inglewood, California, as the project officer in the WS-117L Program Office. I was responsible for the development of the early (low orbiting) signals intelligence (SIGINT) programs of NRO Program A.[2]

The size of the WS-117L Program Office (later to develop into NRO Program A) increased to over twenty people and continued to expand rapidly. The development of a satellite photoreconnaissance capability was given the highest priority, and the classified project was given the name Corona.

*Assigned to the Discoverer Program.* Many organizational changes occurred as the reconnaissance satellite program developed, with the most significant being the creation of the Special Projects Office under General Bob Greer at the Space and Missiles Organization on the west coast. Because of the extremely tight security surrounding the Corona program, a cover program was created to give it the appearance of a benign scientific project. Initially, that cover was called the Discoverer Program, and it was vaguely explained as an Air Force scientific program whose findings would be of value to many related programs. I was assigned to the Discoverer Program Office headed by Colonel Lee Battle.[3] I found this to be an excellent arrangement. The support from this program office was outstanding, and all I had to worry about was the payload.

---

[2] Program A embraced the U.S. Air Force element in the National Reconnaissance Program. It was established in July 1962 and disestablished in December 1992.

[3] C. Lee Battle is a Pioneer of National Reconnaissance (inducted into the Hall of Pioneers, 27 September 2000). See chapter 21 for his recollections.

Figure 9-1. Pioneer John Copley preparing input for this publication at the NRO Headquarters during Pioneer Recollections Day on 26 September 2000. (Photo by Candi Campbell, NRO Visual Design Center.)

*Moving to an Independent SIGINT Office.* One of our initial contractors in California developed several highly innovative payloads. However, its participation came to an end following the student riots for peace in 1961. Our chief project officer, Jim de Broekert, spent an entire night and day personally guarding the safe in which our classified material was stored.[4] Eventually we spirited Jim and the safe to a more secure location. This ended the participation of the contractor in the program, but Jim went on to develop many more payloads and remained a valuable advisor throughout the program.

During this time, all of Program A's SIGINT activities were my responsibility. John O'Connell, who was my very able assistant, joined me while we were still part of the Discoverer program. In November 1962 we determined that SIGINT satellites were developing to the point where they merited a program office of their own. Consequently, SIGINT was activated as part of the Program A Office, and we bid a fond farewell to Colonel Battle and those associated with Discoverer.

**Later Assignments**

I was transferred to Air Force Systems Command in July 1964, and was assigned to a program that the Air Force cancelled several years later. I was fortunate to be reassigned to the National Security Agency (NSA), where I headed a division and was responsible for processing data from NRO Program A satellites.

Our primary objective was to intercept, locate, and identify targets in the Soviet Union. There also was growing emphasis on anti-ballistic missile (ABM) emitters. By 1968 data from these payloads and the follow-on systems had identified early ABM-associated radars, greatly reducing the uncertainty associated with the Soviet strategic threat.

While I was in the SIGINT collection business I received invaluable assistance from the Central Intelligence Agency (CIA), which assigned someone to work with me. The CIA's intelligence inputs were invaluable in designing our payloads. Additionally, I was able to develop a close and mutually beneficial relationship with Howard Lorenzen, who was responsible for the development of the Navy SIGINT Grab satellites.[5]

I worked closely with NSA personnel during my Program A days, and we developed a

[4] James de Broekert is a Pioneer of National Reconnaissance (inducted into the Hall of Pioneers, 27 September 2000). See chapter 10 for his recollections.

[5] Howard O. Lorenzen is a Pioneer of National Reconnaissance (inducted into the Hall of Pioneers, 27 September 2000).

rapport. We worked hard to ensure that data collected by the satellites could be processed and exploited on the ground. I considered it a great privilege to find myself responsible for processing the data that I used to be responsible for collecting. I believe this illustrated the cooperative atmosphere we were able to develop when I was the data collector. For example, one of my first activities when I joined the WS-117L program was to visit NSA and discuss our plans.

### Conclusion

Although many of our early efforts were not as successful as we would have preferred, most of them were part of a learning curve in an area never before charted. As such, they provided the foundation for many of the follow-on SIGINT satellite programs. Later, higher-orbiting satellites carried these systems to a much more sophisticated level. Many of the electronic techniques developed for these programs also provided technology for a variety of later electronic systems.

My assignments with the NRO and NSA in the area of SIGINT satellite development were by far the highlights of my Air Force career. We were in a highly-motivated organization with a sense of the importance of our contributions to the welfare of the United States. The streamlined organization of the NRO made it possible for us to accomplish tasks in weeks that could have taken months or years in military organizations. It also was possible for me to work closely with all the organizations involved in SIGINT satellite operations. Needless to say, I was always very motivated, and it was great to feel every day that we were doing something valuable for our country.

—

**Pioneer Award Presentation and Citation**

Figure 9-2. Pioneer John Copley (second from left) being recognized at the 2000 Pioneer Recognition Ceremony. The pioneer award plaque was presented on 27 September 2000 by DNRO Keith Hall (left) and DCI George Tenet (third from the left). Brigadier General Craig P. Weston (Director, Corporate Operations Office and Chief Information Officer) joined in the presentation. (Photo by Sara Judy, NRO Visual Design Center.)

**John O. Copley, Colonel, USAF**

Colonel John Copley guided the development of Program A signals intelligence satellites from the earliest experiments to the phase-in of constellations that provided broader coverage. His improvements in payload design and ground data processing permitted the Intelligence Community to effectively exploit its signals intelligence satellites.

*Career in National Reconnaissance: 1958-1975*

# James C. de Broekert

James de Broekert's career in national reconnaissance has spanned over four decades. He was a major contributor to the successful development of the first low-earth-orbiting electronic intelligence systems, and later made significant contributions to signals intelligence concept development for several national reconnaissance programs.

## From Campus to Industry: A Career in Signals Intelligence National Reconnaissance[1]

My first job after college was with North American Aviation where I worked on building the displays for the F-100 fighter's radar system, and I also worked on the range tracking loops for tracking airborne targets. These jobs helped develop my knowledge of radar technology.

I was interested in continuing my education, so I applied to Stanford University for graduate school where there was a research assistantship available for someone who was familiar with radar. The Stanford Electronics Laboratories (SEL) wanted to do some development activity with radar. Based on my experience at North American Aviation, I was admitted to Stanford and given the assistantship.

### Stanford and My Introduction to National Reconnaissance

At the time, SEL was involved in national reconnaissance projects, which were new to me. I worked on radar projects, and I also was asked to help design reconnaissance receivers. These receivers were new forms of receivers based on technology that was being developed at Stanford, specifically traveling wave tubes and other components. In the late 1950s transistors were also coming into common use, and I was the first to be given the job at SEL of transitioning the previous vacuum-tube designs into transistorized solid-state designs. We were trying to develop low-power, high-performance circuits.

These projects had considerable technical overlap, especially with regard to satellite design where power, size, and weight are critically important factors. It was fortunate that the projects I was involved with were related to each other. This helped us design and build systems because the knowledge and experience we gained in one project was often transferable to another project.

[1] This section is based on an interview with James de Broekert at the National Reconnaissance Office Headquarters, 26 September 2000.

One of our early efforts involved building a receiver for the Army using items we had on hand at the lab. This receiver was designed to be used for battlefield surveillance. One day when we were testing the system, we picked up a radar signal while overflying a body of water. We assumed it was a ship, but for the next few days we continued to pick up the same signal at the same location out in the water. Finally, we took a small battery-powered receiver in a plane and flew over the body of water. What we found was that an airport had built a dike in the bay and mounted a radar on it.

**Raising the Question of Listening from Space**

At SEL we developed signals intelligence (SIGINT) reconnaissance systems and processes for a number of government customers. The original security caveat for the space work was "Special Handling." Initially, it was just a separate compartment with its own "need to know" access requirements. When the National Reconnaissance Office (NRO) was established, we were given access without receiving additional security indoctrination, because we had been doing national reconnaissance work all along.

Most of the government agencies and contractors involved with the development of electronics intelligence (ELINT) systems and techniques attended regular Technical Advisory Committee (TAC) meetings. One of the functions of these meetings was to provide a forum where people working in the field with the necessary clearances could share information about ongoing projects and review ideas for new concepts and systems.

I met Dr. William Harris at a Stanford TAC meeting in August 1959, and subsequently became involved in space reconnaissance. Harris visited SEL during this TAC meeting, and saw the small battery powered receiver we engineered. We took a subsequent pocket-sized version of that receiver on airline flights and held it up to the window, logging radar signals as we flew. Harris took an interest in our small ELINT receiver and asked me, "Do you think you can put that on a satellite?"

**Applying Signals Intelligence Technology to Satellites**

This was an exciting opportunity for us. Instead of flying at 10,000 or 30,000 feet, we could be up at 100 to 300 miles and have a larger field of view and cover much greater geographical area more rapidly. The challenges were establishing geolocation and intercepting the desired signals from such a great distance. Another challenge was ensuring that the design was adapted to handle the large number of signals that would be intercepted by the satellite. We created a simple model that was used to determine the probability of intercept on the desired radars. The model also determined the interference environment from the other radar signals that might be in the field of view. That was very challenging.

*Of all the payloads we built at Stanford, all but one of them worked perfectly to the end of the vehicle life. One vehicle blew up on launch and went into the water.*

We were enthusiastic and pleased with the outcome of our work. My function was primarily to develop the system concept and to establish the system parameters. I was the team leader, but the payloads were usually built as a one-man project with one technician and perhaps a second support engineer. Everything we built at Stanford was essentially built with stockroom parts. We had a qualified parts list and we knew which parts to avoid, but there were no high reliability parts in any of these systems. We built the flight-ready items in the laboratory, and then put them through the shake and shock fall test and temperature cycling. We did not do thermal vacuum tests in those days because we did not think it was necessary. These were relatively simple systems compared to today's sophisticated payloads, but they did a good job. Of all the payloads we built at Stanford, all but one of them worked perfectly to the end of the vehicle life. One vehicle blew up on launch and went into the water. Consequently, we were generally pleased with the reliability of our systems.

Figure 10-1. Pioneer James de Broekert testing a portable battery-operated receiver in 1958. (Photo courtesy of James de Broekert.)

**Stanford and Student Protests**

In the late 1950s Stanford had various Department of Defense (DoD) contracts with the Army, Navy, and Air Force, as well as contracts with the Central Intelligence Agency (CIA). During the 1960s many students had strong negative attitudes toward Stanford maintaining these relationships.

In April 1969, the student protest movement against universities doing classified research reached a crescendo around the country. For example, a truck bomb loaded with fertilizer was used to blow up the Army Mathematics Research Center at the University of Wisconsin, which killed one researcher. At Stanford the protest was against SEL. The student radicals, together with their outside professional advisors, took over the classified research laboratory by force and held it as "liberated territory" for nine days. They put posters printed in Moscow and China on the walls, including Che Guevara posters. These actions were part of their five-year plan for the revolution, and they were two years into it by then. I watched them carry out their activities as the year unfolded, and I was able to anticipate much of what they would do. They learned how to make Molotov cocktails, burn Reserve Officers' Training Corps (ROTC) buildings, and engage in other types of protest.

The university president at the time, Kenneth Pitzer, who had served as an Atomic Energy Commission Commissioner, supported the student radicals and stated publicly that he did not believe it was proper for the university to collaborate with the evil and amoral United States government. We at SEL were surprised that he made such a statement. Although he made these kinds of statements to the media, he told the people on the campus quite a different story. In the end, he closed the research lab in October 1969 and fired all of us: the staff, the faculty, the graduate students doing doctoral dissertations—everyone. Stanford actually defaulted on contracts and told the government to take its money and get lost! The university washed its hands of classified research. I believe that, to this day, Stanford has no classified research contracts.

**The Transition from Campus to Industry**

The environment that prevailed at Stanford and other universities at this time contributed to my decision to start my own company. Naturally, it was distressing to see SEL falling apart, and I felt like my career was coming to an end. Another group leader and I decided to set up a for-profit company to see if we could continue the same work for the same customers as a private company instead of a university laboratory. We started a company that did systems engineering and built reconnaissance-related hardware. We opened the doors in November 1969, and the company grew to 1,500 people some ten years later. Needless to say, this endeavor was successful.

The company's hardware business expanded, and this presented a corporate conflict of interest with my systems engineering and technical advisory work because the same contractor is prohibited from both manufacturing and evaluating, or recommending products for the government. I left that company in April 1976, worked with a small start-up company, then in April 1978 founded a new company called Advent Systems where we could continue with the systems engineering and advisory work without supplying hardware to the space programs. The Air Force decided to contract with Advent Systems to continue the systems engineering and technical support work.

When I was at Stanford I had no idea that we were going to be thrown off the campus. I also had no idea that I would be involved in starting new companies. In retrospect, even though it was very distressing at the time, it probably was the best thing that could have happened. The closure of the lab gave those of us who worked there an opportunity to continue to make contributions with more freedom and, certainly, with more financial reward.

**Lessons Learned**

I have had the opportunity to be involved in the national reconnaissance field for the past five decades. This experience provided me with some insights and observations that I offer in the hope that they may be useful to current and future professionals in the field. These lessons include: use unbiased study teams for concept development; implement streamlined management; and beware of bureaucratic growth.

*Use Unbiased Study Teams for Concept Development.* During my time at Stanford we were frequently given a technical requirement by the government, and it was our job to develop a solution. Often, we would start from little more than a one-line statement of work. We developed the system concept and then followed it through to development, either by ourselves or by another contractor. During that time period I was quite satisfied with the results of our efforts.

However, in the early 1970s the Air Force and the government decided to do things differently. Rather than having the concept development and systems engineering studies done by study teams or unbiased organizations, the government decided to rely on the big aerospace contractors to do the concept definition and systems engineering work. As a result, companies that may have been motivated primarily by corporate interests and profit often performed studies. The government gave up much of its control when it made that change. The concepts that came out of these studies were not, in my judgment, always the most cost effective solutions to a specific problem. I believe the government and the taxpayers began to lose value for their dollar.

*[I]t is not like in the early days when we once went from chalk on a blackboard to on-orbit operations in eighteen weeks.*

In my opinion, systems concepts should be developed by organizations that do not have a financial interest in supplying the flight hardware.

*Implement Streamlined Management.* As the NRO enters the 21st Century, the organization faces a challenge in terms of streamlining its management structure and processes. Development programs for space reconnaissance systems are frequently quite lengthy—even five to ten years—and often done largely by committees. Even though there is some sense of urgency, it is not like in the early days when we once went from chalk on a blackboard to on-orbit operations in eighteen weeks. That time period included designing, building and testing the receiver. All in eighteen weeks!

With regard to program management, during the early years of national reconnaissance, the financial and technical authority for the government were often concentrated in one

[2] John Copley is a Pioneer of National Reconnaissance (inducted into the NRO Hall of Pioneers on 27 September 2000). See chapter 9 for his recollections.

person. For example, Colonel John Copley, who served on the west coast in the early 1960s, had both the financial and technical authority to make important decisions.[2] If there were a technical problem that needed to be solved, we would propose a systems engineering solution. If he liked it, he would say, "Go do it," and we could immediately begin the work.

*If there were a technical problem that needed to be solved, we would propose a systems engineering solution. If he liked it, he would say, "Go do it," and we could immediately begin the work.*

Another factor that contributed to streamlined management was the fact that we had a long-term contract with a year-by-year renewal. Because we were under contract, we could order parts and begin building things right away. This arrangement provided flexibility, which was another reason the early programs could be operational in short periods of time.

For the purposes of time, efficiency, and cost savings, I believe the NRO should attempt to streamline program management and return to more direct lines of authority and accountability.

*Beware of Bureaucratic Growth.* Seeing the field of national reconnaissance develop over the years has been heartening, but I am concerned about apparent bureaucratic growth. For example, satellite reconnaissance engineering began as a relatively modest enterprise, with a small team of people building systems that were largely experimental. When users saw what they could get out of these systems, they wanted more. Contributing to the improvements in the quantity and quality of the intelligence collected has been quite satisfying for me personally and professionally.

On the other hand, substantial bureaucratic growth has accompanied these successes and program costs have increased dramatically. It now takes hundreds of people to do the work that used to be done by tens of people, or fewer. I have often wondered if there is a way to do a better job of containing costs. But, in fairness, the programs have become far more sophisticated over time so perhaps more people are needed. I suppose that is a trade-off.

Another bureaucratic trend is management by committee. I have seen examples of this process in action, and I wonder how efficient it is when compared with management by a strong, competent leader. For example, in the early days of one particular SIGINT program there was a Lockheed engineer who served as a program manager until 1978. He was an extremely confident and technically competent engineer. He made good decisions on a technical basis, but he was also a good manager, so he made good decisions from a management point of view. While I would not call his management of the program a one-man-show, there was never any question as to whom was running the program. He had the technical and financial authority required to run the program effectively. As a result, the program was efficiently run, with a small number of people doing good work at relatively low cost.

**Secrecy**

Secrecy was paramount in my work, but I did not consider the secrecy a major problem from my family's point of view. My family understood that I could not talk much about what I did. There were certain things I could tell them, and we stopped there. For example, they did not know that my work was related to satellites. When brief conversations did occur, they were limited to airborne reconnaissance since that subject was not classified like space-based reconnaissance. In fact, "airborne reconnaissance" seemed like a good cover, especially since my colleagues and I were also involved with projects in that area.

**Conclusion**

My career in national reconnaissance has provided me with an opportunity to do interesting work while making a contribution to national security. Being a part of the national reconnaissance field over the past five decades has been satisfying, particularly in the sense that I have been in a position to watch the field develop and mature.

I am honored to be recognized as a Pioneer of National Reconnaissance. This recognition has afforded me an opportunity to share with my family some things about the important work that my colleagues and I have been involved with over the years.

—

### My Husband Could be Called "Mr. Security"

Jim could be called "Mr. Security" for all of the work-related information he passed along to me. During the early years of our relationship and marriage I asked specific, pointed questions about what he did during the working day. His responses were always something like, "I can't talk about it;" "It's classified;" "Fine, what did you do?" and so forth. Eventually, I stopped asking. Other than some words such as "radar," "pulse," and "bandwidth" that I would hear mentioned when he spoke with his colleagues, I did not have a clue what his life was like when he left home in the morning.

*If it has been hard on me to not know what he does, it must have been harder for him to close and lock that door when he left work at the end of the day.*

Every now and then, however, something would be in the news about a satellite launch, a foreign military operation, a government employee defecting or being caught spying. I would ask him if he knew anything about it. He would give his usual response ("I can't talk about it" or "it's classified"), but with an additional little smile, which I did not know how to interpret. I vowed I would ask him about them again many years later. I have yet to do that.

The first significant clue I received about his 9-to-5 life came when the book *The Puzzle Palace,* about the National Security Agency, was published in 1983. One of Jim's coworkers loaned the book to me. I found it most interesting, but I understood relatively little of what I read. Some names and events were familiar, having heard them from the news. From reading that book, I added a few more terms to my vocabulary, and I understood a bit more about the tools and techniques of intelligence collection.

Now that some of the work Jim was involved with has been declassified, he has been slightly more forthcoming with information about his other life, and I can better appreciate its importance to our country and to the world. If it has been hard on me to not know what he does, it must have been harder for him to close and lock that door when he left work at the end of the day.

In our case, curiosity did not kill the cat, but it did provide for lively speculation in my mind for almost 33 years. I am grateful for the opportunity to learn from the experts what has been taking place. Perhaps now Jim no longer will have to be so guarded in his responses. Even if he does tell me in detail what he did, I probably will not understand it, so I still will not know!

—

## Pioneer Award Presentation and Citation

Figure 10-2. James de Broekert (second from left) being recognized at the 2000 Pioneer Recognition Ceremony. The pioneer award plaque was presented by DNRO Keith Hall (left) and DCI George Tenet (third from the left). G.W. Mackay (President, Advent Systems) joined in the presentation. (Photo by Sara Judy, NRO Visual Design Center.)

**James C. de Broekert**

Mr. James de Broekert, a contractor with Advent Systems, Inc., in the early 1960s contributed key payload designs for several of Program A's first generation signals intelligence satellites. His pioneering work introduced wideband distributed amplifiers and pulse signal processors.

*Career in National Reconnaissance: 1960-2000*

# Gary S. Geyer

Gary Geyer served as the Air Force Program Manager for a variety of low-earth-orbiting signals and imagery intelligence satellite programs, managing their technical development and improvement. He developed a team-focused managerial approach, new acquisition techniques, and innovative ways to package collection systems, in addition to innumerable technical improvements to these systems. His work provided the nation with vital capabilities essential to our national defense.

## Challenges and Innovation in System Development[1]

I spent most of my Air Force career with the Office of the Secretary of the Air Force for Special Projects (SAFSP).[2] I was fortunate during my career to be able to work on many different satellite reconnaissance systems in various capacities, and to have the opportunity to contribute to the success of these programs that added so much to our national reconnaissance capability. It is important to remember that any accomplishments are really those of a team, most often a group of teams.

### Working at the Satellite Control Facility

I visited southern California about fifteen months before graduating with a Bachelor of Science degree in Electrical Engineering from Ohio State University, where I also received my Air Force commission via the Reserve Officer Training Corps. I wanted my first assignment to be in the engineering area.

After working for General Electric in Evendale, Ohio, during the summer after I graduated, I received orders to report to the Air Force Satellite Contract Facility on 1 September 1966. There was a typo in my orders, and the "Satellite Contract Facility" turned out to be "Headquarters of the Satellite Control Facility (SCF)." I traveled in a new Triumph Spitfire from Ohio to California along with my bride, our dog, and most of our clothes.

We found an apartment in Hermosa near the beach for $115 per month. I reported to the SCF, one of the buildings that housed the Space and Missile Systems Organization,

[1] This section is based on written input that Gary Geyer submitted to the Center for the Study of National Reconnaissance.

[2] The SAFSP was the overt office designator for the National Reconnaissance Office's Program A and was based in Los Angeles, CA. This office also was known as Special Projects (SP).

Figure 11-1. The Satellite Control Facility in Sunnyvale, California in the early 1960s. (Photo courtesy of National Reconnaissance Office Reading Room.)

later the Space Systems Division. After I met my new boss, I was invited to a party at the Officer's Club where I spent most of the night talking to the new commander, Colonel Bill King.[3] The welcome we received that night and over the next few weeks was a significant factor in my decision to make the Air Force a career, along with the fact that the work was fun, and the people were both smart and friendly.

These were interesting times for a new brown bar.[4] Many experienced acquisition types were in Vietnam or supporting that effort, so there were some interesting jobs available. The space business seemed like a mature business at the time, but looking back it really was in its infancy. Much of the equipment, techniques and procedures developed at that time and shortly before set the baseline for the decades to come.

Understandably, we suffered some tough setbacks, including running an antenna feed into an overpass. Also, while carrying equipment from California to the New Hampshire tracking station, we overturned a truck into a ditch on a rain-slicked road in Cleveland. That accident set us back about six months. We also had to keep the existing SCF up and running, with the necessary upgrades to keep the current equipment compatible with evolving satellite requirements. Despite the busy schedule, technical challenges, and pretty strict cost constraints, this was a great assignment to start my career because it allowed me to participate in the development and operations of state-of-the-art technology and systems.

*Hectic Development Schedules, Multiple Projects.* The Air Force was just starting to upgrade the SCF. The Space Ground Link Subsystem (SGLS) had been developed, but not yet deployed.[5] We were deploying the worldwide set of sixty-foot antennas that SGLS would utilize. We were developing and deploying the pulse-code modulation ground stations that would process the received data. The initial facilities for the Satellite Test Center in Sunnyvale had to be expanded, and the "Blue Cube" was being built.[6] The procurement, development, and deployment of these capabilities were a top priority, and they were on an extremely tight schedule.

[3] William G. King (retired as Brigadier General) served as Program A Director from August 1969-March 1971. His attention to detail and quality control in vehicle assembly played a pivotal role in the continuing success of imaging platforms.

[4] "Brown bar" refers to the insignia of a second lieutenant.

[5] The SGLS became, and remains, the standard for command, control and telemetry readout for most spacecraft.

[6] The "Blue Cube" is part of the Satellite Test Center facility in Sunnyvale, CA.

The Defense Support Program was being developed at the same time as well, and there was a requirement for a 20MHz tape recorder. After three rounds of attempts to find a credible builder, the Program Office gave up and went to a more sophisticated downlink modulation to shrink the downlink bandwidth to 1MHz. We never developed the 20MHz recorder. In light of today's super bandwidth, it is difficult to understand what was so challenging about 20MHz. But that was 1967.

*Operations.* Operations also were fun. I missed out on some of the real fun because I did not yet possess the necessary security clearances, but I still had some interesting times such as calibrating the new sixty-foot antennas. At SGLS frequencies, it was not practical to calibrate the far field of deployed antennas using terrestrial sources. Therefore, we had the Test Wing at Wright-Patterson Air Force Base equip a C-135 with a simulated SGLS downlink. We then flew the airplane around the world calibrating the antennas. We flew patterns for eight hours at a time at 40,000 feet through the far field of our deployed antennas. Even at eight hours per day, it still took two or three days to get the job done. Three of us were put on hazardous duty status (the extra $105 per month was great) to coordinate with the flight crew and supervise the operation of the test equipment.

My first assignment at the Test Wing was to calibrate the new antenna at the Indian Ocean Tracking Station. The flight plan to Mauritius Island, which was to serve as our daily base of operations, required refueling at Johannesburg, South Africa. The other two duty officers were African-American, and because of the policy of apartheid, it was decided that I should make the trip to South Africa. It was a lousy reason for a trip around the world.

**Assignment to Special Projects**

My Air Force Special Projects (SP) career almost did not happen. When my orders to report to the Air Force Institute of Technology (AFIT) arrived in August of 1969, I was already working on obtaining a post-AFIT assignment. I stopped by my colonel's office (Jim Heyroth), and explained that I had met some of the Special Projects officers who really seemed to like what they did. I asked him, "What do I have to do to get a post-school assignment in SP?" When I finished eating lunch that day, I walked by the boss' office on the way back to my office. "I understand you want to work for me, young man," bellowed then-Brigadier General Bill King, the Director of SP. While I was at lunch, Colonel Heyroth had called over to SP, probably to the General, and set things in motion. If I had not gone to see Colonel Heyroth, if he had not called General King, and if General King had not acted, I probably would have parted ways with the space business and never returned to that area.

*The Defense Support Program was being developed...and there was a requirement for a 20MHz tape recorder [but] the Program Office went to a more sophisticated downlink modulation... We never developed the 20MHz recorder. In light of today's super bandwidth, it is difficult to understand what was so challenging about 20MHz. But that was 1967.*

When we arrived in Los Angeles in 1966, my wife Jan got a job with SP. I will never forget the day she was briefed. She never said a word, but she was wide-eyed. Over the next 18 months I received dual master's degrees in Aeronautical Engineering (AE) and Electrical Engineering (EE) from the University of Southern California. I exploited a quirk in the system that allowed me to pursue two degrees simultaneously. The Air Force paid for the AE degree, but I knew that I would be going into an EE type of job, so I took all my electives in EE. In the middle of the program I spoke with the EE Dean who agreed to apply my EE electives toward an EE degree, and apply my AE required courses as electives.

I started my SAFSP assignment in early February 1971. I was supposed to work for Lieutenant Colonel Jim Bagwell, but he fired me the morning I arrived. What happened was that there were two openings in SP, and Don Beede arrived the same day I did. Don was authorized for access to the intelligence product, and I was not (in those days, not everybody had access). Consequently, I was reassigned to a component of SP that did launch interface and launch vehicle integration.

**Experience With Launching Satellites**

Initially, I was very disappointed because the job in the Program Office seemed more glamorous. In actuality, it was the best thing that could have happened to me. I ended up working on several different programs, and I worked for some good people like Lieutenant Colonel Jim Everret, Colonel Jack Simonton, and Major (later Major General) Nate Lindsay.[7] My immediate supervisor was Major Jay Starnes, a true national reconnaissance innovator who invented and perfected the National Reconnaissance Office's (NRO) concept of launch vehicle integration. He initiated and developed the concept and procedures for independent checks and balances to assure the successful launch of NRO satellites.

Figure 11-2. Atlas launch, 1960 (Photo courtesy of NRO History Office, likely U.S. Air Force photo.)

I had a principal role in the development of the launch segment for several different programs. In one of these programs, I worked closely with another NRO Pioneer—Pete Wilhelm from the Naval Research Laboratory (NRL).[8] Pete was primarily involved in launching NRL signals intelligence (SIGINT) satellites at Vandenberg Air Force Base (AFB). Pete needed more lift capability for the new, bigger program on which we were working.

My boss at the time, Colonel (later Major General) Hank Stelling, wanted to put the satellite on a Titan III-B out of the Space Launch Complex 4W (SLC-4W). This would have given the program all of the performance it needed, and also helped Hank with the program launch costs. The problem was that the Titan was too expensive. Working together, we decided to use the Atlas vehicles that were being taken off line. They were surplus to Strategic Air Command, and could be refurbished by the launch crew at Vandenberg between launches.

*Assessing Performance Problems.* There were, however, several problems related to performance. First, the only Atlas launch pads were the Advanced Ballistic Reentry System pads in north Vandenberg AFB. If the vehicle were launched in the necessary direction

[7] Jack Simonton served as the Director of the SP SIGINT Programs group from 1972-1974. Before that, he was in charge of SIGINT launch operations, as a launch specialist and an expert in the launch vehicle area. Nathan J. Lindsay (retired as a Major General) served as Program A Director from February 1987-January 1993. Serving as the last full-time director, he continued the tradition of mission success, insuring program integration and co-location of facilities.

[8] Peter G. Wilhelm is a Pioneer of National Reconnaissance (inducted into the Hall of Pioneers, 27 September 2000). See chapter 19 for his recollections.

from that location and had to dogleg back consistent with that inclination, it would have decreased seriously the amount of weight it could deliver to the desired orbit. This option was unacceptable. Another option would have required the evacuation of the Vandenberg AFB housing area every time there was a countdown for this "secret" launch.

Colonel Dave Parrish, the SP Deputy Director, dispatched me to Vandenberg AFB to meet with the Launch Wing Commander, Brigadier General Jessup Lowe, to convince him of this evacuation option. Luckily for me, General Lowe had succeeded General King as the SCF Commander, so I had worked for him and he already knew me. Instead of throwing me out on my butt or laughing in my face, he presented the unclassified briefing to his staff, and then invited me into his classified area. When we were alone, he said something to the effect of, "What the hell is going on here?" I explained the program, the predecessor, the mission, and the schedule. He explained that there was no way we could afford or keep a secure operation if we had to evacuate for each launch. He suggested we modify the existing Thor pad to launch the Atlas out of SLC-3 in south Vandenberg AFB. I returned to Los Angeles with his recommendation.

*...modifying the tower would cost more money than the program had available. I got a call from Colonel McBride telling me to visit the NRO Comptroller's office to ask him for the money. I recall the terror those orders struck in my junior captain's heart.*

*Resolving the Cost Issue.* Colonel Charlie McBride was the SP budget and finance officer at the time. He estimated that modifying the tower would cost more money than the program had available. A few weeks later, I was visiting Pete Wilhelm at NRL when I got a call from Colonel McBride telling me to visit the NRO Comptroller's office the next day and to ask him for the money. I recall the terror that those orders struck in my junior captain's heart. The NRO Comptroller was all-powerful. He was the one person who held the NRO budget together at a time when relatively few people were cleared and authorized for access to this information. His reputation, deserved or not, was that he ate junior officers for lunch, and I was going to ask for significant out-of-cycle funding. I shook through the entire briefing. The NRO comptroller smoked his cigarette and mumbled a few things to people close to him at the table. When I finished speaking I waited for the explosion, when to my astonishment he quietly said, "OK, you got the money." I could not believe it. I later figured out that Colonel McBride had been in there the day before and had greased the skids. He could have told me about that and let me get a good night's sleep, or he could have let me suffer. I guess he thought I needed the experience.

*Returning to the Atlas Solution.* To use the Atlas to inject the payload into the final mission orbit meant that the second stage would be orbital. That was a lot of aluminum to be floating around that far up. Also, we did not have the performance margin needed in the early phases of this new program. We solved both problems at the same time. It was a hot summer day in Washington when we sat down in Pete Wilhelm's office at NRL. The wind from the south was blowing the aroma from the sewage treatment plant into his office. After kicking around a lot of ideas, we settled on a ballistic orbit for the Atlas. The apogee would be at the altitude of the mission orbit. This resulted in the second stage de-orbiting, while increasing the weight delivered to apogee. The only problem was that if nothing were done, the payload also would de-orbit. Pete did some quick calculations and designed the initial concept for an apogee kick motor that would provide the needed delta velocity to insert the payload into the final mission orbit. Not a bad afternoon's work! The apogee kick motor served multiple functions, including fine-tuning the final mission orbit.

Over the next four years, Pete and I, along with our contractor, developed a launch vehicle program that met the program's budget and delivered a reliable product. We cut the fat, but not the meat, from the budgets for traditional integration programs. We were able do this because of the great working relationship Pete and I shared, and the resulting relationship between our teams. The pad and the Atlas were finished before I transferred to a new job, and the first launch was successful. It was really special to watch it using the satellite's sensors from orbit. This was one of the first times we used on-orbit sensors to monitor our own launches.

**Later Special Projects Responsibilities**

I had the opportunity to participate in numerous satellite reconnaissance programs within SAFSP. These assignments provided me with comprehensive experience in the launch and operation of these systems, and represented some of the most exciting times in my career.

I was involved first in the procurement and launch of the Agena upper stage for SIGINT programs. During these years the Agenas performed flawlessly with the exception of one Agena that did not get a chance to operate on-orbit because of a failure of the Atlas booster that resulted in a nine-mile apogee. My office also managed the Agenas for a follow-on program that had an upgraded guidance system, but a different group of officers was responsible for that program. In that program, seven out of eight launches were successful. The second launch failed because of a defect in the new guidance system. They detected, solved, reengineered, rebuilt, retested and successfully launched again in just eighteen months. That was a truly great management and engineering accomplishment that I witnessed, but unfortunately one in which I did not participate.

*Supporting Signals Intelligence.* I also was involved in the successful procurement and launch of two SIGINT programs. I then began working on another program, which unfortunately was canceled. While I was the only NRO interface for this program, the outstanding professionalism, dedication and hard work of several people on the Titan side made this endeavor successful. Another great part of this program was working with my program office contact, fellow NRO Pioneer Forrest Stieg. These benefits made this program interesting, even though it was short-lived. Those years, however, included some tough work and resulted in some revolutionary new contracting and launch procedures. I am sure the working relationship and procedures we developed have been changed and improved, but the success of subsequent NRO launches has been a testimony to our success.

Figure 11-3. President George H.W. Bush. (Photo courtesy of NRO History Office, original source unknown.)

*Interfacing with the North American Aerospace Defense Command.* In 1975, I left Los Angeles and was reassigned within SAFSP. We were operating three satellites and working on several other exciting developments. The system's mission was to collect SIGINT for technical and military applications. While the System Program Office (SPO) did the actual development, the ground station had a significant amount of input on the requirements, interfaces, and training. The North American Aerospace Defense Command (NORAD) became much more interested, and the interface became much more complicated. I was the principal interface with NORAD, working with Major (later Brigadier General) Charlie Bishop in defining the details of that important interface. Bishop was later in charge of intelligence for the Air Force in Europe. I also briefed then-DCI George H.W. Bush on the operations of the system.

**Final Air Force Challenges**

After tours at the Air Command and Staff College and the Pentagon, I returned to SAFSP in Los Angeles. When I arrived, I began working on a program that was just one month old. I was put in charge of the new spacecraft, a major development effort that was made more difficult by the lack of a prime contractor. Air Force SP traditionally selected a prime contractor and let that contractor select its own subcontractors. In this case, the space segment ended up with three associate contractors that hated each other. It was a real management challenge.

Later, I was put in charge of the entire space segment, but the problems with feuding associate contractors remained. One of the contractors was actually working behind the scenes to kill the program, hoping to be able to compete for the contract again. Additionally, the program changed rocket configurations three times, each of which required a re-design of the spacecraft. We started with the initial upper stage, then we were forced to switch to increase performance across the fleet, and finally we ended up with a third option. It is a wonder that the program ever finished. This perhaps was the toughest management job I ever had. Despite schedule delays and funding problems, the program delivered some highly capable satellites.

*This program was the first to send finished intelligence directly to tactical users. This was not only a technical challenge, but an organizational challenge. The National Security Agency wanted all the data to pass through them, and they became both my principal customer and principal adversary to this new and important capability.*

In 1985, General Ralph Jacobson asked me to take over a related program.[9] This was the first of two jobs in which I would follow NRO Pioneer Jon Bryson, completing several of his initiatives.[10] This program was the first to send finished intelligence directly to tactical users. This was not only a technical challenge, but also an organizational challenge. The National Security Agency wanted all the data to pass through them, and they became both my principal customer and principal adversary to this new and important capability.

In 1989, I followed Bryson to take over as SPO Director for another program, and again implemented payload upgrades that Jon started. We successfully operated multiple satellites at the same time. The biggest contribution of this program was the support we provided to operational military commanders.

**Work in Industry**

Upon my retirement from the Air Force in 1992, I went to work for Martin Marietta (later Lockheed Martin). I ran the payload development effort and delivered a payload the year I left. In my last two years, I served as Vice President for Defense Systems, and was responsible for satellite building and operations. We miraculously extended the lives of a number of satellites in orbit, and launched or readied others for launch while developing a new system.

**Conclusion**

I am pleased that I was able to contribute so much to these reconnaissance programs during my career. I had the opportunity to do important work for national security, and I am proud of the success of these programs.

---

[9] Ralph Jacobson (retired as Major General) served as Program A Director from January 1983-February 1987. He played an instrumental role in the genesis of a new reconnaissance program. His background in astronautical engineering and ballistic missiles was critical to the program's success.

[10] Jon H. Bryson is a Pioneer of National Reconnaissance (inducted into the Hall of Pioneers, 27 September 2000). See chapter 8 for his recollections.

—

**Pioneer Award Presentation and Citation**

Figure 11-4. Pioneer Gary Geyer (second from left) being recognized at the 2000 Pioneer Recognition Ceremony. The pioneer award plaque was presented by DNRO Keith Hall (left) and DCI George Tenet (third from the left). Brigadier General Craig P. Weston (Director, Corporate Operations Office and Chief Information Officer) joined in the presentation. (Photo by Sara Judy, NRO Visual Design Center.)

**Gary S. Geyer, Colonel, USAF**

During a career beginning in 1966 with Program A, Colonel Gary Geyer's work resulted in notable improvements in signals intelligence, data processing, and dissemination that permitted the product to reach military and civil users in near real time. He later directed crucial advances in an important overhead imaging satellite program.

*Career in National Reconnaissance: 1966-1999*

# Frederick H. Kaufman

Frederick Kaufman led the TRW team with vision and technical innovation in the development of the concept for an early overhead signals intelligence system. He directed the solution to the problem of launching a flexible, low frequency, mechanical structure on a conventional launch vehicle. He also fostered the development of a liquid apogee injection engine that enhanced the capability of a launch vehicle to boost heavy payloads into geosynchronous orbit. His contributions to systems engineering and management, coupled with creative technology applications, enabled the National Reconnaissance Office to field multiple generations of collection systems that have served the nation for over thirty years.

## My Career in National Reconnaissance[1]

I was the son of an Army Sergeant who was in the Army Engineers. My father was assigned to Schofield Barracks in Hawaii, and I was there on 7 December 1941. That night we went by Pearl Harbor and saw the place on fire, and lit up like it was daytime. Tracers stitched the sky. That intelligence disaster burned an image into my memory that will last forever. As I looked out at the devastation, I remember thinking that if I had the opportunity to do anything to keep that from happening again, I would do my very best.

My memory of 7 December remains with me today. I still have a piece of a Japanese bomb fragment that landed on our roof. When I dug the fragment out of the roof, I noticed that it was shaped like the state of California. This proved to be an ironic coincidence because the *USS California* was one of the ships damaged in the attack, and the California theme continued throughout my life. I was stationed in San Francisco, California, while I was in the Navy; I attended graduate school at the University of California; and I married a California girl.

### Early Influences

After four years at Notre Dame, I went on active duty with the Navy for three years, and I did well with the teams of people with whom I worked. I left the Navy to attend graduate school at the University of California, Berkeley, and I enjoyed that environment in terms of competing with many talented people. I received my master's degree in electrical engineering

[1] This section is based on an interview with Fred Kaufman at the National Reconnaissance Office Headquarters, 26 September 2000.

Figure 12-1. Planes and hangers burning at Wheeler Army Air Field, Oahu, soon after it was attacked the morning of 7 December 1941. Fred Kaufman's home was located just beyond the lower right corner of the photo. (Photo courtesy of U.S. Naval Historical Center.)

Figure 12-2. California-shaped bomb fragment in front of U.S. Navy photograph of damaged *USS California*. The Navy photo is as it appears in the book, *Pearl Harbor: December 7, 1941: America's Darkest Day*, Susan Wels, San Diego: Time Life Books, circa 2001. (Photo courtesy of Fred Kaufman.)

in February 1956, and went to work at Ramo-Woolridge (RW).[2] My thesis advisor, Dr. John Whinnery, consulted for RW and facilitated the initial contact. I worked on the development of missile flight control systems. What more fun can a young man have than blowing up missiles in front of the whole nation! I had found my calling!

**Designing and Building Satellites at TRW**
Following several corporate changes, RW became a group within TRW. I was involved with TRW's Atlas and Titan development programs for the next five years. I also had the opportunity to participate in advanced satellite studies of all kinds, including an opportunity to help Werner Von Braun try to rendezvous and dock two missiles in orbit. I first thought that this was a wild idea but thought, "Let's see what we can do with it." When the time came to work on national reconnaissance programs, I had accumulated knowledge, skills, and abilities that helped me deal with the unique challenges of designing and building new satellite systems.

*The Proposal that Became a National Reconnaissance Office Program.* It was the first day of September 1964, and I had just returned from one of my camping trips in Yosemite. When I went to work, my secretary gave me a note saying that my boss, a TRW Vice President, wanted to see me in his office right after lunch. When I arrived in his office, he handed me an envelope. I opened the envelope, and it contained a Request For Proposal (RFP) letter asking for a study proposal. I had been on vacation for two or three weeks and was away when this letter arrived. During my absence the RFP made its way around the company looking for a leader. Complicating matters was the fact that the letter had a security restriction that meant only ten people in the company were authorized to read its contents. I was put in charge of the effort, which had only three other people. The four of us were expected to do this highly classified proposal by ourselves. Gulp!

[2] Ramo-Woolridge (RW) was a wholly-owned corporation of Thompson Products. Through various reorganizations, RW changed its name to Space Technology Labs (STL), then TRW-STL, before becoming a group within TRW.

*The Government's Objective.* The government planned to select three companies to do the study, and TRW was one of the three. We found out later that the government almost did not pick us because our proposal was not relevant to what it wanted to do. Fortunately, the government included us on the study team, and we went to Washington to the headquarters offices to review the briefing material on what the government wanted. I damn near fainted! I had misjudged severely the scope of what the government wanted this system to do. It was clear that I would need people with antenna expertise. I had not submitted any antenna people for the required access authorization, and now it was too late! One of the first things I had to do when I returned home was to have a meeting with the manager of the antenna department. I told him that he had to teach me all that he knew in three hours. This was only one of many roadblocks we had to overcome.

*Problems in the Company.* To put things in context, several related activities were going on in parallel to our new study. First, an analysis group had completed a study on how to collect the same types of signals. It determined that there was no practical way of doing this except by satellite. I later learned that this study triggered the government to determine that there should be a formal study to evaluate further the use of satellites for collecting this kind of data. I learned this from Lloyd Lauderdale who told me that the National Reconnaissance Office's (NRO) Program B had started those studies to prove the idea was not feasible.[3] Therefore, our study actually was commissioned to prove that we could not collect these signals by satellite. I am glad I did not know this at the time!

Next, there were many things going on in TRW that would impact my study. At the time, the unclassified side of TRW was working on an InTelSat III competition, which was an unclassified international ComSatCorp program. The company was under a great deal of pressure and in a very difficult competitive environment, and there were challenging cost and design requirements. The whole company was caught up in this effort, and when TRW won the competition, management was busy trying to staff the project. At the same time, TRW had been asked to do the system engineering on another NRO program. Almost everyone who had the necessary clearances and access authorizations was assigned to help with the project. As if that were not enough, there was some internal conflict with a group of people in our electronic division that worked on airplane collection systems. That group did not want to lose its business to the satellite side of the company.

With all of these things going on, I was assigned the task to lead the study. I guess I was the only one around who had a chance of getting the study done. Upper management was aware of these other factors and probably believed that this study would not be taken very seriously anyway. Several of the senior managers thought the whole concept was impossible. Fortunately, no one told us. I am pleased to say that the design that we came up with looked a lot like what the NRO later deployed successfully.

*Upper management was aware of these other factors and probably believed that this study would not be taken very seriously anyway. Several of the senior managers thought the whole concept was impossible. Fortunately, no one told us.*

*Initial Government Review.* We had two serious problems at our first government review. First, we had a problem with the launch vehicle, which was no surprise. The slenderness

---

[3] Lloyd Lauderdale is a Pioneer of National Reconnaissance (inducted into the Hall of Pioneers, 27 September 2000). See chapter 13. Program B embraced the CIA satellite reconnaissance element in the National Reconnaissance Program. Program B was established in 1962, disestablished in 1992, and was superceded by functional NRO directorates.

ratio was very high, and we had a very long shroud and very long spacecraft. The bending modes were terrific! Figuring out how to get the payload to survive the ride into orbit was a major challenge. Trying to get our simulation runs to prove we could even do it was almost impossible. Second, we were boxing ourselves in on system capabilities. The government told us what the primary mission was, and we put blinders on and were working the problems strictly within those boundaries.

After the meeting I was out in the parking lot, and Lloyd Lauderdale said, "Fred, I've got to tell you, you're in last place." I said, "Last out of three?" He said, "No, no, you are last out of four. You are doing so badly, we got someone else in to help with the program." I was not looking forward to telling my boss about this. Surprisingly, he did not seem to be too perturbed. In fact, he told me, "If you can't do it, tell the government." I said, "Well, I don't know yet whether I can do it or not, but if we get the next launch simulation runs, and they say we can't do it, then we'll have to throw in our cards. But if the runs do say we can do it, then that's what I am going to say."

*The real mission for our study, whether we were told or not, had to be to look for things that are unexpected. Although satellite reconnaissance programs are sold based on some particular intelligence requirement, it is the unexpected things that can be detected with the systems that really pay for the mission.*

*Origins of an Idea.* As for the mission problem, a very strange thing happened to me a few nights later. I had a vision as I was driving through traffic on the way home. I could see myself standing on the hanger deck of an aircraft carrier looking into this open compartment where access was forbidden because some kind of secret activity was going on inside. Through an opening, I could see there was a lot of electronic equipment with all kinds of wires hanging out, apparently doing some kind of signals intercept work. Driving through traffic I wondered, "Why am I thinking about that? What made me think of standing on the hanger deck?" And then the light came on! The real mission for our study, whether we were told or not, had to be to look for things that are unexpected. Although satellite reconnaissance programs are sold based on some particular intelligence requirement, it is the unexpected things that can be detected with the systems that really pay for the mission. What a realization! I could not wait to get to the office the next morning and expand the mission part of our systems concept. A week later we received the launch simulation runs that confirmed we could survive the trip into orbit.

*Staying in the Competition.* There was just one remaining problem. We were still overweight. If anyone went back to the original briefing they would find that we were honest about it. We said we were about 100 pounds or so overweight, and unless this problem was corrected, it would prevent us from making it into orbit. We listed several possible solutions we could study in the next phase if the government chose us to continue. We definitely would get that negative payload worked off and get the payload within required parameters.

Some of our people thought that telling the government that we were over the weight restrictions was asking for disaster. I chose to just tell it like it was. In the first place, it made it easier on the team because we did not have to do any convolutions to figure out how they would package the design. And secondly, it put them on a constructive path that would fix the problem. One of the key weight trades resulted in the liquid engine apogee injection stage design.

We went into the next briefing just before Christmas in 1964. The TRW Vice President who accompanied me leaned over and asked, "Fred, how's it going? Can you tell how it's going?" I said, "Well, I don't know how we are doing with respect to the rest of the competition but they're interested in what we're saying. So, I know we are going to get another turn at bat." And that is exactly what happened. We turned that program around in a little more than three weeks!

*Selling the Program with a Golden Model.* After that pre-Christmas briefing, we received a message saying that Lloyd Lauderdale wanted a meeting. At the meeting, Lloyd said that he had a set of tasks that he wanted studied. As I reviewed the tasks, I did not see anything that gave me serious heartburn, except that we only had six weeks to complete them! Then he said, "Oh yes, we also want a model. We want a reliable working model of your spacecraft." I thought that was a pretty good challenge. He then said, "We have two constraints. First, you've got three weeks to complete the job. And second, the model has got to fit in a regular safe drawer when stowed. It can't be any bigger than that since we've got to be able to lock it up." I argued with him about the 21-day deadline to build the model but, quite frankly, I had no idea what it would take to get it built.

A contract manager accompanied me to the meeting, and Lloyd insisted that we write out the terms of the contract by hand on a piece of paper. He signed it. His contract officer signed it, and my contract manager signed it. I signed it. We were on the plane heading home when I turned to my contract manager and asked, "Did we execute a real contract with the government?" He said, "We sure did." I said, "Paul, you know that this is not within my authority." He said, "It doesn't matter. It's within mine, and we are legally under contract to deliver this." I said, "Oh, great!"

The head of configurations was John Bennett, who was a very creative man.[4] As I sat on that plane I wondered, "How am I going to tell John what he has to do?" I knew John would do one of two things. He would either get up and walk out of my office and never come back, or he would really get excited and yell and complain and say how impossible everything was. If he did the latter, I would know that everything was going to be fine. Fortunately, John yelled and complained. I just smiled knowing that somehow, with this team, we were going to pull it out. Working almost 24 hours a day for 22 days (we managed to get an extra day), we built the model. I think we should have cast that model in gold because it sold the program and it sold us!

*Working almost 24 hours a day for 22 days, we built the model. I think we should have cast that model in gold because it sold the program and it sold us!*

**Getting TRW to Commit to a Testing Facility**

A few years later, I was working on another National Reconnaissance Office (NRO) RFP. The government chose to compete the various system components separately. This caused somewhat of a problem because the component we wanted to build was going to be competed before the main system components. This meant investing a large amount of money to bid for a comparatively small part of the program without the benefit of knowing the outcome of the other component competitions.

We also were facing corporate and technical problems associated with how TRW would test our proposed concept on the ground. Proper tests required a very large building with expensive air conditioning. Naturally, TRW management was hesitant toward the idea of having to build an expensive new test facility for a limited part of the overall program. However, my team could not proceed without a commitment for the facility. My immediate boss—a Vice President—said he would not approve the facility because it did not make sense to

[4] John Bennett is a Pioneer of National Reconnaissance (inducted into the Hall of Pioneers, 27 September 2000). See chapter 6 for his recollections.

invest that much capital for such small program value. If I was going to get permission for the facility, I had to go up another level in management. So, I went up another level to plead my case. I was told to go back to the contract negotiations and convince the government that it was not really necessary to commit to building the facility until after the system competition had been decided. I knew that this would not work, but upper management insisted that I take this approach.

During the negotiations, we finally got to the facility issue and as expected, the government said that it was not going to award the contract without the facility. Frankly, I understood that. I said I did not have authority to commit to the facility and needed to confer with my boss. The government leader, who was running the competition and negotiation said, "There's the phone!" He walked out of the room, and I called my boss' boss and said, "Well, now is the time to commit to the facility." He moaned and groaned and said, "Well Fred, what do you think?" I said, "In for a penny, in for a pound. We either forget about doing business in this world, or we put up that facility." So he said, "OK, go ahead and commit to it."

A number of years later we had a group visiting from Washington that came for a tour of the facility. I overheard a couple of the people talking. One of them said, "Man, the people who approved this facility sure had guts!" I was behind him, and I thought to myself, "You know, they sure did!"

*Some of the top managers at TRW questioned whether we really wanted to be in this kind of business.... They could not market our systems publicly, or get any public acknowledgement for what we were doing. It was a constant challenge to convince [them] that this was something we really needed to do.*

**Corporate Challenges**

Some of the top managers at TRW questioned whether we really wanted to be in this kind of business, because it was so classified. They could not market our systems publicly, or get any public acknowledgement for what we were doing. My bosses had to deal with many senior corporate managers who were not knowledgeable about our programs, but who had great influence over our promotions and salaries. We also had a number of people who wanted to work in an environment that allowed them to publish articles and discuss their accomplishments at conferences. Some of them wanted public recognition for their work, but this was not possible because of the classification constraints. This constraint sometimes made hiring and retaining individuals difficult. It was a constant challenge to convince the company's upper management that this type of work (high risk and highly classified) was something that we really needed to do.

In the long run, TRW benefited by creating cadres of highly skilled and cleared people whose efforts and ingenuity led to financial success for the company. Because the company was profitable, there was funding available for our internal research and development efforts. These internal efforts benefited both the classified and unclassified sides of the company and helped to attract more talented people. This was an amazing cycle that not only made the company profitable, but also made a significant contribution to the country. Nothing motivates quite like success!

**Career Evolution**

Early in my career a lot of people warned me by saying, "Once you get involved with the skunk works (referring to the domain of highly classified projects), the smell sticks with you forever." I never viewed it as a smell. I viewed it as a set of special skills. Many times, when the company took on new NRO programs, I was chosen to lead them. I was the creative guy who somehow gathered the right set of people to do the job. I developed the skills to deal with people—both customers and management. Also, as I developed a good reputation among the people on my teams, it became easier and easier to lead them. I used to say that I could run those teams by hand signals because they knew how to do what I wanted. They knew they always could get me to listen, and I would understand enough of their problems to appreciate what they were trying to do. People have approached me years later and said that one of the best times of their life was working on those programs. That is such a great feeling!

One time, I had a boss who claimed I had a random path to my career. From his point of view, that was true. I never really had what I would call formal professional goals. If I had a primary goal back then, it was to have fun and work on things that I liked. If it happened to be a national program with great people and great technology, then what more could I ask for? Later in my career, I found myself moving into top management positions, and the fun started to wilt. The financial reward for getting promoted was good, but upper management work was not as much fun. I had more fun leading teams that actually worked on programs of national interest and urgency.

**Working in the Classified World**

Working in a highly-classified environment had its challenges. At TRW, in the early days, we only had two people who could do classified reproduction work. We did not have a separate special access reproduction facility. When it was time to do that classified work, we had to take over the whole facility. The only way the company could make this work was to turn over the facility to us after all of the regular work had been completed. Unfortunately, our reproduction people did not have all of the required clearances and could not have access to the security vault. Someone from my team had to be there to open the vault and store the material. The only fair thing to do was for those of us who had vault access to take turns coming in and overseeing the whole process. We put ourselves on a "call-up" list. We called it "The Watch."

The secretary of the reproduction facility sometimes would call the person on watch in the middle of the night to let him know that the reproduction facility was available, and it was time to come in. If the individual on watch answered the phone first, that was fine. Every once in a while his wife would answer, and that caused trouble (having a strange woman calling her husband at a strange hour). We worked our way around the problems by sharing the burden.

*I used to say that I could run those teams by hand signals because they knew how to do what I wanted. People have approached me years later and said that one of the best times of their life was working on one of those programs.*

In the early days, almost nobody in the company knew what I was doing. One time, one of the customers asked me about my corporate chain of command. I sat there and I thought, "Whom do I report to?" I was not really sure. I looked at the organizational chart and, frankly, no one stood out. I said to the customer, "It's real simple, I report to God." I have thought about that statement over the years, and I still feel the same way. TRW gave me the respect and flexibility necessary to run programs in my own way and without excessive oversight. Where else could a person work and get paid for having so much fun?

I started in the classified world when I was in the Navy. I was made a top secret control officer and crypto security officer on an aircraft carrier. From then on, most of my other jobs (like the ballistic missile program) also required some level of clearance. My wife was used to me not saying much about work.

Working in the classified world is not such a trial on the family. I believe it is the assignments that cause the problems, especially those that take you away from the family for long periods of time. I expect my wife would agree with this statement. I had many projects that took me away for extended periods, including projects for the NRO. At one point, my wife rebelled and decided she was going to go camping without me. She rented a trailer and was determined to take the children on vacation. Thinking of her taking that thing up into the mountains paralyzed me with fear. I decided I had to change my life and start finding ways, at least once a year, to go camping with the family. My wife's rebellion was what really triggered me to change.

**Lessons Learned**

The challenges in my work environment boggle the mind. I never had been asked to do so much with so little and so few. A key to successful management is the ability to transform challenges and difficulties into opportunities for success. I learned how to deal with anxiety, success, failure, and communication to achieve a winning approach.

*If it cannot be drawn, it cannot be built. The opposite is also true. If it can be drawn, it can be built. We insisted on carrying the proof-of-concept down to a low enough level so that the people who had to do the job believed they could do it.*

*Anxiety Can be Constructive.* One thing I noticed when working with talented people was that they often panicked when they reported for duty after getting cleared and were introduced to the full scope of the program and acquired a better picture of what we were trying to do. I could see it on their faces. These were very good people who had successful careers, top college educations, and knew that they were hot stuff. When they found out what we were really trying to do, the first thing that they thought was, "This time I'm going to fail."

One of the things I had to learn early was how to turn that anxiety into constructive action. I would not let anyone tell me what he or she could *not* do. If people said, "Oh I can't do this, or I can't do that," I would not accept it. Instead, I would make them tell me what they could do. I would tell them, "There must be something on this program you can contribute to." That was what I wanted to know and I would tell them to speak with other people on the project. If they were stuck, I would give them a push by telling them where I thought they could begin.

*Success Motivates.* I believed in giving my employees a job that was doable. The success of getting something done motivates people to take on more challenging jobs. I found that people who initially said they could do just a limited amount would end up far exceeding their (and my) expectations. Once I got them hooked, and they understood what the team was trying to accomplish, they gained confidence, and all sorts of wonderful things began to happen. Once in a great while, someone would not rise to the occasion, and we would just have to get him or her off the program. That was rare, and I would say that most of our people reached down deep inside themselves and came up with some major contributions.

Sometimes it was hard to convince people that certain things really are doable. Every job has got to be broken down into something that is doable. As a matter of fact, one of our original trade study concepts was an antenna feed that we never could draw. I learned a lesson from that. If it cannot be drawn, it cannot be built. The opposite is also true. If it can be drawn, it can be built. It can then be tested, evaluated, and analyzed. Everything that we said we were going to do had to be reduced to a job that somebody knew how to do. We insisted on carrying the proof-of-concept down to a low enough level so that the people who had to do the job believed that they could do it.

*Failures are Learning Opportunities.* Another valuable lesson is that we often learn nothing from success. Instead, it took a major failure to teach us some hard lessons. We had to overcome tremendous technical challenges on one of the earlier NRO programs, and we ultimately learned various lessons from the setbacks. For example, we did poorly at program management and customer interface, and we were slow to fix those problems. We also badly mishandled the follow-on program proposal, for which I take a large part of the blame. This episode represented one of the few times that I let management tell me what to design. We were trying to sell an upgraded system instead of what the government really needed. At the time we submitted our proposal, I was convinced we had done a fine job. We lost that proposal, and for good reason. A few years later while I was working a follow-on proposal, I revisited that earlier proposal, and I became physically ill. I saw, through 20/20 hindsight, that we had written a terrible proposal. We gave the government something that was easy for us to sell, rather than what it wanted to buy. I was determined that we would not repeat those mistakes, and I am pleased to say that my management let me do things my way since that time.

*Open Communication Makes Things Happen.* Another lesson I learned (and learned early) was to be, as much as possible, totally open with the customer. The instructions I gave my people were always to tell the government what they were doing, why they were doing it, what their results were, and what their problems were. My people were directed to inquire about any thoughts and concerns the government might have. After listening to the customer, they would come to me about any problems or shortfalls, and I would make the decision on our actions. My staff could not change their direction without permission from me, and they were told to be totally open with the government.

I believe the government respected that approach and, as a result, I am sure we avoided a lot of problems. I believe I earned the respect of not only the people in my company, but also those in the government. Through that approach we maintained an open book as much as we could. I believe that this is an important lesson because contractors naturally tend to want to keep everything back and solve their own problems or hide their failures. In my view, being open with the government is a much better approach. When there is a nice healthy attitude on both sides, a lot of good things happen.

*A Winning Approach Helps Keep Priorities Straight.* When it comes time to sell a program, a contractor's winning approach has three priorities. The first and most important priority is to sell the mission. A contractor must convince the government that the mission is worth doing and that the system concept is credible, along with the schedules and costs—all based on technology readiness. The second priority is to convince government management that your customer and team are the ones to manage the system development. The third priority is to sell your company. To be successful, a contractor must remember these priorities in order to win at all three levels. No one should be selling himself when he should be selling the mission. It is important to keep those priorities straight.

*I saw, through 20/20 hindsight that we had written a really terrible proposal. We gave the government something that was easy for us to sell, rather than what it wanted to buy. I was determined that we would not repeat those mistakes.*

**Conclusion**

I always wanted to be an engineer in order to build things that endured. I admired the Romans for their capabilities because some of their designs and engineering lasted two thousand years. I know my work will not endure that long, but it pleases me that I can still

Figure 12-3. Frederick Kaufman recording his recollections of national reconnaissance at NRO Headquarters, Pioneer Recollections Day, 26 September 2000. (Photo by Candi Campbell, NRO Visual Design Center.)

see Atlas and Titan vehicles in operation at the end of the 20th Century. In this regard, I believe that my contributions are valued. The knowledge that the satellite systems I worked on have evolved in sophistication and continued to be used into the 21st Century is very gratifying to me. These space systems have been valuable in providing important intelligence, as well as in giving U.S. leaders key information that helped them make decisions that were important to protect and advance U.S. national interests.

Over time, I developed two views of the impact of what my talented colleagues and I helped to accomplish. My wife and I were on a three-week trip to Europe prior to the September 2000 Pioneer Recognition Ceremony at the NRO. Each time I visited Europe I was reminded of what I saw in 1951 when I was a young ensign serving on an aircraft carrier in the Mediterranean. Seeing Europe at the end of the 20th Century, in comparison with the years following World War II, is quite remarkable. During the second half of the 20th Century, the U.S. made significant contributions to Western Europe and its people. While there have been disagreements at times, which is natural between friends, the overall record of accomplishment is impressive.

On the other hand, at times it appears that peace and prosperity are being taken for granted. There is a sense that we can have national security without making sacrifices. The sacrifices involve not only funding of national security projects and assigning talented people to work on national reconnaissance programs, but they also require support from national leaders. Additionally, in my opinion the leadership should protect the requirement that these programs operate in secrecy, which over time appears to be getting watered down. For example, I often become very uncomfortable when I see information about reconnaissance capabilities and methods in the media. If we brag about the details of our successes in intelligence collection, we could not only lose the source of that information, but we also could embarrass other nations. In this context, we need not only to celebrate past successes, but also to remain focused and vigilant on present and future challenges.

—

## Living with a Pioneer[5]

I am so pleased with Fred's pioneer award, and I am quite proud of my husband. After all of his hard work, it is just wonderful to recognize his contributions to the country's welfare—even though I'm still not entirely aware of what they were!

[5] This section is based on an interview with Ann Kaufman at the National Reconnaissance Office Headquarters, 26 September 2000.

Fred always wanted to work in engineering. I met him when he was in the Navy in San Francisco. He soon got out and went to graduate school at the University of California, Berkeley. When he was finishing his master's degree and starting to look for a job, missile programs were just on the horizon. They caught his attention and interest, and I do not believe that interest ever has gone away. He always seemed to want to be out there on the frontiers of engineering. It is just a part of who he is.

**Fred's Dedication to his Work**

Fred was so dedicated to his work, and I am convinced he truly enjoyed the challenges. His hours were wild! He often would come home at 2:00 a.m. and get up three hours later to go back to the office. He frequently worked Saturdays and Sundays. I believe that this dedication is commendable, and I suspect that many of the other people who worked on these programs were working the same hours. Fred and his colleagues were more than just mission-oriented. Their dedication was about wanting to do something to help their country. They were proud of their country. The sacrifices that my husband and his team made were in such contrast to the anti-war protests that were going on at the same time. During the 1960s and 1970s, people were stomping on the American flag. When we were growing up nobody did that. Nobody dared!

Despite Fred's wild work schedule, we did try to take yearly family vacations. We usually went camping. Fred believed that it was a good way of getting away from the office and getting some perspective. If he was working on a tough project at the time, he would stop off at a pay telephone and give the office a call. He believed in what he was doing, and he was very faithful to that.

**Pioneer in Training**

When Fred was in graduate school, he always talked to me about what was going on in his classes. As a matter of fact, when he was finishing up his degree, there was one final class that was giving him fits. It was just a one-credit circuits class, and the whole grade was a project. If it worked, he got an "A." If it did not, then he failed. Fred was worried that he was not going to be able to make his project work. He spoke with me about it, and I said, "Why don't you try to do this?" He agreed with my advice, and we completed the whole project. When he took it into class and set it up for the demonstration, it worked. It also blew every fuse in the building! He received his "A."

Figure 12-4. Fred Kaufman and wife Ann standing by the pioneer medallions in the NRO Hall of Pioneers on Pioneer Recollections Day, 26 September 2000. (Photo by Candi Campbell, NRO Visual Design Center.)

**The Security Environment**

As for dealing with the security constraints of Fred's career, they were minor. In our neighborhood, many of the men worked on classified material. When the security clearance people would come around, a neighbor would call and say, "He's over here now for your husband." Everyone just accepted it as a way of life.

My husband gave our daughter a lecture when she went over to Europe as a foreign exchange student. She was going to Berlin, and Fred told her, "You are not to go behind the Iron Curtain (East Germany). You must be careful because you could be a target. They may try to use you to get secrets from me." Of course nothing happened to her while she was away, and she had a wonderful time.

**Why I Never Saw a Launch**

Fred would go down to Florida for two weeks for the launches, but I did not go to a launch at Cape Canaveral because I could not deal with that hot, muggy, humid weather. That hot weather and I did not get along too well. Once, while we were camping at Green River, Utah, we thought there might be an opportunity to see a missile launch from the White Sands missile range. The kids were sleeping at the time, but Fred woke me up in the middle of the night like he had promised. He pointed out into the night and said, "That is where the missile will be." Unfortunately, he was pointing in the wrong direction, so I did not see the launch.

**Conclusion**

We married in 1954. Fred still says, "Ann, it couldn't be all pits because we've been married for 46 years!" He is right. We basically have led a quiet life. He was busy working, and I was busy taking care of our children and "doing my thing." We lived in an environment where other families were in similar situations. The husbands worked on classified projects, and the wives raised the children. For us and other families around us, this was a normal, and certainly a happy, way of life.

—

## Pioneer Award Presentation and Citation

Figure 12-5. Pioneer Fred Kaufman (second from left) being recognized at the 2000 Pioneer Recognition Ceremony. The pioneer award plaque was presented by DNRO Keith Hall (left) and DCI George Tenet (third from the left). Don Winters (Executive Vice-President and Manager, TRW) joined in the presentation. (Photo by Sara Judy, NRO Visual Design Center.)

**Frederick H. Kaufman**

Mr. Frederick Kaufman directed the TRW team that produced two important Program B signals intelligence satellites, including the first communications cross-link system in space. He also conceived and directed development of a liquid-propellant, apogee injection engine that permitted a conventional launch vehicle to boost a heavy payload into a high orbit.

*Career in National Reconnaissance. 1964-1991*

# Lloyd K. Lauderdale

Lloyd Lauderdale was a signals intelligence expert, who worked for the government for the Central Intelligence Agency, as well as for private industry. He made invaluable contributions to improving United States signals intelligence collection capabilities. Lauderdale was honored posthumously as a Pioneer of National Reconnaissance. In this chapter, Virginia Lauderdale, his wife, remembers him as a brilliant and tenacious person whose work helped end the Cold War. In addition, his son, Lloyd Lauderdale, Jr., recalls some of the lessons his father passed on to him. Their recollections follow a brief discussion of Lauderdale's contributions to national reconnaissance.

## Developing A Signals Intelligence Capability from Concept to Launch[1]

Lloyd K. Lauderdale served as the Program Manager for an early Central Intelligence Agency (CIA) signals intelligence (SIGINT) satellite program, that achieved revolutionary advances in the United States ability to perform overhead SIGINT collection. He developed the system from concept to launch and contributed to the development of the system's on-orbit vehicle, ground station, and system engineering. Prior to his Agency service, Lauderdale published various papers and reports on his technical achievements. (See Annex for a list of publications.)

Lloyd Lauderdale died in 1985. On 27 September 2000, the Director of the National Reconnaissance Office (NRO), Keith Hall, and the Director of Central Intelligence, George Tenet, posthumously inducted Dr. Lauderdale as a Pioneer of National Reconnaissance. Lauderdale's wife, Virginia Lauderdale, accepted the award on his behalf.

—

## Helping Future Generations: My Husband's Legacy[2]

My husband believed that his work in national reconnaissance was the best work he did in his career. He was so proud of his NRO projects. He thought of them as his "babies." Lloyd was a brilliant and tenacious man. He was especially tenacious if he believed strongly in

[1] This section is based on information contained in the material submitted with the nomination of Lloyd Lauderdale for selection as a National Reconnaissance Pioneer. We prepared this section in lieu of recollections.

[2] This section is based on an interview with Virginia Lauderdale at the National Reconnaissance Office Headquarters, 26 September 2000.

Figure 13-1. Virginia Lauderdale viewing the display at the NRO Pioneer Hall, Pioneer Recollections Day, 26 September 2000. (Photo by Candi Campbell, NRO Visual Design Center.)

something, or if a question of ethics arose. But he was also very humble, and a very regular guy, despite his brilliance.

My husband was recruited by the CIA, which offered him an attractive opportunity to do some really inventive things. The Agency probably heard of him because he worked on some classified projects at Johns Hopkins University while pursuing his education and teaching.

Lloyd had not been at CIA for long when Bud Wheelon formed the Special Projects Staff and put Jackson B. Maxey in charge.[3] At that time, I also was a CIA employee, and went to work for Jack. Initially, there were just three people on Jack's team: Jack, myself, and Les Dirks.[4] Then came John McMahon as Jack's Deputy.[5] After that, the group slowly enlarged to include my husband, Roy Burks, John Crowley, and others.[6]

I am sorry that Jack Maxey was not recognized in the initial group of Pioneers. He defined the vision of the Special Projects Staff, and had a unique ability to select talent. He did not over-manage, expecting instead that those he hired knew how to get their jobs done.

Lloyd was a typical engineer, seeming distant on occasion. He used to say that the work he was doing, if it turned out successfully, would help our children, our grandchildren, and possibly even further generations. I believe that his claim about the importance of his work turned out to be true, and that it possibly helped bring about the end of the Cold War.

—

[3] Albert D. "Bud" Wheelon is a Pioneer of National Reconnaissance (inducted into the NRO Hall of Pioneers, 27 September 2000). See chapter 42 for his recollections. Jackson D. Maxey served as the Chief of the Systems Analysis Staff in the Directorate of Science and Technology in the early-to mid-1960s.

[4] Leslie C. Dirks served as Program B Director from June 1976-July 1982. He also served as the first director of the Office of Development and Engineering.

[5] John McMahon was a member of the 2000 Pioneer Selection Board. He formerly served as the Deputy Director of Central Intelligence and the President of Lockheed Missiles and Space Company.

[6] Roy Burks is a Pioneer of National Reconnaissance (inducted into the NRO Hall of Pioneers, 27 September 2000). See chapter 22 for his recollections. John J. Crowley is a Pioneer of National Reconnaissance (inducted into the NRO Hall of Pioneers, 27 September 2000). See chapter 26.

## My Father's Advice[7]

I am very proud of my father and his work, the importance of which was reflected first with the CIA's Trailblazer Award, and now as a Pioneer of National Reconnaissance. Hearing about his accomplishments has allowed me to gain a better understanding of what kind of work my father did.

Figure 13-2. Lloyd Lauderdale, Jr. and Virginia Lauderdale at the NRO 40th Anniversary Gala at the Washington Hilton, 27 September 2000. (Photo by Sara Judy, NRO Visual Design Center.)

My father tried to get me interested in subjects like math, which was a challenge for me. He assured me that he had similar problems when he was a boy and that I just had to tough them out. He encouraged me not to give up on anything. My father realized that once I got past a certain point, I would become confident in my capabilities and realize my potential.

—

## Pioneer Award Presentation and Citation

Figure 13-3. Virginia Lauderdale (second from left) wife of pioneer Lloyd Lauderdale. Mrs. Lauderdale received the pioneer award plaque for her husband, who was recognized posthumously at the 2000 Pioneer Recognition Ceremony. The award was presented by DNRO Keith Hall (left) and DCI George Tenet (third from the left). Joanne Isham (Deputy Director for Science and Technology, CIA) joined in the presentation. (Photo by Candi Campbell, NRO Visual Design Center.)

[7] This section is based on an interview with Lloyd Lauderdale, Jr. at the National Reconnaissance Office Headquarters, 26 September 2000.

**Lloyd K. Lauderdale, Ph.D.**

Dr. Lloyd Lauderdale was Program Manager for the CIA Program B team that developed an advanced signals intelligence satellite from concept through first launch. This effort, which included building and staffing an operations center, produced a dramatic improvement in overhead signals collection.

*Career in National Reconnaissance: 1963-1984*

—

## Annex

**Partial List of Publications by Lloyd K. Lauderdale**

"Countermeasures Against Active-Passive Surveillance Systems," The Johns Hopkins University Radiation Laboratory Technical Report No. AF-84, January 1961.

"Considerations Concerning the Selection of an Optimum Jamming Signal for Barrage Jamming of Search Radars," Department of Defense Symposium on Electronic Countermeasures, University of Michigan, October 1957.

"Electronic Countermeasures," Department of Defense sponsored book published by the University of Michigan Institute of Science and Technology under the U.S. Army Signal Corps Contract No. DA-36-039SC 71204.*

"Final Report on the Experimental Radar Jamming Program," The Johns Hopkins University Radiation Laboratory Technical Report No. AF-45, February 1958.

"Nuclear Powered Submarine," NAVPERS 92693, December 1959.

"Optimum Modulations of a Carcinotron Barrage Jammer Operating Against Surveillance Radars," Department of Defense Electronic Warfare Symposium, University of Michigan, September 1959.

"Research and Development in Electronic Countermeasures Techniques and Systems in France and the United Kingdom," The Johns Hopkins University Radiation Laboratory, RL/IMA 60-14, September 1960.

"Spectral Characteristics of Random Modulated Waves," The Johns Hopkins University Radiation Laboratory Technical Report No. AF-94, May 1962.

"Study of Airborne Electronic Countermeasures Techniques for Penetrating a Radar Defense Network," The Johns Hopkins University Radiation Laboratory, RL/IMA 58-7, July 1958.*

*Written in collaboration with one or more authors.*

# Paul W. Mayhew

Paul Mayhew served as TRW's Payload Project Manager and Deputy Program Manager for the development of two vital geosynchronous signals intelligence satellite systems that provided breakthrough collection capabilities to the National Reconnaissance Office. He was part of the team that conceived the initial system, and was TRW's Program Manager and technical lead for the entire follow-on system. These systems produced a new national reconnaissance capability, and established a new direction and scope of satellite reconnaissance and intelligence collection.

## System Integration and Design: Meeting Requirements with Innovation[1]

My career in national reconnaissance focused on the design and development of two vital and precedent-setting signals intelligence (SIGINT) systems. The first system provided the National Reconnaissance Office (NRO) with a new capability, which allowed the monitoring of previously undetectable targets. This was a long-life system adaptable to shifting missions and targets. The second system, which was a follow-on to the first, included performance enhancements in all domains: target coverage, sensitivity, capacity, mission flexibility, dimensional control, security, lifetime, and other technical characteristics. All of the requirements on these systems pushed the state of the art.

Like most engineers, my goals were to solve problems, to create and design systems, and to build something of use and significance, as opposed to acquiring wealth or power. My background was in electronic warfare research and development. My graduate work was in system engineering and analysis. From a professional point of view, my plan was to work in areas of advanced technology, maintain and build technical competency, work with others from whom I could learn, and develop technology that would be beneficial to people.

It was a stroke of luck that I came into contact with Lloyd Lauderdale, the Central Intelligence Agency (CIA) program manager for a new and exciting SIGINT system. Lloyd and I knew each other when he was a graduate student at Johns Hopkins University, and I was in charge of electronic warfare Research and Development (R&D) at Wright Field sponsoring

[1] This section is based on written input that Paul W. Mayhew submitted to the Center for the Study of National Reconnaissance.

his work.[2] This new program provided me with an opportunity to use all of my background in systems design and radio frequency (RF) electronics. My graduate work in system engineering and operations analysis proved invaluable. I applied my knowledge to areas that required new technology and that involved an extremely complex system for a mission never accomplished before. In hindsight, this program and its follow-on program were an engineer's dream because they provided challenges in every direction from mission analysis, system engineering, advances in RF and electronics technologies, and ventures into space.

**Developing Two Signals Intelligence Systems**

These two SIGINT systems were an electrical and mechanical engineer's dream or nightmare, depending on the perspective. We started with nothing except for the mission requirements (efficient operation and high performance) and the constraints (a specific launch vehicle to be used, cost, and schedule). The stringency and difficulty of the requirements demanded that we make several revolutionary breakthroughs in system engineering and in the spacecraft and payload design. These breakthroughs resulted in new and innovative concepts that significantly altered the methods and processes for conducting national reconnaissance. I worked with an innovative, focused team that included TRW contractors and government personnel, and I am proud to have contributed to these important programs.

*Like most engineers, my goals were to solve problems, to create and design systems, and to build something of use and significance, as opposed to acquiring wealth or power.*

In the first program, my role was mission analyst, system architect and engineer, and payload manager. My tasks were to define the payload end-to-end performance, define the subsystems and interfaces with other spacecraft elements, and implement the hardware. I also was responsible for overseeing all other activities in the design process to avoid future problems in the design, test methods and equipment, manufacture, system verification, and on-orbit operation. After delivering the hardware to the spacecraft for final assembly, I was the overall program system engineer, with the task of total system validation and verification.

In the next generation program, I was the proposal manager, system architect and engineer, and program manager. I was responsible for the creation, production, deployment, and operation of a reliable system, including both the spacecraft and ground station segments. This system also required advances in technology and new designs, hardware, devices, and techniques to maximize the payload capability within size and weight constraints. Mission operation had a major influence on the system architecture and the design of the hardware and software.

Although similar to the first program, the follow-on spacecraft configuration enhanced several capabilities. This system took advantage of the knowledge base from its predecessor, and involved many members of the previous design and development team. My major contribution was as system designer, defining the overall configuration and selecting the implementation methods. I can explain my role by summarizing our work in design, payloads, the ground station control software, and system verification.

*Design Concept.* In any spacecraft design, the overall configuration is determined by the payload and its demands in terms of basic geometry, orbit, attitude control, pointing accuracy, command and telemetry, station keeping, stability, power, and temperature control. The system is dominated by frequency coverage and the size of the antenna aperture, which is limited by the weight and shroud capability of the launch vehicle.

[2] Lloyd K. Lauderdale is a Pioneer of National Reconnaissance (inducted into the NRO Hall of Pioneers, 27 September 2000). See chapter 13.

We clearly developed the right architecture. The basic design concept of the first system we developed was as valid at the end of the twentieth century as it was when we conceived it. The system provided great mission flexibility, and enabled the use of the latest evolving signal-processing technology at the ground station to extract the desired signals from the undesired signals. The question was whether we could build such a system using the technology realizable in the then-foreseeable future. This requirement placed great demands on components. The mechanical design of the system was a major achievement. This design was performed by many talented mechanical engineers, and led by John Bennett.[3]

The design process involved all of the technical disciplines that eventually would be needed for the system's detailed design, manufacture, assembly and test. These included mechanical, thermal, dynamic, materials, processes, inspection, quality, manufacturing (including RF and mechanical tests), and the design of the assembly and test facility. Personnel from all of these disciplines participated in the design iteration process, so that everyone knew what the design was and could influence the design to avoid later problems The design of the handling and instrumentation equipment required nearly as much effort and innovation as the system itself. We designed the system and the associated handling and instrumentation equipment to accommodate readily an extension to higher frequencies, closer tolerances, and increased precision of measurement. It was a tribute to the vision of the initial designers.

The spacecraft had an antenna with RF electronics, supported by all the necessary housekeeping utilities including ground support for spacecraft and mission operation. Launch vehicle constraints demanded minimizing weight and power, forcing our design innovation, creativity, and invention. We achieved major advancements in the areas of RF technology, electronics, systems control, and manufacturing and testing methods. Several interesting and sometimes humorous events illustrate the scope of our activities.

*The Payloads.* The overall design of the spacecraft payloads for these two programs was an integrated system of the various segments to maximize performance with minimum weight and power. My particular role and contribution was defining the system architecture for the payload. This architecture included everything from the antenna aperture through the mission data delivery system in the ground segment. It specifically comprised: system and subsystem design and development; production of the flight hardware; verification and testing; and mission operation of the spacecraft.

> *These two SIGINT systems were an electrical and mechanical engineer's dream or nightmare, depending on the perspective.*

Security was the primary design driver for the uplink and downlink between the ground station and the spacecraft. The downlink had to be undetectable by others, and the uplink had to maintain control under the most adverse conditions. These requirements had a profound impact on spacecraft design, ground station design, and ground station location. After a meeting with Lloyd Lauderdale and Bob Hermann, we concluded that the state of encryption technology was not sufficiently advanced for the initial system.[4] So we developed an alternative technique and equipment. Its performance turned out to be outstanding.

About six months into one program, we realized that something was wrong when we had a communication breakdown that resulted from differences in how antenna designers

[3] John T. Bennett is a Pioneer of National Reconnaissance (inducted into the NRO Hall of Pioneers, 27 September 2000). See chapter 6 for his recollections.

[4] Robert J. Hermann served as Director of the National Reconnaissance Office (DNRO) from 1979-1981.

and control designers define coordinate sets. For antenna designers, pointing in the x direction meant distance on the x-axis and rotation about the y-axis. However, for controls designers, pointing in the x direction meant rotation about the x-axis. This disagreement illustrates that what some individuals consider "obvious" is not obvious when integrating different engineering disciplines. Consequently, we spent a significant amount of time and effort ensuring that the various components would function as an integrated system.

We did not want to relearn anything on orbit, so we made a proof-positive polarity test using all final-design components and control hardware. The test went well, and we tested over and over, varying all conditions. Everything moved in the right direction, acquired tracking, and was stable. Thus began an extensive test set validation program.

*We used scale models after we overcame resistance from top management to the placement of a large amount of unsightly "junk" on the roof of the elegant executive building.*

The mission data were transferred to the ground station through a set of transmitters. The transmitter components were built in-house to achieve efficient performance. For one of the initial wave-guide parts fabricated for an engineering model, we gave our mechanical manufacturing division detailed drawings and careful instructions on how to build and gold plate it. With great pride, they placed the pieces on my desk, each a finely polished gleaming yellow gold. They looked beautiful until I looked down the inside of the guide and saw solder globs and drips, rough surfaces and corrosion. Talk about a breakdown in communication! The manufacturing people did not understand that it was the inside that had to be beautiful. One day they asked me why it had to be gold plated. I told them it was cheaper than making it out of solid gold. Fortunately, communication improved greatly after that episode.

*Control System Software.* Attitude control and pointing arrays were tightly integrated in the control system. This was necessary because any action created by the motion of one appendage affected the others, and any unwanted movement affected the spacecraft's pointing. Thus, the structural design components were all inter-related.

This resulted in a control system design that we implemented in the software. At the time of these developments, the background and training of hardware and software people were very different, creating a little understood communication problem. After several rounds of developing flawed software, we began to understand the breakdown. It was necessary for the software and hardware folks to work together as an integrated group for each discipline to understand the others. Everyone needed to understand how the details of the hardware worked and how the details of the software worked. This process produced not only outstanding control software, but also software that really simulated the system and test software that addressed all facets of the integrated hardware/software system. Few people recognize the necessity for this kind of integration of disciplines.

*Testing and System Verification.* We put all the system performance data in a large analytical model, along with thermal models and solar simulation models, all of which were validated by scale models and component testing. Analytic and physical models of portions of the system (both partial and full scale) were built and tested to produce data to validate the analytical tools. Test cases to stress the behavior of the control software were used to verify performance over the range of hardware variation. We put great reliance on analytical tools and simulations that were validated by tests of physical hardware components. We verified the characterization of the behavior of the system across interfaces by full- and partial-scale hardware tests. These tests placed great emphasis early in the program on developing the verification system to qualify and verify the system. The test and verification system itself was a major effort in the program.

At the time of the first program, there were few analytical models and tools available that could represent adequately the details of the antenna design, facilitate design, or evaluate performance. As a result of this lack of precedent, we used a combination of small-scale and full-scale models to evaluate designs. This resulted in a number of interesting modeling and testing experiences including the design of special test facilities to simulate "zero-G" (gravity) and antenna performance measurements.

Subsequently, our analytical tools came a long way, so we could predict accurately the system performance. Scale models validated these tools and verified system performance using full-scale feed-measured performance along with full-scale mechanical measurements of the reflector surface. We used scale models after we overcame resistance from top management to the placement of a large amount of unsightly "junk" on the roof of the elegant executive building. We prevailed—nothing got in our way.

**The Value of End-to-End Responsibility**

I had end-to-end responsibilities for these systems. This end-to-end responsibility forced me to think early about how to define the system and its subsystems, how to operate the system, and how to manufacture and test the system. Incorporating all of these requirements early in the design process avoided subsequent interface problems, manufacturing and process problems, and operational problems. We especially wanted to avoid problems caused by the failure of some component. This method allowed for the early definition of test methods and the design of special electrical and mechanical test equipment.

My experience with these programs was "end-to-end" in two senses: system design and programmatics. System design involved system performance, technology, and optimization of the link from the intercept at the antenna through signal delivery at the ground station. Programmatics involved organizing the work to minimize disruption and miscommunication as progress was made through the sequence of design and development, manufacturing, assembly, test, deployment, and operation. My end-to-end experience profoundly influenced how I thought about working on designs and programs. It provided the opportunity to grow technically and managerially.

We became very aggressive in our creativity, and the design concepts and implementation methods eventually emerged and evolved. I learned that there always is a way, and that way is determination and asking the right questions of the right people. Even though the first system required innovative technology advances in areas where there were no existing methods or hardware, the thought of failure never entered our minds. Open exchange of ideas and teamwork with people of different disciplines (regardless of whether they were government, consultant, or contractor) brought new and better ideas with solid solutions. The TRW/government team believed we could overcome any obstacle—and we did.

*We became very aggressive in our creativity, and the design concepts and implementation methods eventually emerged and evolved. I learned that there always is a way, and that way is determination and asking the right questions of the right people.*

Several aspects of these programs demonstrated the excellence of the team and brought me personal satisfaction. These aspects include the success of the programs; the performance of the system from the spacecraft antenna through the ground station; the fulfillment of mission objectives; and the attainment of a spacecraft life well beyond the original requirements. My involvement with these two programs provided the means for me to develop further my abilities to solve problems, define what is important, determine risks and consequences, and not be intimidated by not knowing right away how to accomplish something. My work led me to new and even more challenging assignments, including my becoming a vice president and general manager.

**Benefits to TRW, the Government, and the Country**

My company, TRW, had many extremely bright, talented, creative people in a wide variety of disciplines. We worked with many intelligent and creative people at the CIA and other government agencies and their consultants. This enabled us to assemble an outstanding team, and we accomplished wonders. The programs were successful and consistently surpassed expectations. We produced new technology and insights into solving systems engineering and operations problems with our teamwork. We integrated different technical disciplines, in particular RF engineering, electronic engineering, and mechanical engineering, and control software design. I derived a great deal of personal satisfaction from the step-by-step achievements culminating in the systems' ultimate deployment, even though very few people knew about these achievements. We knew we did a good job. Although I moved on to assignments with even greater responsibility, this period in my career was the most rewarding.

The experience base of the thousands of people employed by TRW on these NRO projects facilitated major contributions to other systems, both classified and unclassified. The design philosophy implemented on these TRW programs produced designs that exceeded performance requirements, which resulted in a success rate that helped establish TRW as a company of excellence. In turn, this enhanced TRW's ability to acquire new business. These two programs established TRW's advanced technology base for spacecraft and payloads, in particular RF and electronics design that led to many other contracts. These programs contributed to the advancement of knowledge, skills, technology, and processes. (See Table.)

Many of these technology advancements were utilized by TRW in other business sectors of the company, and were spun-off into a variety of military applications. There were commercial applications as well, and in some cases contributed to redefining industry standards. These breakthrough technologies and systems established a new standard for lightweight, high-performance, electronic payloads, making TRW the preeminent payload provider for many space systems. These systems' capabilities provided the government with near-real-time, continuous reconnaissance data that could be acquired by no other means. The overall effectiveness of these programs reduced the need for new systems development, reducing the NRO's costs while preserving performance. Intelligence data collected by these SIGINT systems provided United States leaders the information they needed to deal with potential threats to national security.

**Table. Areas Where NRO Programs Advanced the State of the Art**

| | |
|---|---|
| thermal design | computer simulations for verification |
| structural dynamics | pointing and attitude control |
| propulsion | lightweight thermally stable materials |
| simulated zero-G testing | antenna design |
| design and analysis tools | linear receivers |
| digital encoding | efficient high frequency transmitters |
| composite materials | gallium arsenide technology |
| adaptive control | integrated circuit design and processes |
| analysis tools | digital signal processing |
| simulations | electroform processes |
| precision control | control of flexible structures |
| computer simulations for design mechanisms | |

### My Colleagues and Other Program Support Teams

There are no individual heroes in programs of this size and complexity. These programs drew on many disciplines, skills, and people. It is true that some individuals made greater contributions than others at various times, but our success depended on the support, commitment and creative ideas of each other. We worked interactively to overcome problems. Individual ideas and contributions helped us overcome obstacles, but it took our collective talent to succeed. The key was working interactively to achieve a common objective. Even the most talented people put their egos aside to enable more effective contributions by others. The importance and complexity of the system required that we seek input from all possible sources.

*My involvement with these two programs provided the means for me to develop further my abilities to solve problems, define what is important, determine risks and consequences, and not be intimidated by not knowing right away how to accomplish something.*

As I look back on this experience, perhaps my greatest contribution was in the selection of creative and committed people (e.g., Dr. Marv Stone and Bob Kaemmerer).[5] I was supported by an extremely talented team, and this contributed to an environment of teamwork and trust that enabled this team to bring their ideas and solutions to the table. In addition to our own personnel, a group called the "Tech Panel" also provided valuable support. This panel was comprised of people such as Dr. Dick Garwin, Dr. Bill Perry, and Frank Lehan.[6] Their responsibilities included assessing our possible solutions and alternative sources of information. This panel's talent and wisdom could have brought terror into our hearts. However, from the beginning, its efforts were not to criticize but to help us solve problems, and we accepted it as such. As a result, we were open and honest in presenting the design with all of its capabilities and weaknesses, and we benefited from their knowledge and experience. I highly recommend this process for any program, and the keys are attitude and trust.

Additionally, there was no "we-they" mentality, and new observers in meetings could not distinguish government from contractor based upon interaction. All of the data, good or bad, were distributed to everyone. The government took ownership of the development work along with TRW. This attitude and trust led to our achievement of these highly successful designs with minimum cost and schedule problems. The closeness and interdependence of government and contractor were facilitated by the fact that it was an NRO program, which allowed us to work as an isolated group. It was truly a collective effort. The TRW Vice President, George Solomon, told Director of Central Intelligence, William Casey, that this relationship was the most successful and efficient one he had ever seen.[7]

### Impact of My Work on My Family

The biggest issue I faced when working on classified programs was that I could not tell my family, much less anyone else, what I was doing. As a habit, we just did not discuss my work. At social gatherings, people came to recognize that my work was not a subject of conversation, so we became great listeners of our friends' experiences and successes. My job

---

[5] Marvin S. Stone is a Pioneer of National Reconnaissance (inducted into the NRO Hall of Pioneers, 27 September 2000). See chapter 17 for his recollections.

[6] Richard L. Garwin, Frank W. Lehan, and William J. Perry are Founders of National Reconnaissance. See chapter 3 for Garwin's recollections.

[7] George E. Solomon was then Senior Vice President of TRW's space activities.

Figure 14-1. Paul Mayhew typing his recollections at NRO Headquarters, Pioneer Recollections Day, 26 September 2000. (Photo by Candi Campbell, NRO Visual Design Center.)

was far more exciting than their seemingly boring day-to-day activities, but I refrained from saying that. There were times when our friends and neighbors became suspicious, such as after a visit by the Federal Bureau of Investigation, but even these events became expected. I suspect I was viewed as "that quite weird engineer." My children knew I was involved in engineering, but never asked what I did. When they were asked at school, they merely said, "I don't know."

The job was very demanding—sixty to seventy hours per week was the norm. I worked around the clock on some occasions, and I was gone for months at a time on others. Most major space programs required total commitment, which meant that my wife had to pick up the slack in raising our children. She did that because she understood that my job was important.

**Lessons I Learned**

I learned a number of lessons as a result of my involvement in the design, development, and testing of national reconnaissance systems. Four lessons in particular come to mind that pertain to current and future NRO endeavors. These lessons are in the areas of design, system engineering, professional competence, and teamwork.

*Design for Flexibility.* Overhead collection systems are very expensive and take long periods of time to develop and implement. As a design philosophy, each system should have some overlapping capability with others to facilitate operational flexibility and redundancy, and to minimize the impact of countermeasures. While each system has a specific mission, it should be designed with enough flexibility to accommodate significant changes in its assignment and coverage. There will always be a demand for greater pointing accuracy, resolution, sensitivity, bandwidth, and a dynamic range and larger ratios of payload weight to total weight. Advances in spacecraft technology are necessary to enable the exploitation of advances in payload technology, and should be able to be integrated into existing systems whenever possible.

*System Engineering.* It is interesting that the isolation associated with strict security practices tends to produce integrated and more effective teams. However, this isolation and compartmentation also tends to disconnect the various system elements. This consequence demands a system engineering process that effectively integrates the various program components in order to avoid breakdowns and sub-optimal performance at interfaces. Most breakdowns in large complex systems occur at interfaces, and are caused primarily by ignorance or communication failures—basically misunderstandings. This is a lesson I learned repeatedly.

*Long-Term Government Competence.* The government agencies that procure advanced technology systems must have solid internal technical and managerial abilities. They are needed to acquire systems with superior performance in less time and at lower cost than routine bureaucratic agencies, which have a rotation of staff with no real product ownership. Government technical and managerial competence and long-term responsibility are essential to effective operation with contractors, and clearly impact the success or failure of the program.

*Teamwork.* The number and quality of people I worked with on the two programs illustrates the importance of teamwork, cooperation, and dedication—key factors in the programs' successes. Teamwork between government and contractor produces a better, quicker, cheaper product than an "arms-length" operation focused on blame instead of joint ownership and success. This method requires real skill, leadership, and professionalism from everyone involved, and results in trust and a collective focus on what is needed to achieve success, rather than filling squares.

**Conclusion**

The NRO's mission continues to involve many different overhead systems. It is the nature of NRO programs to be on the cutting edge of technology in order to achieve and maintain a global reconnaissance advantage. Each of these systems provides important capabilities that support national-level decision makers who operate in a climate of shifting threats and information needs. I am proud to have contributed to these NRO and American successes, and I look forward to future advances and achievements in national reconnaissance.

—

## My Husband is Ideally Suited for Secret Work[8]

I am very pleased that the Pioneers of National Reconnaissance are being recognized, because for so long no one knew or talked about their work. I still do not understand the type of work Paul did. I just knew he was doing something important. He did not say that, but I sensed it. I kind of grew into the fact that his work was not to be discussed, and I accepted that.

One reason I accepted the secrecy is that it made it easier for me to find my own happiness. I was not dependent on him and his presence at home to make me happy. It all depends on a person's attitude—if I were one of those people who depended on a daily report of what my husband did, it would have been different. It takes a certain type of spouse to live with this type of situation. It is my understanding that the divorce rate was rather high with families involved with this type of work. It goes back to self-image, having a purpose in life that is not completely dependent on somebody else. People need to have their own agendas.

Figure 14-2. Imogene Mayhew (left) and Nellie Lehan (right), Pioneer Recollections Day, NRO Headquarters, 26 September 2000. (Photo by Candi Campbell, NRO Visual Design Center.)

[8] This section is based on an interview with Imogene Mayhew at the National Reconnaissance Office Headquarters, 26 September 2000.

My husband is ideally suited for intelligence work, and I am ideally suited to accept the consequences of his work. My friends were a very important part of my life, because I am not the type of person who can live in a vacuum or wait at home all day for someone to come home and do something for me. So the secrecy worked out okay—he was a closed-mouthed individual anyway, and naturally a private, quiet person. I may never know what he did, and that is okay, although I would like our children to learn someday.

The only thing that was really bad about his work was the long hours he spent away from home. I thought that he might have a mistress for a while, and I would joke with him about that. I feel that our children missed out a bit, because they never knew what their father did, and could never talk with him about it. Our daughter, the younger child, said to me once, "When someone would ask me 'What does your father do?' I would have to reply 'He goes to work.' If someone would ask, 'Is he a doctor, or a lawyer, or something like that?' I would reply, 'No, he just works.'" Our son, who is four years older, knew that he was involved in space activities, but that was the only thing he knew.

We always had breakfast together as a family with the children, because I thought that was important. Frequently, he would drive them to school. Additionally, our family had opportunities to share in some activities related to Paul's work, such as witnessing launches and things on television. We watched launches together and talked about satellites, just in general terms. Paul and our children would discuss them, especially my husband and our daughter, who majored in Physics at Brigham Young University and later got a security clearance at TRW. Our son started off in computers, and then ended up in lasers or fiber optics, and also followed Paul to TRW.

Our friends thoroughly enjoyed asking Paul to attend social functions, because he did not talk, and he was a very good listener. People always enjoyed talking about themselves and what they did, so he became quite popular because he was a very good listener. I said to him once, "I can't figure this out—I'm the talker and you're the silent one, and you're the one everybody loves!" Everyone would tell me, "He's such a great guy." Our friends did not include his colleagues and work friends, and we basically socialized with his colleagues only during the holidays. It seemed to me that he just wanted to get away from work most of the time.

Anybody with as much dedication and focus as Paul must really enjoy the work. He felt like he was achieving something important. To set a goal, stay focused on it, and achieve it requires work. People have to enjoy their work to put that much effort over the years into it. Paul had a total focus on his work. He was a hard and dedicated worker. He has a way of motivating people, and he never asks others to do what he would not do himself. He works well with people and is very polite to those under him. Paul would say about his long hours, "That is what it takes to get the job done right."

—

## Pioneer Award Presentation and Citation

Figure 14-3. Pioneer Paul Mayhew (second from left) being recognized at the 2000 Pioneer Recognition Ceremony. The pioneer award plaque was presented by DNRO Keith Hall (left) and DCI George Tenet (third from the left). Don Winters (Executive Vice-President and Manager, TRW) joined in the presentation. (Photo by Sara Judy, NRO Visual Design Center.)

**Paul W. Mayhew, Ph.D.**

Dr. Paul Mayhew served as TRW's payload project manager and system engineer for two unprecedented signals intelligence satellite systems. He led development of the payload electronic, mechanical and radio frequency elements from concept to verified designs. His efforts resulted in overhead signals collection systems that proved to be vital national assets.

*Career in National Reconnaissance: 1964-1992*

# Reid D. Mayo

Reid Mayo was a systems engineer and technical director for the Naval Research Laboratory and for the National Reconnaissance Office. He conceived and designed the nation's first operational reconnaissance satellite—the Grab signals intelligence satellite. This satellite represented a significant breakthrough in national reconnaissance capabilities. Mayo later served as project engineer and technical director of the National Reconnaissance Office Program C, which embraced the Navy satellite reconnaissance element in the National Reconnaissance Program.

## Conceiving the World's First Signals Intelligence Satellite

Reid Mayo participated in the Pioneer Recognition Ceremony at the National Reconnaissance Office (NRO) in September 2000, and was eager to contribute his recollections to this publication. Unfortunately, his death occurred before he was able to make that contribution. This chapter is based largely on a 1981 interview with Mayo conducted by David K. Allison, Naval Research Laboratory (NRL) Historian.[1] Mayo's recollections from that interview are included in this chapter, as indicated by indented text.

**Foundations Built During World War II**

Mayo received his draft notice within a month of graduating from high school in June 1943. He chose to join the Navy, where after basic training he initially studied electronics maintenance and repair at the Radio Material School in Illinois. After working and studying there for several months, Mayo was given several options for his next assignment (Washington, DC; Corpus Christi, TX; and San Francisco, CA). He chose the NRL (then called Bellevue) in Washington, and he arrived there in late December 1943. One of the reasons he chose Washington was that he could be relatively close to his father in Toledo, with whom he had recently been reunited.

During World War II the Navy had a desperate need for people who could install and maintain electronic equipment aboard ships, and this was the focus of Mayo's training at NRL. The demand for people trained in electronics was so great that the Navy's admissions standards were somewhat flexible.

[1] In addition, Dr. Ronald Potts of the NRL reviewed this chapter and provided comments.

Figure 15-1. Naval Research Laboratory during World War II. (Photo Courtesy of Naval Research Laboratory.)

"[After boot camp], they put you through an aptitude test, and in those days, with the Navy installing so much electronic gear aboard their ships, right in the World War II heyday, they needed maintenance people desperately. Anyone that could spell algebra, and knew which end of the soldering iron got hot, was automatically a candidate to try out for [electronics] school."

*Training for a Wartime Assignment.* Mayo spent a year at NRL where, among other things, he trained in radio countermeasures with the Special Projects Group. Initially, Mayo was trained for a dangerous mission that involved applying countermeasures against radio-controlled German glide bombs. These bombs already had been responsible for sinking three warships. When the Germans launched a glide bomb at a ship, it was the responsibility of the radio countermeasures team aboard the ship to quickly intercept the radio control signal from the German aircraft that was directing the bomb. Once the the team intercepted the signal, it applied the countermeasure to redirect the bomb away from the ship. If the team did not apply the countermeasure in time, the targeted ship had a good chance of being destroyed.

"The special project training was about six weeks. They kept us a little longer because the first team had experienced operational activity, and they were trying to find out whether the Germans had given up using the device [the glide bomb]. The countermeasures group here at NRL had obtained the schematic and some of the hardware, and they developed a countermeasure for it, which the school was training us how to use. They trained us to do things rather rapidly since a bomb was heading toward you. There was a very deliberate and careful sequence that you had to go through, and it had to be error free."

*A Decision Prior to Deployment.* Prior to being deployed overseas, Mayo married Margaret Ann Blok. Their wedding vows were the first to be exchanged in NRL's new chapel, which was commissioned earlier that day. Mayo's original plan was to propose to Margaret after the war, but the risks associated with what was supposed to be his first overseas assignment caused him to change his plans.

Figure 15-2. The marriage of Reid Dennis Mayo and Margaret Ann Blok, 4 November 1944. (Photo courtesy of Naval Research Laboratory.)

"The lieutenant commander who welcomed us as the second group to go through the course told us that we were a very select group of young men. [He said] most of us would get the Navy Cross posthumously! My plans then, unbeknownst to my fiancée, changed from waiting to get married after the war to getting married as soon as we could, because I did not think there was going to be any after the war [for me]."

*Deployment to the Northern Pacific.* Fortunately for Mayo, the special projects class that preceded his class at NRL had been successful at redirecting all of the German glide bombs. Persuaded of their ineffectiveness because of these countermeasures, the Germans stopped using them.

As a result of this development, Mayo received a different assignment. In December 1944 he was assigned to the Northern Pacific, where he spent most of his time moving from ship to ship providing instruction related to receivers and radar jamming.

"Late on Christmas Eve [in 1944], I had to get on a train and go to Treasure Island [California] to be shipped out to the Pacific. We were sent to temporary duty assignment on each ship, and our officer had a list of the ships in the fleet that had countermeasures equipment installed on them. A few of these ships had [the countermeasures equipment] in the bowels of the ship as ballast—they had not bothered to hook it up! We would ride that ship, with nothing to do, until the next refueling date and then would swing via a 'breeches buoy' highline from the ship we were on to the one delivering mail or delivering fuel. We would walk across the deck on that ship, if it happened to be a tanker, and go on a high-line to the destroyer that was delivering mail. And he would take us to the next ship on our list. This way, we would go from ship to ship.

There were about sixty-five ships in a period of about eight months [on which] we gave extensive onboard training. By standing watch with the ship's company, I was able to tell my maintenance counterpart from each ship how to repair, how to align, and how to make this countermeasure equipment work right. The two operators and I would then stand watch with each ship's operators instructing them on how to differentiate between an image and a real signal, how to determine the signal frequency, and how to tune the jammers.

In the colder parts of our tour, when we would go up north, we would sleep around the electronic equipment because that was the only place you could get warm. We were only on the ship for 4-8 days so they would say, 'You do not need a bunk, and you are not going to be here long enough.' We did not even have a locker. We lived out of a sea bag—an honest to goodness sea bag—and slept on the floor most of the time. We did not get any mail. That was the worst part of it. It was an intensive training tour, [where we] coordinated the countermeasures activity for that segment of the fleet to which we were assigned."

Mayo participated in combat action in the Okinawa campaign and in naval operations that included bombardments of the Japanese homeland. He was aboard a ship in the West Pacific in August 1945 when the atomic bomb was dropped on Hiroshima. When the war ended, Mayo was sent to Hawaii, where he was stationed until December 1945. He was discharged from the Navy on 31 December 1945, and he returned to Washington, DC. Mayo continued his education at George Washington University, where he earned a degree in electrical engineering.

**Starting a Post-War Career at Naval Research Laboratory**

Although jobs were tight in the post-war economy, given his training and operational experience with Navy countermeasures during the war, Mayo was hired by NRL in February 1949 to work in the Countermeasures Branch.

> "I graduated in 1949, and that was a very bleak period to be looking for a job. The economy was down, and there were very few opportunities. If you did not already have a job, chances of getting one were pretty poor. I had, through a personal acquaintance, asked if there was any chance of working at NRL. The answer was, 'No way, they are still getting rid of WWII excess baggage.' I only made arrangements for an NRL interview as a last resort, but when the NRL countermeasures people found out that I had experience in the fleet with equipment they had designed during the war, they very quickly brought in their branch head, Mr. Lorenzen, to carry on with the interview."[2]

*Exploiting German Crystal Video Receiver Technology.* Mayo's first assignment was to design an American equivalent to a captured German radio countermeasures system, specifically the Athos system. Athos was a crystal video radar detector for use by surfaced submarines. Mayo was NRL's principal investigator for crystal video receiver technology, and the scope of his work went well beyond the Athos system. During the 1950s he designed systems for sea, air, and land platforms. His responsibilities included design, fabrication, testing, installation, operational evaluation, and support. In late 1957, the NRL crystal video group produced a detector system for use on submarines. This system proved extremely useful for the U.S. reconnaissance on a major Soviet fleet exercise, and was the NRL's most successful application of crystal video technology to that point.

*The idea was to exploit the German's concept in our own fleet... We would go out into the bay and depart towards Baltimore up towards Tilghman Island and try to see how far we could go and still receive the radars.*

> "My first significant activity [at NRL] was working on applying the German World War II concept of the crystal video receiver to our fleet. Crystal video is a receiving system that detects the radio frequency energy immediately, right after the receiving antenna. Then, the rectangular wave of the detected pulse is amplified in a video amplifier that has a relatively narrow bandwidth and high gain. It gives the advantage of a very simple, low power, small receiver that is quite reliable. It is a receiver that is called a 'wide open' receiver. It looks at all frequencies that are capable of being intercepted by the antenna, or by the band pass filters that are immediately following the antenna. The Germans had used this technique by equipping the lookouts on some of their submarines with something they called the Athos system. It was a crystal video receiver with an antenna mounted to the helmet of the lookout. As the lookout put the binoculars to his eyes and scanned the horizon, his head would turn and he would scan the antenna in the same manner that he directed his binoculars.
>
> The idea was to exploit the [German's] concept in our own fleet, either for surface ships or for submarines. We simulated a system from the concept that the Germans had used on their submarines. We tested it down at Chesapeake Bay mounted on a small landing craft that left the dock near North Beach by the Rod and Reel Club. We would go out into the bay and then depart towards Baltimore up the bay towards Tilghman Island

---

[2] Howard O. Lorenzen is a Pioneer of National Reconnaissance (inducted into the NRO Hall of Pioneers, 27 September 2000).

> and try to see how far we could go and still receive the radars. We could get ranges of 18-20 miles, something like that, with this thing mounted eight or nine feet from the surface of the bay."

*Looking to Space for Potential Signals Intelligence Collection.* As the Cold War rivalry intensified, there was an increasing need for accurate, timely data on Soviet strategic systems. Consequently, there was an immediate and critical need for new collection capabilities and techniques. While it was politically sensitive and technically unproven at the time, space-based systems offered great potential for national reconnaissance.

In October 1957 Russia successfully launched Sputnik I. At the time Mayo was standing watch at NRL's prototype High Frequency Direction Finding system at the Coast Guard station in Alexandria, Virginia, where he tracked the 20MHz signal and helped determine the satellite's orbit using bearing and Doppler measurements. Like many other scientists and engineers at the time, Mayo was motivated by the Sputnik success.

> "It was a very exciting time when Sputnik first went [up]. Everyone was excited by it, and the talk about space was everywhere—everyone was trying to get up to speed. I was anxious to think of how we could do other things from space.
>
> We were handicapped with our crystal video work aboard ship, trying to maximize intercept range in spite of the curvature of the earth. We got some relief when we could put this crystal video set in an aircraft and see, not 18 miles, but several hundred miles. That is a big advantage and if we could get our receiving systems up higher than aircraft, we could see further. So, the opportunity of a wide-open system being useful to look into the heartland of the Soviet Union was pretty obvious."

**Development of an Electronic Intelligence Space Capability—Grab**

The potential exploitation of space for national reconnaissance purposes was just emerging. The Chief of Naval Operations (CNO), Admiral Arleigh Burke, ordered the Navy research community to explore how the Navy might utilize space. The NRL already was involved with the Vanguard satellite program, and the Countermeasures Branch and the Navy Bureau of Aeronautics established a new objective aimed at developing a satellite-based electronic countermeasures system.

*Origins of an Idea.* While refinements to the submarine crystal video detector system were still being made, Mayo conceived an idea of space reconnaissance that would shape the remainder of his career. In March 1958 while stranded by a snowstorm in a restaurant in the mountains of Pennsylvania, Mayo pondered what would happen if the crystal video technologies he had developed for submarines were applied to space reconnaissance. Specifically, could the submarine periscope system be modified so that a solid-state version of this intercept system could be mounted in a twenty-inch solar-powered Vanguard satellite?

*Huddled in the restaurant with his family asleep around him, Mayo began performing calculations on the back of a placemat. He concluded that the extreme height would give the system coverage far greater than that of not only the submarine system, but also of airborne systems.*

> "We thought there might be some benefit in raising the periscope just a little bit…maybe even to orbital altitude. So I did some range calculations to see if truly we could intercept the signals. The calculations showed clearly you could get something up to an altitude of 600 miles."

Huddled in the restaurant with his family asleep around him, Mayo began performing calculations on the back of a placemat. He concluded that the extreme height would give the system coverage that was far greater than that of not only the submarine system, but also of airborne systems. Consequently, the advanced Soviet early warning and height finding radar, Token, could be detected at orbital altitudes. The system's data output could be relayed to small collection stations located around the periphery of the Soviet Union.

Upon returning to NRL, Mayo took his concept to Howard Lorenzen.

> "When I returned to Washington, I brought the placemat and gave it to Mr. Lorenzen. I think that might have given rise to rumors that we've heard on some occasions that the engineers at the Lab had been noted to do very good work on the back of placemats! Not quite as formal as we could be, but innovative nevertheless."

*An Idea Becomes a Signals Intelligence Reconnaissance Satellite.* Lorenzen and other NRL leaders were receptive of the concept, particularly because of the relative simplicity of the system and minimal costs when compared to other space systems under consideration by the Department of Defense (DoD).

*He used a Chrysler emergency brake as a device to clamp on the mast, and parked it so the wind would not turn it. He used a Mack truck steering wheel, which cost $12. He bored out the inside and mounted it to the mast.*

The Navy accepted the proposal. The NRL then proceeded to coordinate the proposal with the DoD, the Intelligence Community, and ultimately President Eisenhower, in order to obtain launch support, funding, and necessary approvals. In August 1959, President Eisenhower approved Mayo's concept for mounting an intercept system on the Vanguard satellite.

The Navy originally named the project NRL Electronic Intelligence Satellite, and later gave it a a classified name, Project Tattletale. By 1959 the Navy placed the project under a tight security control system (Canes), with access limited to fewer than two hundred people. This tightened security followed disclosure in *The New York Times* of the name and some details of Project Tattletale. President Eisenhower terminated Project Tattletale, and subsequent development and interagency coordination proceeded under the name Grab (Galactic Radiation and Background).

In addition to the covert reconnaissance mission, the satellite performed a purely scientific mission to measure solar radiation to help understand the nature and effects of the Sun, and to help predict outages for high-frequency communications. This publicly disclosed scientific payload was known as SolRad.

*Preparing for the First Mission.* There were several challenges to be overcome prior to the first launch.

> "The activity in preparing for this first launch involved two major areas: spacecraft and ground stations for collecting data. In Mr. Lorenzen's branch, under my auspices (with much contractor support), we did the military mission experiment—the electronic intelligence (ELINT) subsystem of the spacecraft.
>
> It looked for a while like the contractor was going to be late in delivery. They had a bright young engineer who was designing the band-pass filter. All he seemed to do was sit at his desk with a slide rule, and we really wanted to see some hardware. This was only a couple months before delivery was due. Finally, he got up from his desk, went into the shop, cut a piece of waveguide, drilled some holes, and soldered some things on it. This became the inter-digital filter using a waveguide cavity that we are

still using today. It was a classic! He made it work, and it worked beautifully.

Under my leadership, the electronic system for the ground interrogation and collection huts was done. Charles W. Price, a staff engineer at NRL Countermeasures Branch, did the mechanical design for the antenna and steering mechanism. He used a Chrysler emergency brake as a device to clamp on the mast, and parked it so the wind would not turn it. He used a Mack truck steering wheel, which only cost $12. He bored out the inside and mounted it on to the mast.

It was imperative before we even approached the launch that we have an isolated area—remote from the Soviet Union—where we could try the system out. We identified a naval communications station in Hawaii, in the center of the island of Oahu. We equipped a hut with not only the interrogation function, but the receiving and recording function as well. This hut was twelve feet long and it had an extra transmitting antenna on top. We had that ready to go as launch time approached."

**The First Launch**

Mayo and his colleagues personally delivered the satellite to the launch facility in Florida. Because of the secrecy of the mission, they maintained a low profile.

"We went down to [Cape] Canaveral and participated in testing out the military subsystem before launch. The transportation of the satellite to Canaveral was a very informal trip in the back of Ed Dix's station wagon! We accompanied it the whole trip. He took it into the hotel with him that night."

"This was all new to me because I had never been at a launch site before. Because of the military mission and my presence in the ELINT community, I was asked to keep a pretty low profile. If any visitor would show up in the area, I would disappear. We were not in the countdown process at all. We were not in any of the pictures. In fact, a rumor existed that we might have had some religious feelings about being photographed."

Figure 15-3. Radio control and interrogation hut in Hawaii. Bottom left to right: Howard O. Lorenzen, CDR Irving E. Willis, and William Edgar Withrow. Middle left to right: PO2 Lee, PO1 Hilbert R. Hubble. Top left to right: Reid D. Mayo, LT John K. Wulfhorst. (Photo courtesy of Naval Research Laboratory.)

The first Grab satellite was launched on 22 June 1960. Mayo and his team returned to Washington, and then on 4 July they traveled to the interrogation and collection

Figure 15-4. First Grab launch from Cape Canaveral, Florida. (Photo courtesy of Naval Research Laboratory.)

hut in Hawaii where they first heard the satellite's output. Initially, there was a brief period of concern that the system was not functioning.

> "The first interrogation did not work. We pushed the button and nothing happened, but 30 or 45 seconds later it did work. We had heart failures in between! We were all huddled inside [the hut] waiting for some sound to occur. When it happened, we all let out a shout of glee that it worked! Then I think most of us then went outside and celebrated with relief. It was really a very, very emotional evening. We turned it on and we could hear radars as far away as Japan—practically the whole Pacific area for a range of 3,500 nautical miles in all directions.
>
> It was not until after the pass, really, that we had a chance to think what had really happened and to go back and listen—when the shouting passed and tranquility prevailed. Not being very skilled operators, we were unable to make much sense out of it except to say, 'There sure is a lot more data there than we expected.'"

The visit by Mayo and his team to Hawaii, and Grab's success, coincided with Hawaii's first celebration of U.S. Independence Day since gaining statehood.

> "It was Hawaii's first 4th of July as an American state, but we pretended that they were celebrating the birth of a new space program!"

**Contributions of Grab**

The Grab technology survived into the 21st Century with applications to multiple signals intelligence (SIGINT) missions. Grab and subsequent SIGINT satellites provided dramatic new capabilities for U.S. intelligence collection.

The intelligence returned by Grab augmented imagery intelligence from the Corona photoreconnaissance satellites and confirmed that the Soviets possessed a robust air defense system. This intelligence provided the Strategic Air Command (SAC) with the information to develop optimum bomber flight paths to the strategic nuclear targets.

> "[The data analysts] were accustomed to looking 50 or 200 miles into the Soviet Union from some vantage point from the ground or from an airplane. Here, suddenly, they were able to see 3,000, 3,500, and 4,000 miles! The magnitude, the massive quantity of signals that were available

Figure 15-5. Galactic Radiation and Background (Grab) satellite. (Photo courtesy of Naval Research Laboratory.)

in one short instance in time, was staggering. They just were unprepared for the millions and millions of pulses they received. The response was a continuous cry of, 'Turn the gain down!' It was one that we were forced to contend with mission after mission after mission."

Not surprisingly after Grab's initial success, NRL was directed to continue satellite developments and improvements. These efforts became a major focus of the Countermeasures Branch for years to come, and for Mayo and his associates they became a full-time job.

Other U.S. SIGINT satellites eventually joined Grab, and by July 1962 all of these operational and planned space reconnaissance systems were placed under the purview of the NRO. The NRL program for which Mayo served as project engineer became part of NRO's Program C. Mayo's work on Navy systems contributed to the successful integration of the Navy's satellite program into the NRO structure. At the time, there was some concern that the Navy satellite program would be excluded from this new organization. However, senior leadership from SAC and the National Security Agency (NSA) appealed to the first NRO Director, Dr. Joseph Charyk, for the continuation of the Navy program. Dr. Charyk was persuaded to include the Navy as a component of the National Reconnaissance Program.[3]

## Conclusion

Mayo continued as project engineer for NRL. He frequently traveled to various U.S. SIGINT stations to supervise installations, evaluate on-orbit systems, and inspire the soldiers, sailors, airmen, and civilian technicians who operated and maintained ground readout systems. They viewed Mayo as a spokesperson for the Chief of Naval Operations and the directors of the NRO and NSA. He helped clarify the nation's needs and what the soldiers, sailors, and airmen could do to help. He stood watch with the crews while they evaluated each new system. He drank tea during quiet moments, and informally instructed them about threat signals and associated weapons systems and about future developments underway at NRL.[4]

Mayo's public career with the NRL, and his classified affiliation with the NRO, continued until his 1981 retirement from the Senior Executive Service (in which he was a charter

[3] The DNRO established Program C on 23 July 1962. Program C embraced the U.S. Navy satellite reconnaissance element in the National Reconnaissance Program and included the Technical Operation Group (made up of representatives from the NRL, Naval Security Group, and NSA) located on the East Coast. Program C was disestablished on 31 December 1992, and was superceded by functional NRO directorates.

[4] Recollections of Dr. Ronald Potts, NRL.

member). Mayo perhaps is remembered best for his "human engineering" mannerisms that inspired and motivated hundreds of soldiers, sailors, airmen, and civilian technicians to be innovative and operate a network of ground stations during the Cold War. The impact of Mayo's human approach is summed up in a reflection on Mayo's career by Dr. Ron Potts, "Reid Mayo directly motivated and inspired us. He turned us on to do great things."

—

**Pioneer Award Presentation and Citation**

Figure 15-6. Pioneer Reid Mayo (second from left) being recognized at the 2000 Pioneer Recognition Ceremony. The pioneer award plaque was presented by DNRO Keith Hall (left) and DCI George Tenet (third from the left). Rear Admiral Rand Fisher, USN, (Director, Communications Systems Acquisition and Operations) joined in the presentation. (Photo by Sara Judy, NRO Visual Design Center.)

**Reid D. Mayo**

Mr. Reid Mayo, at the Naval Research Laboratory, conceived and designed the first Navy signals intelligence satellite, Grab, and later served as project engineer and technical director of Program C.

*Career in National Reconnaissance: 1957-1981*

# Alden V. Munson, Jr.

Alden V. Munson Jr., who was a member of the government and TRW contractor team that defined the automated approach to signals intelligence satellite collection, helped conceive and plan a low-earth-orbit satellite system. Munson advocated a fully automated processing system to expand greatly the volume of reports and their timeliness. In his recollections, he describes the introduction of automated signals intelligence processing and its impact on the tactical use of signal intelligence. Munson also discusses the important aspects of the relationship between the National Reconnaissance Office and its industry partners.

## Automating Signals Intelligence[1]

My undergraduate major was mechanical engineering, and in graduate school I studied automatic control theory. As I was completing my degree programs, I began to focus on computer science. After I left school and went to work at the Aerospace Corporation, I took the substantial coursework in computer science at the University of California at Los Angeles. As a result of these studies and my initial work in industry, I was drawn to signals intelligence with its very challenging computer science problems.

### The Challenge for Automated Electronic Intelligence Processing

In the early days, electronic intelligence (ELINT) analysis was done manually. In the early operational facilities, the ELINT analysts tuned their receivers manually and worked with headsets. They used pencils and pads of paper to write down what they heard through their headsets. It was very much a manual, analyst-intensive process. Later, ELINT processing began using computers, albeit the processing was primitive and very slow.

I believed developing automatic processing, with its flexible architectures, would yield important mission benefits, especially in tactical situations. Initially, there was not much interest in turning ELINT processing over to automated systems, and there was skepticism about automated processes in general. Analysts always had played the primary role in processed ELINT reporting, and that was true at most ELINT facilities. There was a continuous battle to maintain commitment to developing automated, timely, high-volume ELINT

[1] This section is based on an interview with Alden V. Munson, Jr., at the National Reconnaissance Office Headquarters, on 26 September 2000.

processing and reporting. We encountered three challenges: hardware, engineering, and algorithms.

*Developing An Automated Process.* We faced developmental challenges in addition to struggling to gain acceptance of the automated processing concept. First, the hardware available in the early 1970s was not nearly powerful enough to handle automated processing. Second, the unknowns involved in ELINT required us to build a tremendously flexible system. Third, the large volume of information that would result from automating the process demanded the development of complex algorithms.

*There was one certainty about the ELINT environment that we would encounter: no one knew what it would be like.*

*Overcoming Hardware Limitations.* When we started in the 1970s, we were trying to use primitive machines for what were sophisticated algorithms and systems control software. These 1970s computers were primitive when compared with a commercially available laptop computer in the year 2000. In comparison, my current laptop has more memory and probably more processing power than all the computers we had installed around the world in the 1970s combined (including all of the computers at the field sites, training facility, and software maintenance facility).

The automated processing system that we developed was ahead of its time. We had to develop not only the algorithms, but the underlying system software as well. We had to develop a custom data file manager because the available operating system did not provide the kind of flexible data file capability that we needed to handle the huge volumes of data and the multiple tasks of the satellites.

*Building Engineering Flexibility.* There was one certainty about the ELINT environment that we would encounter: no one knew what it would be like. As a result, we needed to build a processing system with a high degree of flexibility. For example, we built very large ranges on the control knobs to tune the system and control the processing flow. In that way, we would have a system that could respond to whatever ELINT environment we might find when we actually deployed the system. This was a new approach in comparison with the earlier systems that tended to be built to specifications based on somewhat rigid assumptions about the operating environments and the ways the systems would be used.

*Developing Complex Algorithms.* Another challenge was developing algorithms for handling large volumes of data while getting precise results. To do ELINT processing, algorithms must look for patterns in the intercept data. Sometimes those patterns can be fairly long, and there has to be enough data in the buffers so that the algorithms can discern the patterns. When the memory on the systems is limited to a total of about a half of a megabyte, algorithms that can detect those patterns require great system complexity. There were people who said it would be impossible to automatically process the volumes of data that an ELINT system might collect. For our team, this sounded like a great challenge. And I believe we met the challenge.

*Success of the Automated Processing System.* Automated processing ultimately changed the paradigm for the use of overhead ELINT data by the military forces. The availability of relatively timely reporting changed the way military commanders and other tactical consumers looked at ELINT data. The system provided the impetus for the development of direct dissemination to military users.

Before automation, there was no need for direct dissemination. Once we demonstrated that automated processing was possible, other programs were encouraged to develop more automated processing as well, and they developed a parallel need for direct dissemination. We saw the results of this in later military operations, when most of the signals intelligence (SIGINT) systems that had tactical reporting missions used direct dissemination.

*Contributions of Many.* I was only one of many people who played roles in bringing the automated process to ELINT. One of my fellow honorees, Jim Morgan, shared the vision for timely ELINT reporting.[2] Jim was a Navy SIGINT enlisted man before he became an officer and senior leader in the Naval Security Group (NSG). The Space Program Office (SPO) directors also understood the importance of automating the ELINT process. Frank Quigley was the director of the SPO during the development and early operational period of the automated system and became a significant mentor for me.[3] Tom Betterton, who succeeded Frank, also shared that vision and brought the system to full realization.[4]

We developed a tremendous sense of camaraderie because we worked so closely, and at a pace and high intensity at which the National Reconnaissance Office (NRO) programs usually operated. To paraphrase Dickens, "it was the best and the worst of times." The work was hard, and that could be the worst of times. But we developed very strong personal bonds with our colleagues, and that made it also the best of times.

**Success and Rewards at TRW**

I had a wonderful career at TRW, beginning with my work on the automated ELINT processing program. It turned out to be a very good business for TRW. At the end of the 20th Century, TRW had the largest experience base and the highest amount of domain knowledge in overhead ELINT, and I am proud to have helped to build that business.

On the psychological level, I have a dual personality. On the one hand, I am suited to work in activities that involve general management. At the same time, I have craftsman instincts—I like to build things.

At TRW, I was involved with the conceptualization of projects as well as management functions, both of which were very fulfilling for me. I had significant management responsibility and ran big enterprises for the company. At the same time, I was able to take personal satisfaction in seeing products and systems in which I was intellectually and emotionally invested come to fruition. I still was very involved in technical invention, concept development, and architecting programs. I was able to maintain my technical involvement in the programs throughout my career.

*The work was hard, and that could be the worst of times. But we developed very strong personal bonds with our colleagues, and that made it also the best of times.*

I have said may times that everyone should have the opportunity to be able to become involved with the development of a program such as the primary one on which I worked for the NRO. It was a phenomenally rewarding experience. I feel fortunate to have been associated with Program C in the old NRO. Program C was the smallest of the three NRO programs and was less formal than the others. When I became involved with Program C in the early 1970s, the corporate culture was more like the NRO had been at its origin ten years earlier. There was little bureaucracy, few constraints, and an ability to work quickly and responsibly to address problems.

**The Government/Industry Relationship**

There are a few things I have learned over the course of my career concerning the relation-

[2] James E. Morgan is a Pioneer of National Reconnaissance (posthumously inducted into the NRO Hall of Pioneers, 27 September 2000).

[3] Captain Frank J. Quigley, USN, was the SPO Director from June 1976 to September 1973.

[4] Rear Admiral Thomas C. Betterton, USN (Ret) was NRO Program C Director from March 1985 to January 1992. He was a member of the 2000 Pioneer Selection Board.

ship between contractors and the government. First, any successful contract must focus on the customer's problem and how to solve it. Second, when the government jumps from contractor to contractor, the agencies suffer because there can be no accumulation and retention of domain knowledge. Finally, a good relationship—and a good product—requires an honest assessment of the true cost of a contract, by both the government and industry.

*Customer Needs Must be the Focus.* Companies that develop data systems must keep in mind that the product has to be relevant to the customer's mission. I always took it upon myself to try to understand the customer's mission domain and the customer's problem. What does the customer's boss want? What is the bigger context in which we are executing these programs? That attempt to understand the customer's needs served me well as I rose in TRW management to eventually run a division doing over $300 million of business each year. I had a reputation among the people who worked for and with me that I could be counted on to understand the problems they were addressing, as opposed to being a numbers-driven or a financially driven manager.

*There are Benefits to Long-Term Relationships Between Government and Industry.* The NRO has a history of long-term relationships with industry. I believe such relationships are absolutely essential, because contractors develop and build domain knowledge when they work on challenging programs. If the NRO were to have a revolving door of suppliers in a particular domain, domain knowledge would erode. The NRO has had relatively long relationships with its suppliers, and I believe that NRO programs have benefited by these relatively long incumbencies. The same is true with aircraft manufacturers, shipbuilders, and the builders of big weapons systems, which also tend to have long incumbencies. Long incumbencies are less common on command and control programs and other information technology intensive systems where there tends to be more of a revolving door of suppliers. I believe it is no accident that the performance record of those organizations doing command and control systems is not uniformly good when compared to performance in other domains. It seems simple enough to just run one set of people out of a job and run another set in when you are building software and command and control systems, but I believe the customers and the nation are paying for this tendency.

*There is Cost for Contracts "On The Cheap."* The unsuccessful missions to Mars in 1999 looks like a good example of how mistakes can be made by trying to do something "on the cheap." When average citizens read that the $165 million mission to Mars was a failure, they probably thought, "You ought to be able to do anything for a $165 million!" I said to my wife Jane, "Well, maybe it was a $300 million mission that the government tried to do for $165 million." Maybe $300 million is what it really would have taken to reduce the risk and increase the certainty of mission success. Trying to do something on the cheap is just going to be painful, difficult, and potentially embarrassing for everyone in the long run.

*I had a reputation among the people who worked for and with me that I could be counted on to understand the problems they were addressing, as opposed to being a numbers-driven or a financially driven manager.*

If a contractor gets a contract that is going to pay five dollars, but has to deliver ten dollars worth of work, there obviously is going to be a problem. Periodically, the government will negotiate contracts for less money than is required to do a credible job. That practice should be avoided because contractors will do crazy things to try to complete the work with the amount of money that is available. The taxpayers do not benefit when this happens, and it is very hard on industry as well.

**Creating an Atmosphere of Commitment**

Successful programs require a dedicated workforce, and management plays an essential role in ensuring that employees are committed and productive. As a manager, I have learned there are many ways to motivate people, and money is often times the least effective. A sense of teamwork and a positive work environment can be very important.

*The Team is a Valuable Resource.* It takes many good people to implement innovative systems. And the leader has to find ways to keep those people feeling challenged, appreciated, and rewarded. It is amazing what people are capable of achieving if they have confidence in the way that they are being led and managed. My advice is to get good people, keep them motivated, give them what they need to do the work, and try to create a sense of common cause. I always tried to lead and manage in such a way that the team members knew they were appreciated and respected so that they could feel good about their own involvement and commitment to the enterprise.

*The early part of my career at TRW was quite intense. My wife jokes that I went to work at TRW one day and came home two years later.*

I would say my greatest success was being able to lead large teams of people to implement innovative and far-thinking visions on a large scale. There is only so much one person can affect. However, effective management can leverage the group's technical skills to create bigger products, bigger systems, and bigger enterprises. Even though I enjoyed the hands-on aspects of my career, I found the greatest success in serving as a leader of the technical enterprise.

*A Positive Work Environment Keeps Good People on the Job.* It is always going to be a fact that the defense business will be competing for people with the non-defense sector. The leaders in government and the defense industry need to create a climate and an environment where they can attract and retain a quality work force. It can be done, but it is not easy. I despair for the government due to the unavoidable constraints that the civil service personnel system places on it. It is most difficult to find and keep new technical talent in the hot fields.

Finding and keeping good people in government service is not so much about loyalty as it is about hard economics. People have a set of choices, and their loyalty is a significant factor at a certain salary range. Employees are unlikely to be loyal to the extent that it would make a substantial difference in their financial benefits. I always have said that talented people could leave their current jobs for a twenty percent increase in compensation. So why does everybody not just do it? Why did I not do it? The answer, I believe, is because if the employer can build an environment where people feel rewarded by what they do, they will not necessarily operate with a mind only for compensation. True, if someone is offered a hundred percent increase in salary, then it is hard for the employee to explain declining the offer to the wife and kids. But in cases where the salaries are close, then money is not the absolute driving factor when people consider where, and with whom, they want to work.

**Family Life**

The early part of my career at TRW was quite intense. My wife jokes that I went to work at TRW one day and came home two years later. It is unfortunate how much family life was sacrificed when we engineers became involved with NRO programs.

The level of commitment, the amount of time away from home, and the amount of energy expended during program development could be very wearing on family life. My wife would sometimes be in bed when I got home. I would go to the kitchen to get the dinner that she had kept warm. Then I would sit on the bed and eat my dinner just so we could have a few minutes of conversation before I just collapsed. I know that many of my colleagues were running at that same pace. The work tended to be very hard on people, and very hard on marriages and other relationships.

Even though the work usually was hard on families, we occasionally were able to include our families in business travel. For example, during the summer of 1976, we were deployed to multiple field sites around the world for several months, and we were able to take our families with us. That helped to build bonds among our families, and these bonds continued to endure and flourish many years after the project ended. The best friends that my wife and I have are those people who we encountered during these experiences. Overall, I believe these unique relationships among the families made for a very rich experience.

**Conclusion**

Much changed during the course of the Cold War. The imperatives that we faced in the 1970s were different from those after the Cold War. During the early days, we had a genuine concern about the Soviet Union's expressed willingness, desire, and capability to project power outside of its borders. The Soviets were building long-range naval weapon systems, long-range aircraft, missiles, and other capabilities necessary for projecting power into distant areas and regions. This new Soviet threat created a clear imperative for the United States to respond.

The NRO and its contractors were a vital part of this response that demanded immediate results. We were allowed to cut through the red tape in order to push for speed and efficiency. We were spared many bureaucratic burdens. I hope this experience in the last 40 years of the 20th Century does not turn out to have been just a unique instant in time, never to be repeated again.

The NRO's ability to cut through bureaucracy has been an invaluable asset. I believe, with a few notable exceptions, the NRO of the 20th Century has had an amazing record of systems development. I hope that the country has not lost that forever. It is likely that the NRO may not be able to respond quite as well when the imperatives are less clear, as they became at the end of the 20th Century. I believe with the end of the Cold War, the NRO became affected by the lack of a clear national imperative. With some good justification, the American public and the Members of Congress felt less threatened by outside forces, and so may have been less supportive of the non-bureaucratic, efficient operating model that had served the NRO so well during the Cold War.

Some might say that in the early days of the Cold War we did too much "saber rattling" at the Soviets, but the fact of the matter is that there was a real threat, and the fear of that threat motivated the U.S. to have a national reconnaissance capability. Cutting edge NRO capabilities provided presidents and other national level decision makers with better information and understanding of U.S. adversaries' capabilities and intentions. We in national reconnaissance provided the information that allowed the U.S. defense community to plan responses to the Soviet threat, while reducing the risk of serious miscalculations or misunderstandings. I believe this allowed the U.S. to be more stable in its responses to the Soviet Union.

*It was easy to lead people during this time because it was easy to explain the imperatives of our mission. It was uncomplicated to convince people that what we were doing was important.*

I found that working on national reconnaissance programs during the Cold War was a unique experience. It was easy to lead people during this time because it was easy to explain the imperatives of our mission. It was uncomplicated to convince people that what we were doing was important. The national reconnaissance assets of this country were helping the world avoid

World War III, which many people thought was inevitable. I believe that those of us involved with national reconnaissance made substantial contributions toward winning the Cold War.

The National Reconnaissance Office, with its wonderful programs, has been one of the most productive elements of this government during the last part of the 20th Century. I have worked in the service of that organization for most of my career, and I am very proud to be able to say that. It is great that security rules now permit us to acknowledge that there is such an organization as the NRO, and that I now can assert that the NRO has built some of the most wondrous machines in history.

—

**Pioneer Award Presentation and Citation**

Figure 16-1. Pioneer Alden Munson (second from left) being recognized at the 2000 Pioneer Recognition Ceremony. The award plaque was presented by DNRO Keith Hall (left) and DCI George Tenet (third from the left). Jon Bryson (Senior Vice President for National Systems, Aerospace) joined in the presentation. (Photo by Sara Judy, NRO Visual Design Center.)

**Alden V. Munson, Jr.**

Mr. Alden Munson, a contractor with the Aerospace Corporation and TRW, conceived and developed an automatic electronic intelligence collection and data processing system that dramatically increased the timeliness of reporting. This represented a significant improvement over previous systems and provided important advances in intelligence support to U.S. military forces in peacetime and war.

*Career in National Reconnaissance: 1967-1994*

# Marvin S. Stone

Marvin Stone provided unique skill and insight to the design, manufacture, and integration of the electronics and radio frequency components of the payload elements of signals intelligence reconnaissance systems. Through his understanding and expertise in systems engineering and management, he was able to bring innovative advanced technologies to bear on the concepts that enabled implementation of the operational systems. The capabilities achieved through his work significantly contributed to meeting the intelligence needs of the nation for over thirty years.

## Successful Management and Personal Fulfillment[1]

I always wanted to be a communication systems engineer. At an early age, I started repairing radios in my father's radio and television store. I also built crystal sets, radios, and other gadgets. In high school, I decided to pursue a career in engineering. I chose electrical engineering with an emphasis in electronics. I subsequently earned graduate degrees in electrical engineering with an emphasis on communication systems.

My strong interest in communications continued as I entered the workforce. I was asked to do some work on one of TRW's signals intelligence (SIGINT) programs. Because I always had been interested in electronics hardware as well as analysis, I found space-based reconnaissance programs, with their advanced electronics content, a match for my personal interests, my background, and my educational training. As a result, I was highly motivated to work on these projects.

### Development of My Focus and Motivation—My First Site Visit

I did not fully appreciate the significance of some of the things that we were doing when I first started to work on SIGINT programs. What really enlightened me was my trip to a ground site. At that time, I was the payload system engineer on a system in development. I was responsible for both spacecraft payload electronics and for certain ground station units that pertained to payload and communication link operations. At the customer's request, I went over to the site to gain a better understanding of mission operations and to see first-hand how certain payload and link hardware was configured.[2]

[1] This section is based on an interview conducted by phone with Marvin Stone on 23 May 2001.
[2] The term "customer" refers to the United States Government and its representatives.

The opportunity to see the system in operation was a real eye-opener. I developed a true appreciation for the work I was involved in and its importance. Furthermore, I understood for the first time how the key payload specifications affected mission performance. Consequently, I became aware of the significance of my work to national security, and I realized that I had to broaden my horizons and channel my energy toward ensuring overall mission success.

**Technical Challenges and Successes**

The SIGINT system that we developed in the late 1970s was a significant improvement over the predecessor system. The new system extended the frequency coverage, was more agile, and had better sensitivity and fidelity. The payload designer's task was to develop a payload electronics suite that provided superior sensitivity and fidelity in the presence of strong environmental interference.

*I became aware of the significance of my work to national security, and I realized that I had to broaden my horizons and channel my energy to ensuring overall mission success.*

There were a number of challenges and obstacles to overcome. At the top of the list was minimizing the size and weight of the payload units without sacrificing system performance. The more weight we added to the payload, the more complex the space vehicle became. To that end, a large number of microelectronics circuits, heretofore not space qualified, were used in many of the designs. Therefore, device reliability was critical to mission success, and a significant effort was put forth in demonstrating adequate component lifetimes.

There were a number of other obstacles related to operating several of the electronics and other system units in a space environment, power consumption, and manufacturability, which we overcame as well. Ultimately, the system met or exceeded all of the key specifications while far exceeding the projected lifetime—an aspect of our legacy that is still of great value.

**The Importance of Team and People**

One of the very positive aspects of our effort was that we functioned as an integral government/contractor team. Our relationship with the customer (i.e., the customer's engineers that we worked with on a day-to-day basis) was excellent. We communicated on a daily basis, and we made decisions in a timely manner.

In addition, we staffed our projects with highly competent people. It was essential that we had an excellent staff in all positions—engineers, technicians, and support staff—given the challenges we faced. One of the key lessons I learned was to recruit and retain top-flight people.

This work environment was conducive to top quality technical work, rapid problem solving, and timely decision making. In other words, we had a "can-do" attitude. I attribute our success to this team effort.

**Impact of My Work on My Family**

Because of the nature of our work, I could not share my career activities with my family. During times of significant activity, namely when we had to meet a major milestone or had to complete a major part of the hardware, it was often hard to attend some family functions. The family often did not understand how consumed I was with work. Of course, those periods were difficult for me as well.

Now that my family members understand the nature of the work I was doing, they have an appreciation for the commitment I made during those years. This understanding seems to have relieved some of my wife and sons' anxiety of not having me around as much as they would have liked. They realize that what I was doing was important.

### Moving into Leadership Roles

At the beginning of my career, I was very satisfied with my communication systems engineering assignments. However, as my career evolved, I sought greater responsibility. I was specifically interested in interacting more with the customer, and in dealing with a broader range of programmatic issues. I wanted a leadership role.

Opportunity knocked in the late 1970s, and I was appointed head of payload systems engineering on our new SIGINT system development. Later, I assumed the role of payload project manager. I thrived on the growing responsibility and was excited about my work. I enjoyed the management aspects of the job, as well as the expanded technical content. I met my career goal of transitioning from an individual technical contributor to a leadership position.

### Conclusion

I am very proud of the work I did in national reconnaissance, and I am grateful for having had the opportunity to contribute to these important programs. I believe my strong technical background and my drive to succeed were the key factors in my successful career. I enjoyed all aspects of my work, including the technical challenges and the programmatics.

Worldwide communications, methods, and equipment will continue to grow in complexity and sophistication. Thus, the technical requirements of 21st century SIGINT systems will be even more demanding than those of their predecessors. I am hopeful that the payload system engineering design tools, and the design and manufacturing processes that we developed will be of value to this generation of SIGINT engineers.

—

## Pioneer Award Presentation and Citation

Figure 17-1. Pioneer Marvin Stone (second from the left) being recognized at the 2000 Pioneer Recognition Ceremony. The pioneer award plaque was presented by DNRO Keith Hall (left) and DCI George Tenet (third from the left). Don Winters (Executive Vice-President and Manager, TRW) joined in the presentation. (Photo by Sara Judy, NRO Visual Design Center.)

**Marvin S. Stone, Ph.D.**

Dr. Marvin Stone served as a TRW payload systems engineer on Program B electronic intelligence satellite reconnaissance programs. He brought numerous critical innovative advanced electronic and radio frequency technologies from development through flight production. His contributions yielded operational systems that revolutionized multiple generations of overhead signals intelligence collection.

*Career in National Reconnaissance: 1968-1988*

# Don F. Tang

Don Tang served as President of Lockheed's Space Systems Division after holding the positions of Chief Systems Engineer and Program Manager for the National Reconnaissance Office imagery intelligence and geosynchronous signals intelligence satellite programs. He led the way in fiscally sound technical management of national reconnaissance programs, and provided the impetus for the transition of Program A systems from testing to production.

## Success with a Technical Approach [1]

It was truly an accident that I ended up in my line of work, developing classified satellite programs. I broadened my experience during my initial years at Lockheed, and I gained exposure to various areas of system engineering. I never was drawn to any particular specialty, and it was by accident rather than by design that I ended up in the national reconnaissance business.

I came out of school in the early 1960s with the intention of working for several different companies before deciding where to settle and in which field of engineering I wanted to work. On my quest to determine where to establish myself, I worked for Link Aviation and then Lockheed. I ended up working at Lockheed for 34 years. Lockheed was a good company for me because it provided me with the opportunity to have a variety of different technical experiences.

I started out as an engineer, but I immediately recognized that I needed to move into management in order to advance my career. This was the area that I enjoyed the most, and also the one in which I most excelled. The best position I had was as Program Manager for one of the signals intelligence (SIGINT) programs. I always enjoyed working with people. I had around 100 people working for me, and I knew everyone by name, as well as each person's contributions to the program. That experience could not have been more enjoyable, and I did that for four or five years.

### A Team Approach at Lockheed

The government-contractor relationship was great because we all were in it together. There was trust between the different groups, and the relationship was successful and cooperative,

[1] This section is based on an interview with Don Tang at the National Reconnaissance Office Headquarters, 26 September 2000.

not adversarial. Contractors are not perfect, and our National Reconnaissance Office (NRO) Program A and B counterparts were quick to tell us when we were not perfect. At the same time we all were trying to achieve the same objective, and they always gave us the opportunity to earn back any lost fees.

There was minimal oversight, and this environment required us to perform well. The ability to work in an environment without strict oversight and scrutiny was a privilege, not a right. By the end of the 20th century, there was much more oversight of the program development teams, and it became more difficult to get things done.

*The fact that my family did not know what I was doing, or sometimes where I was traveling, caused occasional complications in my life.*

I believe the NRO of the early 21st century may have lost the pattern of the successful team relationships that we had in the early days. I expect this development was a result of the bureaucracy that often has a tendency to creep into organizations. This usually leads to bureaucratic oversight that hinders progress. Also, I found that the rapid rotation of personnel in Program A resulted in overall inefficiency. People rotated among jobs too often to maintain true professionalism, continuity, and stability. I believe that continuity of personnel contributes to the success of these types of programs.

### The Challenge of Communication and Security

An enduring problem that I faced was misunderstanding, or a lack of understanding, because of security. Part of program management was getting people motivated when they did not understand the entire mission. It was not always an easy task to make the satellite mission clear. However, when we talked to our people in plain terms, "This is the SIGINT collector, and this is the radar or communications data we will intercept with it," or showed them a satellite image, they could appreciate better the value of the system they were developing.

Figure 18-1. Photograph of Don Tang taken in 1994. (Photo Courtesy of Don Tang.)

Because of security restrictions, our contractor teams did not always have all of the information about the systems they were building. We did not always have a big enough imagination in terms of how our systems might be used, and what capabilities and qualities they should provide. Therefore, we built the spacecraft and its computer system without knowing at the time how they ultimately were going to be used. That made system engineering and development even more difficult.

### Security and the Family

The fact that my family did not know what I was doing, or sometimes where I was traveling, caused occasional complications in my life. I believe the need for secrecy required a bit of trust on the part of my wife and family. But they recognized that the work I was doing was important. I could not take members of my family into my immediate work environment, but at least I could show them the rest of the company. This was the middle ground in terms of my family's knowledge about my work. I expect working in the classified world became easier by the end of the 20th century. In the early days, the security people were stricter. However, in being strict they also were very successful in their jobs.

**Conclusion**

My most significant contribution to national reconnaissance was my technical approach to program management and the successes it produced. The contribution that allowed me initially to succeed in this field was my ability to bring experimental systems into operation through refinement and development. I believe system engineering provides a very disciplined approach to solving both technical and management problems, and the principles of system engineering should be applied often. There was not much resistance to my ideas at Lockheed because engineers by nature are accustomed to having things be very exact. At the same time, our government clients accepted my approach because we produced systems that worked.

I have no regrets about the field in which I specialized, and I only have positive memories of Lockheed. It was very enjoyable for me to work in that environment. I felt a strong sense of pride knowing the importance of the work I was doing, even though I could not talk about it outside of work. As a program manager and systems engineer, I believe my teams and I achieved many accomplishments, and I am proud that my work continued to be applicable to programs into the 21st century. My program management skills also helped provide profits for Lockheed, which was an important consideration for the company. More importantly, the systems we developed provided an advantage for our national policymakers that helped them prevent and win wars.

—

## Pioneer Award Presentation and Citation

Figure 18-2. Pioneer Don F. Tang (second from the left) being recognized at the 2000 Pioneer Recognition Ceremony. The pioneer award plaque was presented by DNRO Keith Hall (left) and DCI George Tenet (third from the left). Dr. Vance Coffman (President and CEO, Lockheed Martin) joined in the presentation. (Photo by Sara Judy, NRO Visual Design Center.)

**Don F. Tang**

Mr. Don Tang, while serving as a Lockheed spacecraft engineer on a Program A signals intelligence satellite, established a "collection scale" for determining what signals technically could be collected at an affordable cost. This allowed payload design and experimentation to move to production with confidence, and served as a paradigm for other NRO programs.

*Career in National Reconnaissance: 1960-1995*

# Peter G. Wilhelm

In a career with the Naval Research Laboratory that began in 1959, Peter Wilhelm was responsible for many technological advances that significantly improved the reconnaissance capabilities of early electronic intelligence satellites. In his recollections, Wilhelm discusses some of the highlights of his career at the Naval Research Laboratory, and explains the national security importance of maintaining robust government laboratories.

## Cutting Edge Work at the Naval Research Laboratory[1]

Sputnik was launched on 4 October 1957, about four months after I graduated from college. The thought of Russian hardware orbiting over the United States (U.S.) was scary. I recall 4 October was a Saturday and, even though I am not very religious, I was in church the next day. I was that afraid of Sputnik!

I remember when we were working on our first electronic intelligence (ELINT) satellite —galactic radiation and background (Grab)—I was worried that the Soviet satellites and rockets were so big and our satellite was this tiny thing.[2] I wondered how we were ever going to catch up to the Soviets because they could throw huge amounts of weight into orbit, and everything we did had to be lightweight and low power.

Looking back, if we had not been as scared as we were, I do not believe we—as a country —would have put as much effort into developing the new technologies as we did. We felt like the underdog, and I believe that was probably good for our psyche. We concluded that we could not overpower them, so we had to beat them with smarter systems and technology.

### Introduction to the Naval Research Laboratory

My wife graduated from college a year before I did, and she accepted a teaching position in Chicago, so that is where I looked for my first job after college. In 1957, I was hired as an electrical engineer with Stewart Warner Electronics. Stewart Warner had a contract with the Navy to redesign the UPM-70, a Navy radar test set for submarines. The contract stipulated that the equipment that we built had to be tested at the Naval Research Laboratory (NRL), so the company sent me to the Lab in Washington, DC.

[1] This chapter is based on an interview with Peter Wilhelm at the National Reconnaissance Office Headquarters, 26 September 2000.

[2] The term ELINT refers primarily to searching for radar emitters.

Figure 19-1. The Soviet Sputnik satellite. (Photo courtesy of NRO History Office, original source unknown.)

In March 1959, I accepted a position in the NRL Electronic Division. Late that year I was offered a position in the Lab's Satellite Techniques Branch. It took me a nanosecond to say yes to that offer!

Before I joined the Satellite Techniques Branch team, the team had been working on the Vanguard program, the first U.S. satellite program. The Vanguard program's legacy began with experiments by the NRL using captured German V-2 rockets. By the early 1950s, NRL had developed the Viking, which was the first large guided rocket and represented a significant improvement over the V-2. The NRL used the Viking to conduct atmospheric probes in space. By the late 1950s, NRL had developed the Vanguard, a multi-staged rocket whose development was based on the NRL's work on the Viking.

In October 1958, when the National Aeronautics and Space Administration (NASA) commenced operations, all of the NRL equipment and people who had been dedicated to Vanguard were transferred to NASA.[3] All NRL employees who had been working on Vanguard automatically became NASA employees—the whole Vanguard team was transferred en masse. If an NRL employee who was connected with the Vanguard program wanted to stay at NRL, that individual had to find another job at the Lab. This upheaval made it a very interesting time at the Satellite Techniques Branch when I arrived. There were only four or five people left in the whole branch. The NRL essentially had to recreate its capability to build the satellites for the Navy.

After Vanguard was transferred to NASA, there were a few NRL people who recognized that space was still going to be important to the Navy and to national security. They believed that space had a military intelligence aspect that needed to be developed, and that the U.S. involvement in space should not be entirely the dominion of a civilian space program. It was those people who came back to the Lab.

**Solar Radiation and Galactic Radiation and Background**

One of my first assignments at the Lab was to test the payload timer of the Grab satellite. Project GRAB combined a classified electronics intelligence (ELINT) mission with an unclassified solar radiation (SolRad) measurement mission. ELINT instrumentation was designed to collect RF pulses from Soviet air defense radars and transpond data to ground stations girding the Soviet Union. SolRad was an open, scientific endeavor to understand solar physics and how the sun interacts with the earth to cause radio fade-out. It also served as a cover mission for GRAB's ELINT operations.

For SolRad, we developed the first solid-state memory system. Prior to that we tried

[3] The National Aeronautical and Space Act created NASA in 1958. NASA was charged with the responsibility for all non-military space programs, hence the transfer of Vanguard from NRL to NASA.

something similar to a magnetic tape recorder, but we had to start and stop it all the time. The mechanical devices were not very good. The playbacks were not synchronized, and they were all jittery. We could not "lock up" on a signal.

Because solid-state devices and magnetic cores had been developed, we were able to build a solid-state memory. It was not very large, but it had no moving parts—it was completely electronic. We used the solid-state memory for SolRad. We also used it to sample the attitude when we were trying to figure out what was causing oscillations in the gravity gradient attitude control systems.

For many years SolRad and ELINT instrumentation flew together, but I could see the day coming when the two programs were going to have to go their separate ways. Because the SolRad was a spin-stabilized satellite, it needed the equatorial band of the satellite to scan across the face of the sun. We developed an attitude control system that would always keep the spin access perpendicular to the satellite sun line. That system was fine for SolRad, but we wanted to cover more frequencies for the purposes of reconnaissance. The solar radiation community wanted to concentrate on aiming their sensors toward the sun, but the Intelligence Community wanted to concentrate on directing the antennas toward the earth. The two programs went their separate ways.

**Atlas vs. Titan**

I developed a technique for launching futute satellites that could achieve high orbits by using a refurbished Atlas ICBM. The Atlas could not get into orbit by itself, but it could get very close. By stacking another solid-rocket on top of it, it could reach orbital altitudes. I was an electrical engineer, and I did not understand too much about orbitology in those days. But I taught myself, and I was able to do all the calculations.

Program A (Air Force) of the National Reconnaissance Office (NRO) provided launch support for Program C (Navy). In the early 1970s, Program A was in the process of transitioning from the Thorad to the Titan III, in order to orbit bigger and more capable satellites. Program C also needed a more powerful booster, but not as big and expensive as the Titan III. The NRO Staff asked Program C for alternatives to its latest satellite design proposals, "to reduce costs," and expected Program C to reduce weight or launch fewer satellites. Instead, I conceived of a way to do more for less.

Figure 19-2. The Galactic Radiation and Background (Grab) satellite. (Photo courtesy of the Naval Research Laboratory.)

I went back to what I had learned from the navigation satellites, and I did the calculations. If we used an Atlas to launch into the orbit, we needed only a very small rocket within the satellite itself, what we wound up calling a "multiple satellite dispenser (MSD)." I did the calculations,

[4] Program A was the Air Force satellite reconnaissance element in the NRO. The Director of the NRO established Program A on 23 July 1962, and it was disestablished on 31 December 1992.

and I convinced my boss, Howard Lorenzen, that MSD would work.[5] The Titan would have cost $25 million, compared with the Atlas at $6 million. That was a $19 million savings on every launch! My boss and I went to "4C1000," what we called the NRO offices in the Pentagon at the time. I got up in front of several of the NRO people and started describing our concept. General Bradburn jumped up and said, "No, no. That is not going to work! What in the heck are you talking about! No way! You can't make orbit with that vehicle."[6] I said, "Sir, I think you can." He said, "Well you're not an expert! I want my people to check this out! This briefing is over!" He was very upset. He said, "I don't want to hear anymore of this!"

*He said, "I can't get over this! My whole career I have tried to make the rocket bigger and bigger, and you guys are making things smaller and smaller!"*

The comptroller said, "All right, I want everyone back here in one week. General, you get your experts to look over this proposal." As Howard and I walked out, he said to me, "You had better be right." The next week we were all back together, and General Bradburn said to me, "Well, congratulations. My people tell me that you are absolutely right. It will work."

**A Lesson—The Value of Government Labs**

The mode of doing business in the aerospace industry changed considerably at the end of the 1990s. There became a much greater emphasis on industry endeavors at the cost of the government laboratories. I believe that we have forgotten some of the past lessons we learned about the value of government labs. I have kept trying to remind people of the value of the government labs, and that some of our greatest successes were the results of work done there.

A good example is the Global Positioning System (GPS) program. The technology for GPS was developed in a government lab, because there was no real commercial interest in that technology. Development of GPS first required years of work at the NRL, and then a government/industry team was formed to produce the satellites that transitioned the technology out of the laboratory. Today, GPS is a huge success. Not only is it beneficial to the military, but it is also a worldwide business. Initially though, GPS had to start within the government because it was not a moneymaker.

The idea that everything can be done commercially is flawed. Some things simply have no commercial application. Take signals intelligence (SIGINT) for instance. There is no commercial application for SIGINT. Imagery intelligence is different. At the start of the 21st Century, there appears to be a fledgling market for space imagery, but it is not clear that satellite imagery will make a lot of money.

Industry tries to get maximum market share. It also is much better at lobbying on Capitol Hill than the government labs are. Industry lobbyists go to the Hill and tell Congress, "You don't need to spend money in research and development (R&D) in the government. Spend it with us." Industry leaders will admit privately that they need the technology that comes out of government labs. The government labs are just not effective at marketing their own technology.

---

[5] Howard O. Lorenzen is a Pioneer of National Reconnaissance (inducted into the NRO Hall of Pioneers, 27 September 2000)

[6] Brigadier General David D. Bradburn, USAF, was the Director of the National Reconnaissance Office Staff, Office of the Secretary of the Air Force, Washington, DC, where he served from 1971 until 1973. General Bradburn later served as Director, Program A (the NRO Air Force satellite reconnaissance element).

Research and development investment can be at risk in the commercial sector. Industry has to watch its stock prices and make investors happy. This became even more of a factor during the 1990s when Wall Street began to dominate thinking. If a company's stock price was not rapidly increasing, the company was in trouble. As a result, companies cut back on their R&D and did not spend much money on developing core technology. They may have spent it on demonstrations, but they tended not to spend it on technology. That is shortsighted, but then Wall Street can be very shortsighted.

Figure 19-3. Werner Von Braun examines the television camera for NASA's Apollo 15 lunar landing mission. (Photo courtesy of the National Aeronautics and Space Administration.)

There is some R&D that is inherently government. It is that R&D that is high technology and has a long development cycle. If the U.S. is going to stay on the cutting-edge and be better than other countries, it is going to have to work on those long-term projects. It is dangerous to be shortsighted because some new technologies are going to take many years to mature, and they could give us the next giant leap forward. The U.S. was way out in front of any other nation, but we began to lose this advantage at the end of the 20th Century. I would like to see an examination of a better way to develop new technology and new programs by creating more of a partnership between industry and government laboratories.

### Werner Von Braun

One of my great memories is meeting Werner Von Braun when he visited the NRL in the early 1970s. Von Braun had been one of my heroes since I was in junior high school. We gave him a tour of the whole laboratory, and I took him into the clean room where we had some of the micro-thrusters. I showed him how, over the years, we had managed to make them smaller and smaller. He was sitting at the microscope, adjusting it and shaking his head. He said, "I can't get over this! My whole career I have tried to make the rocket bigger and bigger, and you guys are making things smaller and smaller!"

Figure 19-4. Werner Von Braun with NRL staff standing behind the SolRad High Satellite during Von Braun's visit to NRL in the Fall of 1975. Pictured (l-r) are Peter Wilhelm, Werner Von Braun, Harry Dornbrand, and Jim Winkler. (Photo courtesy of Peter Wilhelm, likely NRL photograph.)

As we sat in the clean room together I told him, "You are one of my heroes! I would like to have your autograph." He was embarrassed and said, "I'll tell you what, you give me your autograph, and I'll give you mine." I still have his autograph. It was a great experience.

### A Rewarding and Gratifying Career

I will never forget when we launched the first satellite with a transmitter that I had built with my own

Figure 19-5. Pioneer Peter Wilhelm during interview session, Pioneer Recollection Day, NRO Headquarters, 26 September 2000. (Photo by Candi Campbell, NRO VIsual Design Center.)

hands. When the satellite launched in Florida, I was back at NRL working on a satellite that would be launched only a few months later. Knowing where the satellite would come over, I set up an antenna on the roof of our building. It was such a thrill when I heard the transmitted signal. Just being there on the cutting-edge of this technology and then seeing the constant improvements was very gratifying.

One lesson I have learned over the course of my career is that nothing worth doing is easy. Almost every good idea requires convincing others that it is worth doing, especially if it costs a good bit of money to do it. If someone is really convinced they have a good idea, they should not be easily discouraged. They should just slug away and not be discouraged when it does not work perfectly the first time, because it will not! People will make mistakes, but that is all part of life.

I absolutely never regretted that things in my life turned out the way they did. In fact, I know I was just damn lucky. It was not through any great career planning on my part, for sure. It just happened. Some young people do a lot of what they call "career planning." I just dreamed, and when an opportunity came up that seemed to fit the dream, I grabbed it. If I had the chance to do it all again, I cannot think of anything I would change. I probably would do it all the same again.

There is a saying about "standing on the shoulders of giants" that applies to my career. I accept all the recognition that I have received as an endorsement of the quality of the whole program, and so it reflects well on the rest of the lab. I feel very fortunate to have had the opportunities I had in my career. I believe it definitely has been unique. I was there at the right time, and I just had a set of experiences that no one else had. It has also been personally rewarding. I have worked with some great people. I had a very general goal in my career: I just wanted to be a part of making important contributions.

—

## Pioneer Award Presentation and Citation

Figure 19-6. Pioneer Peter Wilhelm (second from left) being recognized at the 2000 Pioneer Recognition Ceremony. The pioneer award plaque was presented by DNRO Keith Hall (left) and DCI George Tenet (third from the left). Rear Admiral Rand Fisher, USN, Director of Communications Systems Acquisition and Operations, joined in the presentation. (Photo by Sara Judy, NRO Visual Design Center.)

**Peter G. Wilhelm**

As the chief spacecraft engineer at the Naval Research Laboratory, Mr. Peter Wilhelm invented new techniques and devices that added capabilities and improved performance of signals intelligence satellites.

*Career in National Reconnaissance: 1959-2000*

# Robert W. Yundt

Robert Yundt served as the Air Force Director of the Signals Intelligence Project Office in Program A of the National Reconnaissance Office, where he advanced the development and operation of signals intelligence and electronic intelligence satellites. An expert in low-orbiting satellites, he is considered by many as the father of low-orbiting signal intelligence satellites of the big, multi-purpose type. Under his direction, the National Reconnaissance Office placed into orbit satellites with longer lifetimes and greater location accuracy. Through his work, Yundt helped the Intelligence Community learn about Soviet air defenses at a critical time during the Cold War.

## A Brief Memory of a Space and Missile Observation System Launch[1]

Perhaps the most distinct and fond memory I have of my national reconnaissance work is the last Space and Missile Observation System (SAMOS) launch in which I participated. The SAMOS was an early satellite development program that emphasized read-out photo satellite systems, either by use of television transmissions or use of video recording tape. One SAMOS photo satellite project attempted to develop a high-resolution film-return system, while other SAMOS projects were involved with electronic intelligence satellites.

Figure 20-1. Robert Yundt enjoying retirement. (Photo courtesy of Robert Yundt.)

The launch, I best recall, occurred before I retired from the Air Force in 1966 and joined TRW. The morning was consumed with countdown, and we encountered an amazingly small number of problems. I gathered my staff and we went to San Jose, California, just in time to witness the first

[1] This section is based on written input that Robert W. Yundt submitted to the Center for the Study of National Reconnaissance.

crossing of the satellite's track over Kodiak.[2] The crossing confirmed that all deployments had taken place, and that we had an operational satellite. Before I left the station, I congratulated the Lockheed team—they were a great bunch of people.

—

## Pioneer Award Presentation and Citation

Figure 20-2. Robert Yundt, Jr. (second from left), son of pioneer Robert W. Yundt, received the pioneer award plaque for his father at the 2000 Pioneer Recognition Ceremony. The award was presented by DNRO Keith Hall (left) and DCI George Tenet (third from the left). Brigadier General Craig P. Weston (Director, Corporate Operations Office and Chief Information Officer, NRO) joined in the presentation. (Photo by Sara Judy, NRO Visual Design Center.)

**Robert W. Yundt, Colonel, USAF**

Colonel Robert Yundt directed the Signals Intelligence Project Office in Program A. He introduced a new, long-life, multi-purpose signals intelligence satellite. This program became a template for other programs that followed, and a cornerstone of the National Reconnaissance Office's early electronic intelligence collection efforts in support of the Strategic Air Command.

*Career in National Reconnaissance: 1960-1966*

[2] Kodiak is an island in the North Pacific near the base of the Alaska Peninsula. It is the location of one of the remote tracking stations that supported the SAMOS program. Other stations were located in Alaska, Oregon, Hawaii, New Hampshire, Iowa, and California.

# Section 3 IMINT Pioneers

## Changing the World of Imagery Intelligence and Mapping Through Satellite Reconnaissance

Martin K. Gordon, NIMA Historian

During World War II, aerial reconnaissance—specifically visual observation and photographic imagery—was utilized to support military intelligence operations such as locating and identifying targets for aerial bombardment and bomb damage assessment. However, reconnaissance was generally limited to supporting military objectives. The advent of nuclear weapons at the end of World War II, and the onset of the Cold War, dramatically changed the international and strategic environment.

This post-World War II environment, largely defined by superpower competition, contributed to the development of the new concept of peacetime strategic reconnaissance. Applying the lessons of Pearl Harbor, a cadre of scientists, engineers, and military officers—many of whom were associated with the Army Air Forces' Aeronautical Photographic Laboratory—articulated this new doctrine. This doctrine was intended to help provide advance warning of surprise attack. Periodic, high-altitude overflight of an adversary's territory would be conducted for the purposes of collecting photographic intelligence and mapping data. This concept of peacetime strategic reconnaissance was approved by the Eisenhower administration, and it became national policy.

### Historical Context for post-World War II Photoreconnaissance

The Soviet Union was the primary target for strategic photoreconnaissance during the post-World War II era, particularly after the Soviets tested an atomic bomb in 1949. However, as collection requirements increased and technical capabilities improved (in terms of both imaging technology and collection platforms), reconnaissance overflights were conducted against an increasing number of nations. Initially, converted military aircraft (e.g., RB-47s and RB-45s) were employed for reconnaissance missions, but these aircraft were susceptible to radar and interception, which added considerable risk given the political and military sensitivity of these missions. Consequently, high-altitude aircraft (e.g., U-2) and balloons were developed for these reconnaissance missions, but these platforms proved to be only near-term, limited solutions.

President Eisenhower and his technical advisers recognized that eventually the Soviets would develop the means to detect and shoot down the high-flying U-2, so an alternative collection method was required to mitigate the risks associated with overflight of Soviet territory. The Soviets' vulnerability to these unauthorized overflights led them to accelerate development and deployment of countermeasures, including surface-to-air missiles (SAM) and aircraft capable of intercepting the U-2.

**Satellites as a Revolution in Photoreconnaissance for Intelligence and Mapping**

In July 1955, President Eisenhower announced plans to launch "small, unmanned, Earth-circling satellites as part of United States participation in the International Geophysical Year," scheduled between July 1957 and December 1958. His statement avoided reference to the underlying purpose of the enterprise, which was to establish the principle in international law of "freedom of space" with all that implied for the use of space for national reconnaissance.

Photoreconnaissance satellites changed the world of intelligence, and Project Corona was among the first and most successful of these early photoreconnaissance satellite programs. Corona became operational in August 1960, and this system opened new worlds to imagery analysts and decision makers with its delivery of comprehensive photographic coverage.

Corona employed a film-recovery system, and the first film capsule recovered successfully contained photographic images of more than 1,650,000 square miles of Soviet territory. By June 1964, Corona had photographed all known Soviet intercontinental ballistic missile (ICBM) sites. At the same time, Corona eliminated the need for risky and politically sensitive aerial overflight of the Soviet Union.[1]

American decision makers at the highest political and military levels were able to identify and monitor the deployment and location of Soviet offensive and defensive weapons systems. This intelligence mitigated some of the risks associated with national security and arms control decisions as a result of the increased accuracy of the information that it made available. For example, the monitoring of Soviet nuclear and military research and development test sites provided advance knowledge of systems under development. Similarly, the tracking of arms shipments provided knowledge of Soviet arms transfers.

These photoreconnaissance satellite systems also enabled President Eisenhower and a few others authorized for access to this information to know with certainty that Soviet missile and bomber strengths were not present in the numbers alleged by the Soviets. Consequently, they could plan American defense expenditures with the confidence that they were providing the strategic force levels required for deterrence and national defense. The long-term effects of this change probably were best illustrated in 1972 when the Strategic Arms Limitation Treaty (SALT I) recognized and accepted "national technical means" (i.e., reconnaissance satellites) as the method of treaty verification.

Photoreconnaissance satellites, with their increasing reliability and their continuous improvement in image quantity and quality, changed forever not only imagery intelligence, but also cartography. This new capability facilitated the replacement of the narrow-strip

---

[1] McDonald (1997, pp. 25-74).

photography acquired by aircraft, and also gave the mapping community an opportunity to update the obsolete maps they were using.

Project Argon is an example of how photoreconnaissance satellites had an impact on mapping. Argon used special cameras and different satellite orbits on the systems six successful flights to advance greatly the former Army Map Service's geodetic accuracy and global mapping efforts. The flow of Argon imagery helped improve targeting accuracy that greatly reduced the circular error of probability for ICBMs.

**Contribution of Pioneers**

The success of satellite imagery reconnaissance operations, both in terms of the spacecraft and sensors, is a story of talented people harnessing their knowledge, experience, and creativity for important national purposes. They advanced the limits of technology as they continuously improved and refined the technologies and systems they designed, developed, and operated.

The imagery products that resulted from these Pioneers' efforts made an enormous contribution to U.S. national security and international stability. Their systems provided critical information to President Eisenhower and all of his successors. This information reduced the risks and uncertainty associated with many important political, military, and economic decisions, often when vital U.S. national interests were at stake.

The recollections of the Imagery Intelligence National Reconnaissance Pioneers in this section reflect a story of overcoming significant obstacles and constantly pushing the boundaries of technical innovation. It is a story of their achievements—namely the creation of important new intelligence sources and methods that revolutionized the collection, analysis, and dissemination of imagery intelligence.

—

## References

McDonald, Robert A., ed. (1997). *Corona Between the Sun and the Earth: The First NRO Reconnaissance Eye in Space.* Bethesda, MD: American Society for Photogrammetry and Remote Sensing.

# C. Lee Battle, Jr.

C. Lee Battle, Jr. was the Air Force director of the Discoverer/Corona Program, the nation's first satellite imagery reconnaissance program. From 1958 to 1963 he led the government-contractor team that developed and operated the Corona system, giving the country vital overhead intelligence of the Soviet Union and other denied territories. "Battle's Laws" of program management established standards for subsequent programs that were developed by the National Reconnaissance Office.

## The Laws and Principles of Program Management[1]

As I look back on my national reconnaissance career, the aspect that stands out as the most significant is my doctrine of the laws and principles of program management. I believe that my compilation of principles—including techniques for external interaction, internal personnel management, and technical advancement—is my most significant contribution to national reconnaissance. As such, it is fitting that my recollections include these lists, and I have included them as annexes to this chapter.

My teams and I operated more or less along the guidelines described in these two annexes, which state the laws and principles essentially in their order of importance. The key points of what came to be known as "Battle's Laws" (see Annex 1) are to select a small group of good people, demand quality performance, focus on mission accomplishment, and do not waste time on busywork. Other important components of Battle's Laws are a close working relationship between program office and contractors, rigorous analysis and correction of all failures, and constant forward progress.

My Principles of Program Management (see Annex 2) expand Battle's Laws to include more details about how to direct a program. Some of the main points include the importance of dealing with organization, planning, and schedules; focusing on the technical side of a problem; examining mistakes immediately and thoroughly, and avoiding paperwork and useless management exercises.

I am proud that "Battle's Laws" continue to be recognized and employed by many at the National Reconnaissance Office and in the Intelligence Community, and I believe that my management doctrine still can be applied to program development. In addition to helping

[1] This section is based on written information that C. Lee Battle, Jr. submitted to the Center for the Study of National Reconnaissance.

establish the nation's satellite imaging capability by directing the Corona program, I am pleased to leave behind a legacy that still can contribute to the improvement of our national reconnaissance capability.

Figure 21-1. Colonel C. Lee Battle (far right) standing behind a recovered Discoverer/Corona bucket. Also in the picture (from left to right) are: General T.D. White (Air Force Chief of Staff), Major General Osmond "Ozzie" Ritland, and Lieutenant General Bernard Shriever (Air Force Systems Command), circa August 15, 1960. (Photo courtesy of NRO History Office, original source unknown.)

—

**Pioneer Award Citation**

**C. Lee Battle, Jr., Colonel, USAF**

As the Air Force manager of the Corona Program, Colonel Lee Battle directed the government-contractor team that produced, launched, and operated the world's first satellite film recovery system. Corona provided U.S. leaders with comprehensive and repetitive images of all the states in the Sino-Soviet Bloc.

*Career in National Reconnaissance: 1958-1963*

## Annex 1

**Battle's Laws—5 September 1961**

1. Keep the program office small and quick-reacting at all cost.
2. Exercise extreme care in selecting people, then rely heavily on their personal abilities.
3. Make the greatest possible use of supporting organizations. You have to make unreasonable demands to make sure of this support.
4. Cut out all unnecessary paper work.
5. Control the contractor by personal contact. Each person in the Program Office has a particular set of contractor contacts.
6. Hit all flight and checkout failures hard. A fault uncorrected now will come back to haunt you.
7. Rely strongly on contractor technical recommendations, once the Program Office has performed its function of making sure the contractor has given the problem sufficient effort.
8. Don't over-communicate with higher headquarters.
9. Don't make a Federal case out of it if your fiscal budget seems too low. These matters usually take care of themselves.
10. Don't look back, history never repeats itself.

*A fault uncorrected now will come back to haunt you.*

## Annex 2

**Principles of Program Management**

1. Be schedule-oriented.
   a. If you don't start that way, you will end that way anyhow.
   b. Haste does not make as much waste as foot-dragging in this business.
   c. Decision time is critical.
   d. A tight schedule avoids letting anyone off the hook.
   e. Early launch testing shortens time to fix.
   f. Only in the Program Office does schedule motivation exist.
2. A good Program Office is oriented to the technical side of the problem.
   Don't kid yourself—the program payoffs come upon results. That means when it works.
3. Recognize the contractor's role and live with it.
   a. The contractor is in the driver's seat, technically speaking.
   b. At same time, make sure you hold the contractor fully responsible.
   c. The Program Office concentrates on evaluating the amount of effort and quality of people, and on problems you think are important.
4. The direct personal involvement of all Program Office members is vital.
   a. The Program Director must be held personally responsible for all aspects. (This automatically becomes the case.)
   b. In turn, the Program Director holds individuals under him in same status, etc.
5. (Corollary to 4) The Program Office *must* remain small. Parkinson can kill you.
6. (Corollary to 5) Use all other offices you can. Apply principle 4 to this. Always make unreasonable demands.
7. Never ask for help. You might get it.
8. Comply promptly with all report requirements in the most meager fashion that will pass inspection.
9. Restatement of 8. Don't over-communicate with higher headquarters.
10. Financially, it is the same story. You have to live with the contractor. Never let him get behind in keeping a finger on his status.
    Incurring unpredicted overruns is bad, but overrunning without knowing it is disastrous.

*A good Program Office is oriented to the technical side of the problem.*

11. Troubles: hit them hard and instantly.
    a. Unfixed troubles will bite you again.
    b. There is no such thing as a random failure.
    c. Personnel mistakes are far more frequent than design defects.
12. System integration is very important.
13. Insist that all principles herein apply to all contractor activities.
14. Don't generate paperwork. There are plenty of people willing to do this for you.
15. Committees are the world's most useless activity. Avoid "let's-have-a-meeting-ers" like you avoid poison.
    a. They never accomplish anything.
    b. There is always some individual who has the responsibility for doing what the committee thinks it is doing.
16. Management surveys are punitive. Recognize it and employ them (if ever) accordingly.
17. Examine closely the tie between the Home Office and the field. There's many a slip here.

# A. Roy Burks

A. Roy Burks was the Central Intelligence Agency's Technical Director of the Corona program, the nation's first imagery intelligence satellite system. Throughout his career, Burks faced and conquered both technical and managerial challenges. His efforts at integrating government, military, and contractor personnel were a primary factor in the success of this vital program, which solidified the nation's satellite imaging capability.

## Three Decades with the NRO[1]

My national reconnaissance career gave me the opportunity to contribute to various activities of national importance over the three decades I served in the government. I participated in both imagery intelligence (IMINT) and signals intelligence (SIGINT) satellite programs, and essentially witnessed the birth of the United States (U.S.) national reconnaissance capability. It all started with my first assignments after completing my military service.

### How I Became Involved in National Reconnaissance

I joined the Central Intelligence Agency (CIA) in 1956 as a member of the Junior Officer Training (JOT) Program. After my training, I became a member of CIA's Office of Scientific Intelligence (OSI). I worked at Headquarters with the OSI's Guided Missile Division, and later as a member of the Directorate of Intelligence Guided Missile Task Force. In 1960, I was assigned to the OSI team overseas, where I assisted with the successful collection of main beam SA-2 radar intercepts, which I hand delivered to Headquarters. The Agency was anxious to have precision power measurements of SA-2 radars. The CIA was developing the A-12 Oxcart aircraft which was the first of the so-called stealth designs, and the designers needed radar data.[2] It was through my OSI activities that I first met Jackson Maxey and Lloyd Lauderdale, two brilliant CIA analysts.[3]

---

[1] This section is based on A. Roy Burks' written input as amplified by an interview conducted at the National Reconnaissance Office (NRO) Headquarters, 26 September 2000.

[2] The A-12 was the successor to the U-2 program. The program's codeword was Oxcart, and the military variant became the SR-71.

[3] Lloyd Lauderdale is a Pioneer of National Reconnaissance (inducted into the Hall of Pioneers, 27 September 2000). See chapter 13 for his recollections.

Figure 22-1. Lockheed Advanced Projects (AP) "Skunkworks" building, East Palo Alto, California (Photo courtesy of A. Roy Burks, original source unknown.)

After my return from overseas to CIA Headquarters in 1964, I was assigned to the Defensive Systems Division of OSI. I was working on the analysis of Antiballistic Missile (ABM) systems when I was called to Maxey's office. Maxey was then the Chief of the Special Project Staff (SPS). This group was a part of the Directorate of Science and Technology (DS&T) headed by Albert D. "Bud" Wheelon.[4] Wheelon was responsible for the U-2, Oxcart, and CIA satellite reconnaissance systems. He also was responsible for all CIA scientific intelligence analysis and photo interpretation.

As an OSI intelligence analyst working on missile programs, I worked regularly with photoreconnaissance products, including Talent-Keyhole controlled imagery.[5] However, I did not have the appropriate access authorization for satellite operational details. Maxey briefed me into the operational system and gave me a complete overview of the Corona program. He described the unplanned vacancy the CIA had because of the sudden reassignment of Lieutenant Colonel Vern Webb.[6] Maxey asked if I would be willing to accept a position on the Corona program, and if I could catch the afternoon flight to California because a launch was scheduled the following day. I answered yes to both questions and spent the next three years of my life in East Palo Alto, California, as the Technical Director of the Corona program.

**The Corona Program**

My various assignments at CIA allowed me to witness the technical and managerial developments in the Corona program, and the use of imaging in intelligence analysis. I believe that my greatest contribution to the program was my management of the Corona Improvement Program. Because of my experience with Corona throughout most of the years of its operation, I also had the opportunity to contribute to the historical documentation of the program and, eventually, its declassification.

*The Beginning of Corona and the CIA's Involvement in Reconnaissance.* The Air Force and CIA began the Corona program as a joint activity, following up on the success of the cooperative U-2 program. When the President approved Corona in 1958, the CIA ended up

[4] Albert D. "Bud" Wheelon is a Pioneer of National Reconnaissance (inducted into the Hall of Pioneers, 27 September 2000). See chapter 42 for his recollections.

[5] Talent-Keyhole is a Sensitive Compartmentalized Information control system that protects technical data used in collection tasking, imagery or signals processing/exploitation techniques for collected data, and intelligence products derived from overhead reconnaissance programs.

[6] Lieutenant Colonel Vernard Webb, USAF, was involved in the operations of several NRO IMINT programs during the 1960s. He also served as CIA's Chief of Satellite Operations on the west coast, and the Deputy Chief of Operations at the CIA's Advanced Projects Facility.

managing the payload and film-recovery pieces of the program, while the Air Force kept the launch activities. The success of Corona led to a great deal more responsibility for the CIA in national reconnaissance activities. The CIA later took on the development of other satellite reconnaissance programs, both SIGINT and IMINT.

When the Russians launched Sputnik and their manned space flight, the National Aeronautics and Space Administration was established and assigned the U.S. manned space mission. It was then that some senior Department of Defense officials wanted the Air Force to reestablish its responsibility for aircraft and satellite reconnaissance activities, including Corona.

From my experience in the Army at Redstone Arsenal, I was not surprised by the Air Force attempt to regain responsibility for U.S. satellite reconnaissance programs.[7] During the 1950s, the Air Force and the Army fought over the control of medium-range missiles. The Air Force's top priority in the 1950s was the development of medium-to-long-range ballistic missiles, to the detriment of the WS-117L satellite reconnaissance program they also were developing.

The Air Force-CIA battle into which I was sent in my assignment in California was the struggle for satellite control. The reason the Air Force did not win the battle for control over all U.S. satellite programs was that some individuals—such as Edwin Land, Dick Garwin, John McCone, Jackson Maxey, and Bud Wheelon—believed it was good for the country to have the CIA remain involved in these intelligence collection programs.[8] Land believed that this helped to ensure that National Reconnaissance Office (NRO) programs remained responsive to all intelligence needs. I believe that the Director of the National Reconnaissance Office (DNRO) at that time, Brockway McMillan, thought it was a good idea to have the Air Force resume full control for all satellite reconnaissance.[9] However, his successor, Alexander Flax, apparently did not share his views.[10] In 1966, Flax signed the first formal agreement on NRO responsibilities, which established CIA as a partner in the National Reconnaissance Program.

*Early Days in Palo Alto.* Upon my arrival in East Palo Alto, where Lockheed's "Skunkworks" was located, I encountered some of the effects of the Air Force-CIA conflict. For instance, the Air Force raised an issue about my availability on a twenty-four hour basis. During orbit operations, we worked twelve-hour shifts. But we also traveled between the Satellite Test Center in Sunnyvale, and the Payload Integration Facility in East Palo Alto, when the satellite was not being tasked. The Air Force said we needed to be available by communication during this travel. To accommodate their desire, we leased a Hertz vehicle with a mobile phone.

Shortly after my arrival, during launch preparation for one system, the Air Force raised an issue about the reentry vehicle heat shield having exceeded its initially approved one-year shelf life. At that time, the NRO had a requirement for Corona to maintain twelve complete payload systems in storage at all times, and to be able to launch each payload within thirty days. The actual launch rate was approximately one system per month. I scheduled a meeting at General Electric (GE) in Philadelphia, with personnel from GE, Lockheed, and Aerospace in attendance. We examined the age history of successfully recovered heat shields. Based on this study, the contractor extended the shelf life to twenty-seven months.

---

[7] The Air Force initially had full responsibility for satellite reconnaissance programs.

[8] Edwin Land is a Founder of National Reconnaissance, see chapter 4. Richard Garwin is a Founder of National Reconnaissance, see chapter 3 for his recollections. John A. McCone served as the Director of Central Intelligence from 1961 to 1965.

[9] Brockway McMillan served as NRO Director from 1963-1965. He presided as Corona continued its successful operations, and vigorously promoted the development of a second generation of reconnaissance satellites. He was a strong advocate for maintaining the NRO as the primary agency in space reconnaissance.

[10] Alexander H. Flax served as NRO Director from 1965-1969.

Figure 22-2. Arthur Blackwell (Photo courtesy of Anne Blackwell.)

The launch continued on schedule and both recovery vehicles (RV) were successfully recovered. Arthur Blackwell, a highly qualified engineer, worked for the Aerospace Corporation and represented the Air Force in the heat shield study. I recruited Art to be my deputy many years later in my CIA career. Art was the first African-American to achieve super grade status at the CIA.

Another early memory I have of my time in Palo Alto occurred in August 1965 during the riots and fires in the Watts district of Los Angeles. Rumors of potential riots in East Palo Alto were widespread. Our concern, of course, was for the twelve Corona payloads held in storage at the East Palo Alto facility. Should we move them? If so, how could this be done rapidly, yet quietly? To what extent should the guard force be prepared to use firearms? With as little fanfare as possible, we moved most of the payloads to a Lockheed Sunnyvale warehouse until the threat subsided. We also began a study at that time of a potential move of the entire facility. The issue of secrecy versus physical security began to be examined for the first time.

*Corona Improvement Program—My Greatest Contribution.* If I had to pick my one crowning achievement or one special moment that I have had within the NRO, it would be the Corona Improvement Program where I worked as a member of the Special Project Staff.[11] I say this for several reasons. The program rebuilt the relationship between the Air Force and the CIA, it significantly improved the system's resolution, it introduced the use of color and infrared film, and it improved the mapping capability of the Corona system.

When I joined SPS, most of the staff was studying ways to develop a follow-on for Corona. Maxey had recruited John McMahon as his deputy, and hired John Crowley as the CIA Corona Program Manager.[12] Crowley, and the Chief of Operations, plus a security officer, a secretary, and I comprised essentially the entire CIA Corona staff.

Gene Fubini, who was a good friend of John Crowley, suggested to Crowley that too much attention was being given to the Corona follow-on, and too little effort was being made to improve Corona's performance.[13] In the Spring of 1965, John tasked me to develop a Corona Improvement Plan that could be presented to the DNRO. My first action was to request that each of the Corona contractors assign an individual to work with me as a member of a systems staff. Lockheed, Itek (the camera contractor), and General Electric (the reentry vehicle contractor) supplied representatives. All of these men were exceptional officers and quickly integrated into a team.

I also met with Air Force personnel in Los Angeles. I had Crowley's support for this meeting, which was a big boost because of his friends in the Office of the Secretary of the Air Force for Special Projects (SAFSP). Colonel Paul Heran and his staff agreed to cooperate fully in the Corona Improvement Program.[14] We had a proposal before Flax within months. It turned out to be the case for Corona "KH-4B," the last and most successful version of the system in terms of both resolution and success rate.

---

[11] See Annex 1 for a listing of some of the modifications included in the Corona Improvement Program.

[12] John McMahon served as the Deputy Director of Central Intelligence and later as the President of Lockheed Martin Corporation. John Crowley is a Pioneer of National Reconnaissance, (inducted into the Hall of Pioneers, 27 September 2000). See chapter 26 for his recollections.

[13] Eugene Fubini served as the DoD Director of Defense Research & Engineering (DDR&E).

[14] Colonel Paul Heran served as the Air Force Manager of the Corona Program Office, 1963-1967.

The Itek Corporation, under the direction of a brilliant young photo scientist named Robert Kohler, carried out experiments for the Corona Improvement Program on the use of different films and techniques.[15] Kohler later joined the CIA and introduced photographic edge measurement and edge sharpening tools into the overhead imaging community. Later he would hold senior operational and management positions at the NRO before returning to industry as a senior executive in 1985. He continued through the first decade of the 21st Century to serve as a senior advisor to the DNRO and the Director of Central Intelligence (DCI).

The first of the KH-4B systems was launched on 15 September 1967. The cost of all payload modifications including the camera, RV, system modifications, engineering and qualification testing was significantly less than what was estimated in our briefing to the DNRO. Much of the credit for the cost underrun goes to Frank Madden and the GE reentry vehicle Program Manager.[16] Frank was the Itek Program Manager for the KH-4B camera, and was the Itek Systems Engineer for the KH-1 through KH-4A systems. It was a very good group of people to work with, from the CIA, Air Force, and contractor community. Much of the credit for reestablishing a cooperative relationship within the NRO belongs to John Crowley.[17]

**Documenting Corona's History and Declassification**

I had the opportunity to document some of my activities in Corona history. I was a member of the Army Reserve, and often did my active duty tours with the SAFSP in Los Angeles. In 1972, General Lew Allen was Director of SAFSP.[18] Lew called me into his office during one of my active duty tours and said, "Roy, the Corona Program was really this nation's first satellite program. History will not look kindly on us if we do not preserve some part of it for future generations." He asked if I would take the lead in preserving a portion of the Corona history.

I discussed Allen's request with my management, John McMahon and Carl Duckett. They gave me approval to write a history

Figure 22-3. Corona KH-4B camera on display at the National Air and Space Museum, Washington, DC, Summer 2001. (Photo courtesy of Timothy J. McDonald.)

[15] Robert Kohler is a Pioneer of National Reconnaissance (inducted into the Hall of Pioneers, 27 September 2000). See chapter 28 for his recollections.

[16] Francis J. Madden is a Pioneer of National Reconnaissance (inducted into the Hall of Pioneers, 27 September 2000). See chapter 32 for his recollections.

[17] In 1965, Maxey left CIA and Wheelon appointed Crowley to replace him as the head of SPS, with John McMahon as his deputy. Crowley played a unique and pivotal role in the NRP at a time when it was on the verge of self-destruction. His cool head and negotiation skills brought the CIA and the Air Force back into a partnership agreement that endured for the next several decades.

[18] Lew Allen, Jr. (retired as General) served as Program A Director, April 1971–January 1973.

of the program, produce a film about the program, and preserve the last two reentry vehicles and the KH-4B engineering model for use as a museum display.[19]

The history turned out to be a five-volume document that the Air Force had to publish for us because of its classification.[20] The CIA film staff produced the film, which was called *A Point in Time*. The film was screened at the Corona Declassification Conference in 1995, and portions of the film also have been used commercially. As a result of General Allen's foresight, we were able to preserve Corona's history in written, visual, and exhibit form so future generations would be able to learn about the achievements of the Corona program.

*As a dedicated family man, Crowley recognized the need for a father to be close to his children. Because my children were on the west coast, he said he would respect my wishes.*

In the 1990s, I again had the opportunity to work with the NRO as a member of the Classification Review Task Force that recommended declassifying the imagery from the Corona satellite program, and the U-2, A-12 and SR-71 aerial photoreconnaissance programs.

**The Corona Follow-on**

As I stated earlier, most of CIA SPS was working on a follow-on system to Corona. Both the CIA and the Air Force had similar studies underway to meet the Intelligence Community's requirements. Director Flax decided that the Corona follow-on program would be a joint CIA-Air Force program.

The NRO then held a series of industry competitions for the Corona follow-on system. Separate contracts would be awarded for the payload (camera), the recovery vehicle, and the orbital support system. The CIA would procure the payload, and the Air Force would procure the other subsystems, but both organizations participated in the proposal evaluations. I served as a member of the proposal evaluation team for the recovery vehicle.

At this time, I was scheduled for an overseas assignment, but a medical condition caused the cancellation of that assignment. Instead I helped set up a CIA liaison office at the payload contractor's facility. After the liaison office was established, I returned to CIA Headquarters in Washington, DC, to begin work as member of a CIA project team supervising the development of the payload system. However, integration activities would be performed on the west coast. I spoke with Crowley about returning to the west coast to serve as the Test and Integration Manager.

At the same time, Crowley had received approval from the NRO for OSP to begin a study of a different future photoreconnaissance system. He offered me the choice of managing the study, which clearly would represent the future of the NRO, or the position of Test and Integration Manager on the west coast. As a dedicated family man, Crowley recognized the need for a father to be close to his children. Because my children were on the west coast, he said he would respect my wishes. I selected the west coast assignment, even though I realized the huge potential of the new study.

I served on the west coast, and this was a time of excellent cooperation between CIA and its contractors, and between CIA and the Air Force. We were experiencing technical

---

[19] Carl E. Duckett served as CIA DDS&T from January 1963-January 1967. He also served as Program B Director, January 1963—May 1976.

[20] This history has since been declassified. Kevin Ruffner of the CIA History Staff used portions of the history to produce the unclassified CIA monograph, *Corona: America's First Satellite Program.* Robert A. McDonald also drew on Roy Burks' historical research when he compiled the 1997 American Society for Photogrammetry and Remote Sensing book, *Corona: Between the Sun and the Earth: The First NRO Reconnaissance Eye in Space.*

successes and were meeting program requirements. The one "dark side" was cost, as cost per launch increased by an order of magnitude from Corona to this follow-on system.

**Signals Intelligence Support**

During my NRO career, I had the opportunity to broaden my experience and move from a focus in IMINT to SIGINT satellite development. Lloyd Lauderdale had proposed a complex and ambitious space initiative for the NRO. Lloyd worked tirelessly to develop the system requirements and to convert them into a viable operational concept. Almost single-handedly he developed the program plan and obtained approval. This program ultimately represented another huge NRO success story.

During the late 1960s, Lloyd Lauderdale regularly visited with his contractor team on the west coast. Because we had both worked as analysts on ABM and other related technologies, we would meet during his west coast visits to discuss our mutual interests. After John Crowley and Lloyd Lauderdale succeeded in getting an NRO approval for the program, Lloyd asked me if I would be willing to go to the ground station as the first Chief Engineer for site activation. I accepted the position.

I began attending site activation meetings at the contractor's facility. Although I had taken on this new SIGINT role, I continued to work on the Corona program. We achieved two successful launches of KH-4B Corona systems, and I was still conducting the camera "buyoffs" monthly in Boston, and the RV "acceptances" in Philadelphia. I also attended the Corona performance evaluation team post-mission evaluations at the National Photographic Interpretation Center (NPIC) in Washington, DC.

My double duty came to an end in for medical reasons in April 1968, and I returned to CIA Headquarters in July 1968 to work in the Corona follow-on. Before I retired from the CIA in 1984, I had an opportunity to return to SIGINT matters when I became CIA Director of SIGINT Operations (1981-1984). In this position, I had oversight for a number of NRO mission operations.

**The Impact of My Work on My Family**

How did my family handle the demands of my career? Not always well, and I ended up getting divorced once. The work undoubtedly put some pressure on my colleagues and their families. The secrecy that was involved, the inability to talk about their work at home, and the inability of the people who worked at the Corona ground facilities even to have their families know where they were working, were all things that created problems on the "home front."

*I believe what the NRO has done in supporting openness and declassification—when there no longer is a need for secrecy—is quite positive.*

After the Corona program was declassified, the children and grandchildren of many of the people who worked on Corona have been interested in finding out what their fathers did. That includes my children, who have been quite interested in visiting the Smithsonian, watching the film, and reading the books. So I believe what the NRO has done in supporting openness and declassification—when there no longer is a need for secrecy—is quite positive.

**Reflections on the Past and Lessons for the Future**

My work on Corona and other national reconnaissance programs provided me with both managerial and technical experience with IMINT and SIGINT systems and operations. I witnessed many successes and failures that taught me lessons that are relevant to NRO operations and programs into the 21st Century. These five lessons are: failure gives lessons for

Figure 22-4. Pioneer A. Roy Burks during the interview session at NRO Headquarters, Pioneer Recollections Day, 26 September 2000. (Photo by Candi Campbell, NRO Visual Design Center.)

success; the value of a sense of urgency; cost must be viewed in context; support to natural disasters is one non-conventional mission; and the value of NRO's "old program" organizational framework.

*Failures Gives Lessons for Success.* It would be very useful for current NRO personnel to look back at the Corona statement of work (SOW) and compare it to present statements of work. The Corona SOW is one and a half pages long. The SOW effectively read, "Go build a satellite reconnaissance system. Take pictures that have this resolution, and recover the film by catching it in the air. Complete your job by next year." The Corona schedule, from beginning to first launch, was one year.

It is hard to believe that the first launch did take place within one year. If anyone were to attempt to do what we did on that kind of schedule, with a statement of work that was one and a half pages long, the likelihood of success probably would be low. So it is not too difficult to look back and say, "My goodness, we should have expected to have problems!" While we certainly had problems, the important thing was that we stuck it out and learned from our mistakes. By correcting the mistake on the next flight, we avoided the same mistakes. Once we learned to do that Corona's success rate was as high as that of the NRO's programs at the start of the 21st Century. For some reason, however, the success of launch vehicles has not gone up much in the NRO's forty years, in contrast to the improved reliability of the satellites.

*The Value of a Sense of Urgency.* I still believe that the NRO is the country's best acquisition organization. The post-Cold War NRO has much more stringent requirements than we had back in 1960. Perhaps this is because safety and reliability are more important to Congress and other oversight authorities, which did not exist at that time. These people ask, "Well, why is the NRO having the same problems that it did forty years ago? I can understand those problems when the U.S. never had launched a missile or a space vehicle before, but now that forty years have gone by—during which the U.S. has been to the moon and back—why is the space community still having problems?"

The stringent requirements of the 21st Century result in cost increases, and cause the NRO not to be able to do things as quickly as it otherwise could. If the NRO had a really urgent need to do things, I believe it still would be possible to move quickly. The trouble is that in the post-Cold War environment, there is no longer a perception that we have an urgent threat that requires immediate priority. Instead, risk management becomes a higher priority than the urgency of the mission.

*Cost Must be Viewed in Context.* The idea of cost management became extremely important after the end of the Cold War, especially because the NRO no longer launches at the rate it did back in the early days. If the NRO launched a system every month, the unit cost of launch vehicles would not be as high. But when the NRO launches one vehicle per year, the cost to keep the contractors in business to produce the launch vehicles has to be higher.

For example, in the 1970s we produced a number of studies in which for cost reasons we proposed to launch intelligence payloads using the Minuteman III training launches that occurred monthly. With Lockheed and Program B, we proposed to use these launches for some small intelligence payloads. The proposal was to put an intelligence payload on top of one of the two training payloads, and a dummy warhead on the other. At launch time, the decision would be made to launch Missile A or B. In that case, the launch vehicle would cost nothing, and we would only have to pay for the cost of the payload. It is not the same payload weight as the Titan III-D, but some small payloads and the right kinds of orbit selections could solve some significant intelligence problems. To sum up, if someone insists on no failures and only one launch every couple of years, the cost per launch will be very high.

*Support to Natural Disasters is one non-Conventional Mission.* Since the time I worked in the national reconnaissance field, I have expanded my involvement in exploring the possibilities for non-conventional missions for reconnaissance systems. President Clinton issued Executive Order 13151 on 27 April 2000 for a Global Disaster Information Network. I have been a member of a working group on natural disasters, which is part of the Senate Caucus on Natural Disasters. In addition, I serve as a member of the Board of Directors of Western Disaster Center in California.

One of the biggest problems for the natural disaster community (both in the United States and globally) is that there are too many organizations involved, and no one is fully in charge. The NRO hoped to establish procedures to support natural disasters such as the forest fires that occurred in 2000. To date there has been only limited success, and local cooperation is necessary to establish an effective disaster mitigation system.

*The Value of NRO's "Old Program" Organizational Framework.* I worked in the days when the NRO was organized into four groups: Program A (Air Force), Program B (CIA), Program C (Navy), and Program D (Airborne). I am probably in the minority in making the argument that in spite of some competition that took place between the Air Force, CIA, and Navy, the previous organization was superior to the meshed IMINT/SIGINT/COMM orientation that exists in the year 2000. The reason I say this is because I believe the NRO has lost some of the support it had from each of these agencies prior to meshing them into these stovepipes. At the beginning of the 21st Century, the Air Force and Navy are building systems that look almost competitive to NRO systems. Additionally, I do not believe the CIA provides the same sort of senior-level support to the NRO that it did back in the 1960s and 1970s. Similarly, the CIA and NRO do not maintain as close a link in the research and engineering arena as they did in the past.

**Conclusion**

There is no question that NRO operations during the Cold War were more clearly defined than operations in the post-Cold War period, because during the Cold War we had an overarching strategic threat. It was crucial to justify the systems that were being developed with an intelligence requirement. Having worked in the CIA's Missile Task Force during the time of the supposed missile gap, I can tell you that we desperately needed hard information on Soviet missile and bomber strength. This information requirement justified the national investment in high-risk reconnaissance programs like Corona.

Aside from my personal satisfaction of having worked on intelligence programs that I know were extremely important during the Cold War, I worked with some of the best technical talent in the world and I made many friends.[21] Additionally, the Corona Advanced Project Staff holds an annual reunion. My wife, my children and their spouses, and I attended

[21] John Crowley, Lloyd Lauderdale, Jackson Maxey, and Art Blackwell are deceased, but all remained good friends of Burks' until their deaths. Robert Kohler, Frank Madden, Bud Wheelon, and John McMahon all remain friends with whom Burks continues to correspond.

the reunion in February 2000 at the Hiller Museum in San Carlos, California. I saw many old and treasured friends from the days I spent at the "Skunkworks" in East Palo Alto. These people should be proud because of our collective achievements. My experiences with the NRO will be treasured throughout my life.

—

## Pioneer Award Presentation and Citation

Figure 22-5. Pioneer A. Roy Burks (second from left) being recognized at the 2000 Pioneer Recognition Ceremony. The pioneer award plaque was presented by DNRO Keith Hall (left) and DCI George Tenet (third from the left). Joanne Isham (Deputy Director for Science and Technology, CIA) joined in the presentation. (Photo by Sara Judy, NRO Visual Design Center.)

**A. Roy Burks**

Mr. Roy Burks served as CIA Technical Director of Program B's Corona Program. He played a pivotal role in the development and improvements of the Corona cameras and film recovery system. His skillful integration of Air Force, Central Intelligence Agency, and contractor teams ensured the long-term success of our nation's first space imaging system.

*Career in National Reconnaissance: 1965-1995*

—

## Annex

### The Corona Improvement Program

For the Corona Improvement Program, the Air Force lengthened the Thor booster tank to provide increased orbit payload capability. Developed and executed between 1965-1967, this improvement program modified the Corona stellar index camera to provide a 3-inch focal length, Argon-type index lens and dual stellar cameras. The program called for the main Corona panoramic camera to be modified to:

- improve photographic performance by removing reciprocating scan arms and reducing vibration from moving components,
- improve the velocity-over-height (Image Motion Compensation) match to reduce image smearing,
- improve photographic scale by accommodating proper camera cycling rates to enable altitudes as low as 80 nautical miles (the J-1 camera operating altitude was 100 nautical miles),
- eliminate camera failure due to the film's pulling out guide rails (an occasional problem with the J-1 camera system),
- provide better calibration of the panorama image with precision holes drilled into the guide rails,
- improve exposure control through variable slit selection to get greater performance at low sun angles (as opposed to the J-1, which had only a single exposure for the entire mission),
- enable use of ultra-thin base film, yielding a 50 percent increase in area coverage with no increase in weight, and
- allow handling of different film types and split film loads through addition of an in-flight changeable filter and a film change detector.

# Frank S. Buzard

Colonel Frank Buzard worked on the Corona satellite photoreconnaissance program, and was the System Program Director for the Corona follow-on program. He organized and managed the program from the identification of its requirements to the first successful launch and operation. Buzard focused on a small Program Office with a few highly motivated and talented people, good rapport with key personnel on the government-contractor team, and a strong system engineering integration with an open exchange of information. This approach contributed to establishing a record of consecutive successful launches and facilitated the development of a new approach to monitoring arms control treaties.

## A Doer, Not an Administrator[1]

The 1960s was an exciting time for launching satellites, and we knew we were doing something important. The work was never dull or routine. After working for six years in the Air Force Security Service, where I could not say anything about my work, it was a nice change to be able to talk to family and friends about my work—launching satellites and operating them on orbit—even if I could not say much about the nature of the payload.[2]

### Working on the Corona Program

The Corona program and "Battle's Laws" were my introduction to the world of satellite programs. As far as I am concerned, the program had three heroes: Richard Bissell from the Central Intelligence Agency (CIA), Lee Battle from the Air Force, and Jim Plummer from Lockheed. As much as anyone, these three men made the program a success.[3] The thing that stands out in my mind about the Corona program is Battle's insistence that there is no such thing as a random failure, and that every failure has a cause. Consequently, we had to

---

[1] This section is based on an interview with Frank S. Buzard at the National Reconnaissance Office Headquarters, 26 September 2000.

[2] The Air Force Security Service later was integrated into the Air Force Intelligence Command, which subsequently became the Air Intelligence Agency.

[3] Richard M. Bissell, Jr., served as the CIA Corona Program Manager. He previously served as the Assistant Director of the CIA, and the U-2 Project Director. C. Lee Battle is a Pioneer of National Reconnaissance (inducted into the Hall of Pioneers, 27 September 2000). See chapter 21 for his recollections and a version of "Battle's Laws." James W. Plummer served as Director of the National Reconnaissance Office from 1973-1976.

find and fix every error. He had a disdain for management and bureaucracy, and he did not want anyone "messing" with his programs. Everyone working with him had to be productive, which mostly meant engineers and scientists. Administrators had to look elsewhere for something to do. This was a theme throughout my work in national reconnaissance. Namely, I always enjoyed being a "doer" rather than an administrator.

*Keeping the Program Office Small.* One reason for this relative lack of bureaucracy and administration was the small size of the Corona program office. The maximum number of people we ever had assigned there during my five years was 23, and that included the procurement staff. The first year we had three people—Battle, Bill Johnson, and myself.[4] Two more people were added the next year, and more arrived after the first successful film recovery.

*Minimizing Paperwork.* There was very little documentation in our office about what occurred in the Corona program. That made it a very interesting place to work at the time. Everyone concentrated on getting the job done with an absolute minimum of busy work, such as reports. When the National Reconnaissance Office (NRO) was established, it was a very non-bureaucratic organization, which made it a great working environment. One consequence of this disdain for administrative work is the almost total absence of documentation and reports prepared by the Corona program office. Unfortunately, the lack of records and reports made it quite difficult 20 years later for the people who were trying to document the histories of those programs.

Simon ("Si") Ramo of Ramo-Woolridge wrote about one instance when he was riding in an airplane with Secretary of Defense Charles Wilson. Wilson, sitting with a big stack of papers on his lap, said, "My staff thinks that you ought to document all decisions that you make in reports like this. That way, you can find out what happened in case of failure." Ramo replied, "I don't want to be associated with a well-documented failure. If we succeed, nobody will care, and if we fail, it would not make any difference." Essentially, that meant, "To hell with the documentation," and that was how things worked!

Figure 23-1. Aerial recovery apparatus in the rear of a C-130 aircraft in position for retrieval of a Corona capsule. (Photo courtesy of NRO History Office.)

*Dealing with Failures.* The challenges we faced before we achieved mission success ranged from specific technical design questions to broad management issues. We responded to failures with positive action, and we obtained truthful analyses.

[4] Albert "Bill" Johnson served as payload recovery manager in the Corona Program Office.

*Responding with Positive Action.* On 19 August 1960, we successfully recovered the first Corona bucket with film. However, it was a tough road to that first success on the thirteenth launch. Our first ten launches all had failures of one sort or another. Many of these were human or procedural errors. Some errors were caused by a lack of basic knowledge. A problem we experienced with the film is a good example. On the earliest launches, the film kept breaking. It turned out that the standard film was acetate-based, and it crumbled like autumn leaves in the harsh space environment. Eastman Kodak learned to coat the photographic emulsion on a polyester base, and that solved the problem.

On our eleventh launch, the Agena performed well, and the whole load of film moved through the camera into the recovery bucket. However, the recovery vehicle (RV) separated from the Agena and just disappeared.

*Wilson said, "My staff thinks you ought to document all decisions you make in reports like this. That way, you can find out what happened in case of failure." Ramo replied, "I don't want to be associated with a well-documented failure. If we succeed, nobody will care, and if we fail, it would not make any difference."*

This failure made a major redesign imperative. First, we decided to replace the solid spin/de-spin rockets with a cold gas spin system. We used these rockets to spin and stabilize the RV before the retrorocket was fired, and then de-spin it afterward so it would survive re-entry. The new system was much simpler, and we could test it before launch. Second, we decided to ignore the conventional assumption that electronic components would burn up if we left them operating through the ionization layer. We put a primitive telemetry transmitter and tracking beacon on board so that we could monitor what occurred between RV separation from the Agena and emergence from the ionization layer. Our solution solved that problem.

*Getting Thoughtful Analyses.* I believe that a critical component of the system development process is that the failure analysis has to be open and honest. That means that the government should not threaten the contractor with a financial penalty when it investigates a failure, because then the contractor will be more likely to conceal things. We wanted the contractors to be completely honest, to stand up and say, "This is what happened, and this is how it happened." In order to get that kind of cooperation from the contractor, there must be trust that the government will have the maturity to say to the contractor, "Okay, you messed up that time, but we are not going to penalize you."

One such situation occurred when the Douglas Aircraft Company experienced a failure on a launch. As I recall, a technician put a block-one system in a block-two vehicle, and the wires were connected one way instead of being connected another way. This caused the vehicle to fly wrong. Douglas confessed immediately to the error, and there was no problem on our end. However, later it came to Secretary of Defense Robert McNamara's attention, and he wanted to punish Douglas financially for the error. He sent out a message, which I received in the Corona program office. He requested details about the event. Battle told me not to send him that information, so I did not even answer the message.

Shortly thereafter, an Air Force major visited us from the Pentagon. He was instructed to talk to Battle to find out what happened so that they could charge Douglas for this mistake. Battle would not even speak to him, and I told the major, "You are beating a dead horse; we are not going to tell you. We are not going to penalize Douglas because the minute we do, they will stop cooperating on finding the cause of the error." Punishing the contractor makes a good failure investigation impossible. The purpose of such an investigation should be to look for the truth, not to look for whom to punish.

**Developing a Corona Follow-on Program**

The initial motivation for the development of a Corona follow-on system arose in 1964 and 1965, when the government began to think that Corona had reached its performance limit. The government wanted to improve its satellite capabilities, and two competing programs emerged. There was a back-and-forth exchange, which was disruptive to those of us still working in the Corona program. A single design was chosen and duties assigned, with overall responsibility for the system assigned to the Air Force. I went back to the Secretary of the Air Force for Special Projects (SAFSP) office in Los Angeles and became the System Program Director.

*I did everything possible to make sure the entire system worked on the first launch. I did not want to find myself in the position of having to inform DNRO Flax that I just blew tens of millions of dollars on the first launch.*

When I was managing the Corona follow-on program, I tried to keep the program office at about the same size as the Corona program office and run it along the same lines. I was largely successful at doing that, and I ran the program office with a minimum of administrative work. Our contracts were straightforward and contained a minimal number of clauses. The only reports we had to write to the Director of the National Reconnaissance Office (DNRO) were for the financial reports by Charlie McBride or Jimmie Hill stating what we had done that month.[5] Those reports really got things done. I usually made the decisions myself in Los Angeles, as opposed to asking DNRO Flax to make decisions in Washington, DC.[6] This approach significantly simplified the decision making process.

Although I used the Corona program as a model when possible, there were a number of challenges that were specific to this program. Consequently, a tailored approach was necessary to meet certain program and mission objectives in the areas of requirements, source selection, design, systems engineering, and launch.

*The Requirements.* Because I grew up in the Corona program, I understood the problems of a weight-limited spacecraft design. As a result, I developed seven requirements from the outset of the program:

1. The design had to have an adequate weight margin so the designers would not have to make high-risk decisions to save weight.
2. There had to be a very strong systems engineering organization and focus.
3. There could be no single-point failure to avoid a catastrophic on-orbit failure.
4. The design had to be highly redundant with high reliability.
5. There had to be a comprehensive test program before shipment to the launch base, with an all-systems test and all subsystems functioning at the production facility.
6. The vehicle had to have multiple recovery buckets.
7. The system had to have an adequate growth capability in its design to accommodate longer on-orbit lifetime without an extensive redesign of the system.

These requirements helped ensure that the system would not only meet its initial mission objectives, but also could be flexible enough to accommodate additional missions as those requirements emerged.

*Source Selection Challenges.* The source selection was a very difficult time, because when we were conducting our source selection, the Request for Proposal for the spacecraft needed to accommodate two very different design proposals for the sensor subsystem. We originally solicited proposals from two companies, and later we asked three other companies to

---

[5] Jimmie D. Hill served 29 years in the NRO at all levels, 14 of them as Deputy Director of the NRO.
[6] Alexander H. Flax served as DNRO from 1965-1969.

submit proposals. One company chose not to bid. One proposal was like a NASA program, very elaborate and very expensive. Another proposal was unfocused, perhaps because the company had never done business with the NRO. We eventually narrowed the selection down to two companies, and made our final selection soon after.

For the recovery portion of the program, we solicited proposals from three companies. One company chose not to bid, and another submitted a very bad proposal. For several years, this latter company had been trying to change the recovery bucket on the Corona program and use a new material for the heat shield. I, among others, opposed the idea for two reasons: the existing material always had worked, and the Corona program was about to come to an end. The company's proposal personnel apparently thought that I was prejudiced against their recovery material, so when they submitted their proposal for the Corona follow-on, they proposed the same heat shield material that they had been telling us for three years was no good. I think they figured that I would influence the source selection board to help them win, but I did not. The third company won the competition, which made the other company very unhappy.

*Initial Design Challenges.* We spent a year negotiating the interface between the satellite vehicle and the sensor subsystem. The reason was that there were two primary proposals for the satellite vehicle, and we could not indicate to the sensor subsystem program office and contractor which bidder was going to win. The source selection had not been determined, and as a result, the sensor team whose proposal already had been approved had to begin its work without knowing the overall system design and interface.

When the spacecraft contractor was selected, we began the long and often contentious negotiations between all the participants in the program, including SAFSP, CIA and the various contractors. I set up a formal interface working group on each of the major disciplines (e.g., electrical power, thermal, structural, command and control, and test and operations). Each of the participants had representatives on each of the working groups. The groups were chaired by my representatives and co-chaired by the Sensor Subsystem Program Office. These groups met regularly (at least once a month initially) to work on specific agenda items. We kept formal minutes, and assigned specific action items. Once a month we held a program managers meeting, frequently at a contractor facility, to review and approve the progress of the working groups.

I thought we could resolve the interface differences in 90 days, but it took us a year. The week before Memorial Day weekend in 1968, we had a major meeting in Los Angeles to resolve the final differences. After we had worked almost all week without final agreement, I told everyone, "Okay, we have fiddled around long enough on this. If you do not come to some agreement, we are going to work all weekend and nobody can go home for the holiday." Because the meeting was in Los Angeles, and most of the participants were from out of town, we finally reached an agreement. But it was a hard struggle, and it caused a major delay in the program.

*Strong Systems Engineering Approach.* We had a complete and free exchange of information across all interfaces, so that a change in one subsystem could be evaluated for its effect on the other subsystems and the system as a whole. The satellite vehicle, the sensor subsystem, and the recovery vehicles were designed and manufactured by different contractors in different cities. We had a thermal vacuum chamber that simulated the space environment, and a second vacuum chamber with an optical test capability so the sensor could be given a final check at the satellite vehicle contractor's plant prior to shipment to the launch base.

We used the factory-to-pad concept, meaning that when the vehicle was shipped to the base, it went directly to the pad to be launched in 15 days. That meant that we had to perform the final tests of the overall system at the prime contractor's location. To save time, we brought in the Vandenberg Air Force launch personnel for them to observe and participate so that we would not have to repeat the systems test at Vandenberg.

We faced many problems during the development stage: a thermal control design problem, parachute development problems, and a sensor subsystem problem. In the case of the sensor subsystem, we could replace the component and retest rather than ship the whole sensor back to the contractor, because we had an optical capability in one vacuum chamber.

Despite these difficulties, I was less troubled by failures during tests on the ground than I would have been if the failures had occurred during launch and deployment. I did not want failures while on orbit. I did everything possible to make sure the entire system worked on the first launch. Obviously, I did not want to find myself in the position of having to inform DNRO Flax that I just blew tens of millions of dollars on the first launch.

*Some people worried that we would spread hydrogen chloride ions all over southern California. On the day of the launch, we went out and put litmus paper all around the launch site. People thought we were nuts...*

*First Launch.* On our first launch, we faced a major problem when the temperature in the battery compartment did not level off as predicted. If the batteries became too hot, one of them might explode and disable the others. We had an on-board auxiliary battery that would let us operate for one day. Unfortunately, once we switched to auxiliary, we could not switch back.

We learned a lot about battery characteristics that night and the next morning. The batteries became hot when they were being charged by the solar cells. Similarly, the batteries became hot when they were discharging their power to operate the satellite and the sensor. If we let the batteries rest between charging and discharging, they would cool down. However, if we let the charge in the batteries get too low, we would not have enough power to switch to the auxiliary battery, if necessary.

At our 4:00 p.m. decision meeting on launch day plus one, most of the experts wanted to switch to the auxiliary battery. I decided that was not the way to go. I said, "No, we will stick with what we have. Do not switch." Fortunately my boss, General Lew Allen, said, "Buzz, you are the boss, that is what we will do." After operating a few days in a rather restricted mode, the batteries cooled down. Our operations team learned more about the system, and gradually began unlimited operation of the sensor. It turned out later that the cause of the problem was contamination during launch from the solid rocket blast, which deposited debris on the satellite vehicle surface and almost completely negated our very sophisticated thermal design.

There was also an issue with the reefing line on the parachute. The first film recovery bucket fell into the water because the chute did not deploy fully. We retrieved the bucket from the water. We lost another recovery bucket because the parachute did not work. The cause was that, unbeknownst to us, the vendor who made the device changed the material that cut the reefing line. The cold on-orbit temperature affected the new material the contractor had used. As a result, the parachute would not work within the time limit for the operation. The contractor did not anticipate that would happen, and the parachute did not deploy.

We also ran into an environmental question during the first launch. Some people were worried that we would spread hydrogen chloride ions all over southern California. On the day of the launch, we went out and put litmus paper all around the launch site. People thought we were nuts, and they thought we were even crazier when we went back and picked it all up. But we did not have any environmental problems with the launch.

These are examples, just from the first launch, of the importance of having a diverse team of highly qualified experts on hand that knew the hardware design and on-orbit operations. Throughout the life of the program, these individuals met all kinds of on-orbit emergencies with clever, innovative ways of making the system work despite any anomalies.

**Impact on Family Life**

Surprisingly, the primary effect of work on my family and personal life was that it provided great stability. The main reason was that I had long tours: five years in the Discoverer/Corona office at the Air Force Ballistic Missiles Division in Los Angeles, three years in 4C1000, six years at SAFSP back in Los Angeles, and then retirement.[7] This type of experience probably differs from the experiences of many of my colleagues during those years. Their assignments usually were not as long. If I had been reassigned more frequently, like many of my colleagues and friends, balancing my work and personal life might have been more difficult. In actuality, the stability contributed to my success in working on and managing these satellite programs.

**Comparing the National Reconnaissance Office of Past and Present**

Sometimes I would like to return to the early days of national reconnaissance. I do not have insight into what happened in the post-Cold War NRO, but I often think back to the beginning. The programs were kept small and non-bureaucratic, we concentrated on the work, and we did not worry about anything but the success of the program. There was no busy work, and there were no voluminous reports. When we had a failure in the Corona program, Battle (the Air Force program manager) and Plummer (the Lockheed program manager) went from Los Angeles to Washington, DC, to meet with DNRO Charyk and Bissell (the CIA program manager) to discuss what was going on.[8] That was it. There were no elaborate meetings, reviews, or outside agencies to interfere. The newly formed NRO was a small group of dedicated people doing something they knew was important.

The fact that our work did not receive public attention did not bother us. As a matter of fact, we were delighted to be out of the public eye. I have an intense distrust of the ability of congressional personnel to keep secrets. I witnessed too many instances when someone shared information with congressional personnel, even though the recipient never had to sign a security agreement. Those individuals would claim, "We were elected by the people, so we do not need to do that, but we won't tell." However, they often leaked information when it was convenient.

I spent three years in 4C1000, but I was not particularly happy there. I was not a headquarters guy. I was a doer who was more comfortable out in the field. The fun was to be found in the thick of the action. However, if I had to be at the Pentagon, being at the NRO was preferable to being on the Air Staff. Back in those days, we spent more money and attention on hardware than on people. After the Cold War ended, it seems that more money was being spent on administrative functions than on the actual systems. I believe that the trend should have been in the opposite direction.

*Surprisingly, the primary effect of work on my family and personal life was that it provided great stability.*

Frankly, I am aghast at the publicity surrounding the NRO activities at the beginning of the 21st Century. The attention probably makes it very difficult to do business. I have spoken with individuals who worked for me and are still working in the national reconnaissance field.[9] They said the complexity of the management arrangements and the rules and regulations make it almost impossible to get anything done. One example they cited was the non-direct chain of command from the DNRO to the program directors to the contractors.

---

[7] 4C1000 is the room number of the NRO's Pentagon offices. It served as NRO Headquarters until the mid-1990s.

[8] Dr. Joseph V. Charyk served as the first DNRO from 1961-1963.

[9] See Annex for a summary of what they see as operating principles for program management.

During Corona, I was responsible for submitting reports for Colonel Battle to the higher authorities. Battle told me my objective was to get a "C-minus" on the report. He would say, "Just get it in on time. If you get over a C-minus on the report, you are spending too much time on it. You are supposed to spend your time working, not reporting." Perhaps we were living in a much simpler time, but I suspect that in the post-Cold War era the administrative procedures at the NRO have become overly complicated.

### Conclusion

We pushed the state-of-the art in national reconnaissance across new frontiers. Some of the programs' innovative developments were multiple re-entry vehicles, generous weight margins, high system redundancy to avoid on-orbit failures, and rigorous system testing. I am proud to have been a part of these efforts, and to have been associated with the dedicated and talented individuals with whom I worked.

It was a tremendous privilege to work with such outstanding people throughout my career. In addition to working with intelligent and dedicated people, the reasons I enjoyed my work in national reconnaissance were that the work was fun, it was important, and it was never routine or dull. There was no need for motivation. We were not practicing what we would do in case of a war—we were actually doing something.

—

## Taking the Bad with the Good of My Husband's Career[10]

My husband, Buzz, first got involved with classified work for national reconnaissance because of his education and training. The Air Force sent him to the University of Illinois for graduate study in mathematics. The objective was to make him and other young officers knowledgeable in computers, which just were beginning to emerge. Because of his background in physics, chemistry, and mathematics, along with his training in computers, the Air Force assigned him to this new program that turned out to be Corona.

Figure 23-2. Pat Buzard being interviewed on Pioneer Recollections Day, NRO Headquarters, 26 September 2000. (Photo by Candi Campbell, NRO Visual Design Center.)

### The Problems of Secrecy

Most of our neighbors had regular nine-to-five jobs. Buzz was assigned to the Air Force Security Service before he went into space and missile defense work, so I was used to not asking questions about these things, although sometimes unwillingly. On one occasion, I overheard something during morning coffee with friends. Later that day, I mentioned what I heard to my

[10] This section is based on an interview with Pat Buzard at the National Reconnaissance Office Headquarters, 26 September 2000.

husband. His reaction, because it was classified, was, "Where did you hear that?" Instances like this were inconvenient, and sometimes stressful.

The one time I really had problems with Buzz's work came when he was still in the security business. One of our children became very ill. He was behind the Iron Curtain, and I was unable to contact him. I did not know where he was or what he was doing. One of his bosses finally contacted him, and he came to the hospital by plane. It was scary! Because he often was out of contact, I learned not to depend on him for things like this. In fact, I believe women in my position became self-sufficient, even though we were not in the workforce as most women were by the start of the 21st Century. One of my sons told me, "Mom, you have been a liberated woman all your life, you just did not know it."

One complaint throughout the years was that Air Force promotions often ignored those who could not talk about their work. I tried once to explain it to an officer, when he had posed the question "Why is the Air Force losing so many of these young officers who are resigning?" I answered, "Because they are not getting promoted." He responded, "But these officers know how important their work is!" I explained to him what went on in the family. "The problem is that the wives do not know how important the work is. Their husbands are unable to talk and brag about their work. The only recognition that the wives can see is a promotion. When there are no promotions, the wives convince their husbands to resign."

**Long Hours and Personal Sacrifice**

When Buzz worked on the Corona program and it had its first launch, he was gone for 48 hours! I often worried, because he would say "I will call you if I am going to stay really late." The next day, he would say, "It got to be midnight, and I did not want to wake you." I was worried that he would be so tired that he would have an accident on the way home. Instead, he would return the following evening like a zombie. But those are the hours that they kept, seven days a week, ten hours a day. Needless to say, I was in charge of things at home.

The men who worked under Buzz used to be out of town on the job for weeks at a time. Buzz tried to insist, although he found it difficult because of the Air Force rules, that his subordinates call home at government expense once a day. He felt that this was very important. He had to fight the communications people, but he finally got that allowance through. Unfortunately, some of those families ended up separated anyway.

**The Benefits of Openness**

I definitely feel free now that I know and can talk about something related to Buzz's work, even though he worked on a program that has not yet been declassified and I still do not know everything he did. Although our friends and family knew that my husband had something to do with launching missiles and things like that, they never knew anything about the photoreconnaissance part of his work. Now Buzz has been able to tell them some of the stories.

Just after Corona was declassified, Buzz and I gathered our family, including grandchildren, at a family dinner party. Buzz said, "Well, I just wanted to explain a few things to you about my work." He described some of the things he did that are now unclassified, and they were startled. Their eyes popped! Previously, the fact that Dad was away on business travel was nothing new to them, and they were not overly curious, possibly because he was away so much.

Our children and grandchildren were very impressed by the videos and the other information we received after the Corona declassification. The first time we visited the Corona exhibit at the Smithsonian, we saw a man with his children looking at the exhibit. Buzz decided to tell them all about the program. He was like a regular tour guide. It is so nice for Buzz and his colleagues to be able to do that because they could not say anything about their work for so long.

## Conclusion

When Buzz and I attended high school and college reunions in St. Louis, where I grew up and went to school, we had very little in common with classmates who had not had the opportunity to travel, live in different places, and have the wonderful experiences that we had. All in all, I believe that Buzz's rewarding work and the chance to make many friends and see the world through our travels provided us with a fabulous life.

—

## Pioneer Award Presentation and Citation

Figure 23-3. Pioneer Frank Buzard (second from left) being recognized at the 2000 Pioneer Recognition Ceremony. The pioneer award plaque was presented by DNRO Keith Hall (left) and DCI George Tenet (third from the left). Brigadier General Craig P. Weston (Director, Corporate Operations Office and Chief Information Officer, NRO) joined in the presentation. (Photo by Sara Judy, NRO Visual Design Center.)

—

**Frank S. Buzard, Colonel**

In 1966, Colonel Frank Buzard became the Director of a Program A imaging satellite program described as "the most complex electro-mechanical device ever placed in orbit," establishing a record of consecutive successful launches. His efforts helped overhead imaging achieve a new level of sophistication and made possible the monitoring of the SALT I Arms Control Treaty of 1972.

*Career in National Reconnaissance: 1960-1972*

## Annex—Operating Principles for Program Management[11]

I asked two people who worked for me for their ideas on the principles of program management and systems development. This annex is a summary of their views.

People. Hire good people, match their skills with the job, and delegate appropriately. Flat, small-as-possible organizations work the best. I would rather be undermanned than overmanned. There is nothing wrong with Battle's Laws[12]!

Organize for Teamwork. Arrange a hierarchy of working groups so that every problem has a home and parallel conflicting solutions are avoided. Keep interference from the "management by virtue" folks to a minimum. The highest-level working group must contain all contractor and government executives who can commit their organizations to the agreed-upon actions.

Informal Organization. The "informal" organization that develops over time is usually more effective than the "formal" organization. Understand and exploit it.

External Interface. Watch and define your external interfaces, particularly with other government agencies. Any attention you get from these interfaces is usually not good. It is better to underpromise than overpromise.

First Step. Spend a lot of time in the beginning getting a mutually-agreed program concept. Assure that all participants are committed to the success of the entire enterprise, not just the individual parts.

Contractors. Work with the contractors, not against them. Government managers and contractors need to work as a team. Esprit de corps is a good thing. Do not expect the product to be produced as proposed. It will never happen. In fact, it is not always good for the contractor to do what was proposed, because requirements, schedules, and costs change. It is the government's job to manage those factors the best it can.

Profit. Profit is not a dirty word! A contractor's success is the government's success. Contractually incentivize the contractors creatively to get to the overall end product (not to lower level steps). Reward them well when they do well, and poorly when they do poorly.

The Real Work. Understand how the "real work" on the project gets done. An end capability is what the System Program Office wants, and it takes hardware and software and people to do that. Paper, reportage schemes, specifications, and plans are a method of getting to the capability (the product)—they are not an end unto themselves!

---

[11] This section is based on written input that Frank Buzard submitted via e-mail to the Center for the Study of National Reconnaissance.

[12] C. Lee Battle is a Pioneer of National Reconnaissance (inducted into the Hall of Pioneers, 27 September 2000). See chapter 21 for his recollections, which list and describe his rules and principles of program management.

# Cornelius W. Chambers

Cornelius Chambers provided key operations support to the Corona program, and contributed to solving many vehicle anomalies and operational issues. He also provided the cornerstone design for the vehicle protective measures architecture in support of a near-real-time photoreconnaissance satellite program. Chambers reflects on his experience in developing this protective measures design during the early development of space reconnaissance technologies.

## Protective Measure Designs as a Response to Risk Analysis[1]

In 1957 I graduated from the University of Arizona with a degree in mechanical engineering. After graduation, I worked for Hughes Aircraft for a few years in its missile program. In 1960 I joined Lockheed Missile and Space. I went to Lockheed specifically to work on satellites. Satellite technology was a new and exciting field at the time, and I wanted to be a part of it.

My primary contribution to national reconnaissance was the development of a protective measures design for satellite operations and fault detection. The design synthesis of the protective measures architecture required an approach with multiple levels of fault detection, corrective action, and ground monitoring. It was challenging technically, managerially, and politically, and we faced strong resistance from several different directions.

### The Setting for a Computer Revolution

During the early 1970s we used FORTRAN for programming, and we relied on large mainframe computers with punch cards and batch processing.[2] We submitted runs, and if the run did not have a high priority we received the output the next day. It was a computing environment defined by mainframe computers, but nothing like what became available by the end of the 20th Century. The prospect of a flight computer that performed on-orbit diagnoses of the health of vehicle components was unimaginable during the early days of satellite reconnaissance.

---

[1] This section is based on Cornelius Chambers' written input as amplified by interviews on 26 September 2000 at the National Reconnaissance Office Headquarters and 14 February 2002 by phone.

[2] "FORmula TRANslator. The first and still the most widely used [high-level] programming language for numerical and scientific [and engineering] applications. The original versions lacked recursive procedures and block structure, and had a line-oriented syntax in which certain columns had special significance. There have been a great many versions." (Free On-line Dictionary of Computing ©)

Flight computers that were reloadable on orbit never had been used before, and many people on both the government and contractor teams were skeptical about whether this approach would succeed. Reloadable onboard computers permitted us to make programming changes remotely from the ground. They involved software rather than permanent "firmware." We had many fears about onboard problems that could cause an early termination of a satellite mission.

We spent a great deal of time selling our concept, because it was a new way of thinking and doing things. People were more comfortable with burned-in, fixed memories, or firmware. In contrast, by the year 2000, with powerful personal computers and reprogrammable microprocessors, a protective measures design is the norm.

**Developing a Protective Measures Design**

With the advent of far more complex satellite designs that incorporated passive redundancy, the government had to develop alternate approaches to spacecraft control and operations. The complexity of the new generation of satellites would not permit detailed analysis and troubleshooting problems using only existing technology and methods. Fortunately, I was in the right place at the right time. I was a systems engineer with operational on-orbit satellite experience. The development of a successful protective measures design depended on proper risk analysis from a systems engineer's perspective.

*The importance of this systems engineering approach—especially the definition of the primary goals—is difficult to overstate. When there is no goal in mind, then any path will do.*

Our systems engineering approach started with defining the threats to the satellite vehicle. This approach required prodigious risk assessments: fault detection, fault definition, response definition, reconfiguration definition, and recovery. The success of the protective measures design had a significant impact on overall vehicle life. Specifically, it extended vehicle life and reduced overall system costs. It also had an impact on reducing the time "out of service," permitting the vehicle to return to its primary mission in a timely manner when a fault occurred.

The importance of this systems engineering approach—especially the definition of the primary goals—is difficult to overstate. It probably is best illustrated by the experience of the white hare in the classic, *Alice in Wonderland.* When there is no goal in mind, then any path will do. (See the Annex for some of the steps used to provide an organized approach to this serious problem.)

**Working Relationships**

We also had fun. I remember one incident that involved a test engineer who was a friend of mine. I would tease him about his apparent inability to speak without gesturing with his hands. During one pre-launch phase, I was in Washington, DC and he was at Vandenberg where he conducted pre-launch diagnostics. At that time, secure communications were awful, so we were using Teletype to communicate. He found some problem in his pre-launch diagnostics and sent a Teletype to me trying to explain the situation. I sent him back a message in which I jokingly told him that I could not understand what he was saying, because I could not see his hands flying around. What I did not know was that the Teletype had gone to every location involved in the launch: the generals at the launch base, Lockheed-Martin, General Electric, the ground station, and even the Pentagon. It was very embarrassing for me!

**Impact of My Work on Personal Life and Family**

I was never able to tell my family what I was doing or where I was going. My wife understood the need for secrecy, because she was involved in the Corona program. Nevertheless, my absences still were difficult for her. One year in the late 1970s, I was on travel for over 200 days. We were building a house at the time, and I was hardly ever there. There is no question that my career had a very significant impact on my wife.

I believe these circumstances were difficult for my children while they were growing up. I know I did not have enough interaction with them. Early in my career, I did not work as many hours, so I was able to spend more time with my older children, my two sons. I coached their Little League and soccer teams, and I was active in their swim club. I felt that I had a relationship with my sons. In contrast, my daughter was about eight years younger and I was not able to spend as much time with her as I should have, and I am sorry for that.

When my son Greg was older, he surmised the type of work I was doing. At the time, Greg was a student at the University of California at Santa Barbara, which is located about thirty minutes from Vandenberg AFB. Once, I had to be at Vandenberg for a launch, so while I was in the area I took Greg out to dinner. The next day, Greg was at the beach with some friends when they saw a satellite launch from Vandenberg, and then explode. He called his mother that evening, and she told him I was on the way home. The next morning's headlines were all about the "spy" satellite that had exploded the previous day. Greg put two and two together and figured out that I somehow was involved with that satellite

**Conclusion**

Looking back, I am extremely impressed with the talent of my colleagues in every area. These individuals pushed the state-of-the-art in terms of design and technology. Furthermore, they were risk takers. Yes, risk takers! However, they did their homework and consequently understood the magnitude of the risks. The risk analyses they performed resulted in backup contingency plans.

I believe that after the Cold War, the system engineering and risk taking mentality of most NRO personnel (both government and contractor) diminished. For one thing, the organization became burdened by too much oversight. My perception is that the norm became to "play it safe." Unfortunately, playing it safe denies the benefits inherent in the detailed risk analyses required of alternative plans.

*I believe that after the Cold War, the system engineering and risk taking mentality of most NRO personnel (both government and contractor) diminished. For one thing, the organization became burdened by too much oversight. My perception is that the norm became to "play it safe."*

It was easy to stay motivated during the early days of national reconnaissance in the midst of the Cold War. We were doing challenging things that had not been done before, and that were important for the government and the country. The Intelligence Community benefited from the high quality intelligence products that our satellites were able to produce. Every area of our programs required a state-of-the-art approach. We had to find solutions that were far removed from the technologies of the time, as well as from what had been done before.

Overall, my experiences in national reconnaissance were tremendously satisfying. I loved the challenge and I feel as though I contributed to something of major importance. I believe the intelligence collected by national reconnaissance contributed to ending the Cold War and promoting the relative peace that the United States enjoyed as it entered the 21st century. It was just dumb luck that I was there, but I am proud to have been a part of it.

—

**Presentation of Pioneer Award and Citation**

Figure 24-1. Pioneer Cornelius Chambers (second from the left) being recognized at the 2000 Pioneer Recognition Ceremony. The pioneer award was presented by DNRO Keith Hall (left), and DCI George Tenet (third from the left). Dr. Vance Coffman (President and CEO, Lockheed Martin Corporation) joined the presentation. (Photo by Sara Judy, NRO Visual Design Center.)

**Cornelius W. "Connie" Chambers**

Mr. Cornelius Chambers contributed flight "protective measures" adopted for use on most NRO satellites. A contractor with Lockheed, he integrated satellite subsystems in a novel approach to on-board fault detection. This procedure permitted autonomous re-configuration, and in extremis, "safe mode" operation that became the cornerstone of flight control after 1975.

*Career in National Reconnaissance: 1962-1994*

—

## Annex

**Risk Analysis Steps Comprising the Protective Measures Design**

| *Steps* | *Description* |
|---|---|
| Define potential threats | Vehicle Threats (e.g., loss of electrical power, loss of propellant, inability to communicate with the satellite, inability to utilize redundant hardware). |
| Define internal or external potential faults | Identify potential functional faults. |
| Define timeliness | Define timeliness required for corrective action. |
| Conduct failure analysis | Provide detailed definition of hardware/software that could manifest into a functional fault. |
| Detect faults | Define a reliable method (sometimes triply redundant with majority voting logic), including outer loop method to detect multiple faults and protect against unforeseen faults. |
| Define corrective action | Define corrective action required, and determine need for autonomous reconfiguration (allowing the vehicle to go into safe mode for diagnosis and reconfiguration). |
| Define flags and diagnostics | Define flags and diagnostics to allow ground personnel quickly to determine fault sources and the corrective action required. |
| Establish reaction procedures | Establish quick and efficient reaction procedures for ground. |

# Robert H. Crotser

Robert Crotser was Lockheed's Business Manager for the Corona program and for the first electro-optical satellite program. His management skills contributed to the delivery of a major National Reconnaissance Office program being delivered on time and under cost. His method of team building to solve unique intelligence problems was widely respected and adopted across the Central Intelligence Agency and Department of Defense.

## Ensuring Mission Success from Behind the Scenes[1]

My principles of business management are quite simple. I believe team building and communication are key qualities in any cost and schedule management program, through all phases of the development and acquisition process. Throughout my career, I tried to make the most of my relationships with the government and subcontractors in order to help make these programs successful.

### Supporting Corona

I joined Lockheed in 1957, and I began working on the Polaris program. I was the engineering personnel coordinator, and I was responsible for hiring and processing of personnel. At the time I joined the program there were about 1,000-1,500 people, and by the time I left in 1960 there were over 4,000, so obviously I did a lot of hiring.

One day one of our engineers was moved out of Lockheed's Missile and Space Division in a hush-hush manner and without anybody's knowledge. As the personnel coordinator, I approached this bashful person, and I raised hell because I wanted to know what was going on. Essentially I was told to shut my mouth, and that this matter was being handled by the front office. I was impressed by the way this hush-hush program could get things done, so I fiddled around and found out who was working there. I discovered that the guy who hired me into Lockheed was working there. I went to speak with him, and I said that I would like to be a part of that organization. He processed the Personnel Security Questionnaire (PSQ) for me, I was cleared, and then I learned that I had just signed on to the Corona photoreconnaissance satellite program.

[1] This section is based on interviews with Robert Crotser at the National Reconnaissance Office Headquarters on 26 September 2000, and by telephone on 19 September 2001.

They were obviously very far ahead of the game in terms of technology, film systems, and things like that, but what intrigued me from the management perspective was their ability to get things done without going through all of the cumbersome bureaucratic procedures. For example, if they wanted some person to be added to the program they processed a PSQ on him (sometimes without his knowledge), and when he was authorized program access they would go tap him on the shoulder and say, "Come with me." Best of all, they got away with it because senior management was totally supportive. I loved that.

*As the prime contractor, I believed it was important also to make the subcontractors part of the team. When I treated people like they were important to the mission, they respected and trusted me.*

My support for Corona was on the personnel rather than technical side of the program. I was the personnel manager as well as the business manager. Given the secrecy of the program there were some interesting challenges, such as creating "cover" arrangements to protect sensitive operations. We had to ensure those individuals' associations with the highly classified and sensitive operations were concealed.

**Critical Elements of Team Building: Communication and Motivation**

The Central Intelligence Agency (CIA) always was very good at motivating people, and they made sure to get the contractors involved. As the prime contractor, I believed it was important also to make the subcontractors part of the team. When I treated people like they were important to the mission, they respected and trusted me. Another important part of the teambuilding process was educating the customer (the CIA), and I had the opportunity to help in that education. The Agency organized a weeklong program management course for Program B personnel.[2] The Agency picked its bright, young engineers and sent them to the course. Various company representatives would talk about program management, and I frequently gave them my "How To Be A Good Customer" presentation.

Sometimes getting the job done required creative management, and I did a lot of things with the customer that were highly unusual. If I saw something that needed to be done, I did it regardless of what the policy was or whether it was contrary to normal procedures. Policies and procedures were secondary to me when compared with the success of the program. We never lost sight of what we were trying to accomplish, and did not let anything get in the way.

One example of this occurred in 1957 or 1958 when part of Lockheed was based at a different facility in another county. This facility had a closed union shop and all of our hourly employees had to be unionized, which of course was impossible for Lockheed employees. So we had an interesting problem to sort out from a practical and legal point of view. During this time we were so small that everyone on the Lockheed team at this facility performed multiple jobs. In one instance when the non-Lockheed personnel went on strike, I ran the switchboard and reproduction machine for a day. Everyone pulled together and did what had to be done, which contributed a lot to camaraderie among the people. Working together in that way was also a great motivational tool.

Working with the customer did not always go smoothly, but we managed to resolve our differences. Sometimes we would have knock-down, drag-out arguments with the customer, and then at the end of the day we would go out and party together. We were united by the common sense of purpose and we all did what was in the best interest of the program as we saw it. This approach caused me to be trusted by the customer, and they even gave me an

[2] Program B embraced the CIA satellite reconnaissance element in the National Reconnaissance Program and consisted of the Office of Development and Engineering in the Directorate of Science and Technology.

Agency badge that gave me free access to the whole place. As far as I know, in those days I was the only non-agency employee that had that privilege.

Lockheed management also was committed to the success of these NRO programs and it often put the needs of the program ahead of corporate gain. I could go to the President of Lockheed and tell him that we were going to do something that was not in the company's best interest from a profit perspective, and he would say, "No problem, we support it." The program was the chief priority, not making a buck.

### Life with Secrecy

The secrecy of the work did not have as much of an impact on my life as it did on a lot of other people, but perhaps that is because I took a different approach to dealing with it than many. My whole family knew the type of work I did, but they did not know what it was for or specifics about the project. I would tell people that I worked for Lockheed and that I was involved with contracting, budget controls, and personnel issues, without going into further details or suggesting it was related to covert projects. Some colleagues would say, "I can't tell my wife anything about what I do." I would reply, "You're a design engineer—tell her you're designing a spacecraft or just that you're designing something. You don't have to say that you're working on covert projects." In my opinion, the security should have given people better guidance about how to answer questions without drawing attention. I believe that by their secrecy and evasiveness many people actually brought more attention to themselves. The problem probably could have been avoided more often than it was.

### Successful Program Management

The most critical element of program management is putting together a plan that lays out the objectives, identifies the factors for meeting the objectives, and provides a realistic implementation schedule with contingencies built into it. The plan also needs to estimate program costs, keeping in mind that people are about 90% of all program costs. Hardware and materiel costs are secondary in comparison to the costs associated with people. Once the plan is in place, it is important to monitor status over time. By following all of these steps, things should go well.

Another important component of program management is what I call the shoe leather program. Every day when I was at Lockheed, I physically went out into the work areas to talk to people. I was amazed how people appreciate the fact that someone shows concern and interest for them and their work. Additionally, when we had monthly program review meetings when the customer would come out for several days, I insisted that the customer take part of the day and visit a different part of the area each time to talk with people. The employees felt energized by having the customer visit them and show an interest in who they were. It was one way we let our people know that we appreciated them, and it worked like a charm for us.

### Conclusion

When I retired in February 1989, I was the Vice President and Assistant General Manager of Space at Lockheed with $4 billion in annual business from our various space customers. In hindsight, I believe I accomplished more than I ever thought I could. I reached levels at Lockheed that I would have never dreamed I could reach. The greatest thrill of my career was seeing the electro-optical system brought on board in 1976, and watching the first images arrive at the ground station in near-real-time. That accomplishment clearly illustrated that I had made a significant contribution through my years of work in business and program management.

**Pioneer Award Presentation and Citation**

Figure 25-1. Pioneer Robert Crotser (second from the left) being recognized at the 2000 Pioneer Recognition Ceremony, 27 September 2000. The pioneer award plaque was presented by the DNRO Keith Hall (left) and the DCI George Tenet (third from the left). Dr. Vance Coffman (President and CEO, Lockheed Martin) joined in the presentation. (Photo by Sara Judy, Visual Design Center.)

**Robert H. Crotser**

Selected as Lockheed's business manager for the Program B electro-optical imaging satellite, Mr. Robert Crotser conceived and put in place the cost and schedule controls that permitted this vital program to be completed on schedule within acceptable budget allocation. His handbook on cost and schedule management remains a standard reference in spacecraft acquisition.

*Career in National Reconnaissance: 1960-1989*

# John J. Crowley

John Crowley served as the Director of the Central Intelligence Agency's Office of Special Projects. He reunited the National Reconnaissance Office during a critical period in the Cold War by brokering agreements between Programs A and B. He successfully managed the approval and initial development of four of the nation's most successful overhead reconnaissance programs. Crowley was honored posthumously as a Pioneer of National Reconnaissance. In this chapter, Gladys Crowley, his wife, describes the pride and satisfaction he took in his work throughout his career, and how that led to his success. Her recollections follow a brief discussion of Crowley's contributions to national reconnaissance.

## Reuniting the National Reconnaissance Office[1]

John Crowley, as the Director of the Central Intelligence Agency's (CIA) Office of Special Projects (OSP), was instrumental in obtaining approvals for the Corona Improvement Program, and three other imagery intelligence and signals intelligence satellite reconnaissance programs. He assumed command of the OSP at a time when the national reconnaissance roles of the CIA and the Department of Defense were undefined. Because of his leadership and management skills, the Corona Improvement Program proceeded, and development of three other systems advanced.

Crowley was among the most respected individuals in government during the late 1960s. He had served both in the Department of Defense and in private industry prior to joining the CIA. He was instrumental in establishing a strong CIA-Air Force partnership that ultimately endured. There was unnecessary confrontation that reached the senior levels of the U.S. Government with the Deputy Director of Defense and the Director of Central Intelligence expending huge amounts of their time and energy on keeping the National Reconnaissance Office (NRO) focused.

Fellow Pioneer Roy Burks recalled that "Much of the credit for reestablishing a cooperative relationship within the NRO belongs to John Crowley. ...Crowley played a unique and very pivotal role in the National Reconnaissance Program (NRP) at a time when it was on

[1] This section is based on information contained in the material submitted with the nomination of John J. Crowley for selection as a National Reconnaissance Pioneer. We prepared this section in lieu of recollections.

the verge of self-destruction. His cool head and negotiation skills brought the CIA and the Air Force back into a partnership agreement that endured for the next several decades. Without a seasoned and respected professional like Crowley in charge of the CIA/OSP, it is unlikely that the NRO's many significant advances from 1965-1975 would have been possible. He was a very seasoned manager, had excellent contacts throughout the Department of Defense and was highly respected for his integrity and managerial skills."

*John's greatest strengths were his abilities to learn and understand new disciplines, organize projects, and bring about cooperation among his staff in order to achieve the greatest possible results.*

Crowley died in 1984. On 27 September 2000, the Director of the National Reconnaissance Office (DNRO Keith Hall) and the Director of Central Intelligence (DCI George Tenet) posthumously inducted Crowley as a Pioneer of National Reconnaissance. Crowley's wife, Gladys Crowley, accepted the award on his behalf. The citation on the Pioneer award plaque illustrates the significance of his leadership and direction of United States imagery reconnaissance programs.

—

## Pride and Satisfaction in My Husband's Work[2]

John's CIA career ended in 1970 after six years of service with the Agency. It seems to me now that every position John held during his career was another step in preparing him for his final employment at the CIA, where his last position was the Director of the Office of Special Projects.

Although John took great pride in each position he held, usually working his way up to a leadership position, I believe that he viewed his role at the CIA as a true fulfillment of many of his life's goals. He had never found a better group of well-educated, disciplined, dedicated group of colleagues, whose goals were to achieve success in their projects–not for themselves, but for the nation.

Figure 26-1. Gladys Crowley (bottom center) surrounded by family and friends, left to right include Eldrena Engleson (nephew's wife), Ruth Ann Crowley (daughter), Leo Engleson (nephew), John Crowley (son), and Cay Crowley (daughter-in-law). (Photo by Candi Campbell, NRO Visual Design Center.)

[2] This section is based on written input from Gladys Crowley, wife of John J. Crowley.

I was his partner since 1936, and I always was aware of what he was doing. However, we always practiced the routine, "I won't tell you, so don't you ask" where his CIA work was concerned. As far as the impact of his secret work on our two children, they were in college during those years, so it did not matter that they knew nothing of the details of his work, and only that he worked for the CIA.

John's greatest strengths were his abilities to learn and understand new disciplines, organize projects, and bring about cooperation among his staff in order to achieve the greatest possible results. A good example of John's impression on others took place after the 2000 Pioneer Recognition Ceremony at the NRO Headquarters. A gentleman whom I did not recognize approached me. He was a current employee of the NRO, and was attending the ceremony. When he found out that I was in attendance, he sought me out in order to tell me what a lasting impact John had on his life, and that John had given him the sense of direction he needed.

I know much more about John's work now than I did before he was recognized by the NRO, and this knowledge gives me a sense of both satisfaction and pride. John would be very pleased about the continued importance of his legacy.

—

**Pioneer Award Presentation and Citation**

Figure 26-2. Gladys Crowley, (second from left) wife of pioneer John Crowley. Mrs. Crowley received the pioneer award plaque for her husband who was recognized posthumously at the 2000 Pioneer Recognition Ceremony. The award was presented by DNRO Keith Hall (left) and DCI George Tenet (third from the left). Joanne Isham (Deputy Director for Science and Technology, CIA) joined in the presentation. (Photo by Sara Judy, NRO Visual Design Center.)

**John J. Crowley**

In 1965, Mr. John Crowley became chief of Program B's Office of Special Projects, responsible for the CIA's overhead reconnaissance programs. He directed the successful development of follow-on imaging satellite programs and a crucial signals intelligence program, and is credited with establishing a true partnership between the CIA and Defense Department elements of the NRO.

*Career in National Reconnaissance: 1965-1975*

# Thomas O. Haig

Thomas Haig directed the program that developed the National Reconnaissance Office meteorological satellite system. This system provided accurate and timely cloud cover forecasts for imaging photoreconnaissance satellites, starting with Corona. His previous Air Force assignments prepared him for this work, and included projects such as tracking and measuring weather conditions by radar and radio, developing reconnaissance balloons for project Genetrix, and dsigning ground station components for several satellite programs. Haig's meteorological satellite system significantly increased the quantity of cloud-free pictures and virtually eliminated the need to re-photograph targets. He met strict scheduling, cost, and operational requirements, and developed techniques that influenced all future meteorological satellite programs. His Air Force successes allowed him later to develop the world's first global satellite weather system.

## Meteorology and National Security[1]

I first became involved with the military when World War II brought me into the Army Air Corps, and I almost immediately became involved in meteorology. This turned out to be my specialty area of national reconnaissance. I volunteered to be an aviation cadet, but I flunked the eye exam and was sent to the University of California, Los Angeles (UCLA), where I first learned to be a meteorologist (and where I met my future wife). This unique set of circumstances eventually led to a career in meteorology and satellite development.

### My First Military Assignments

Newly commissioned, I learned to fix radiosonde ground stations, single cylinder gas engine power units, and primitive hydrogen generators at the Meteorology Instrumentation School at Fort Monmouth, New Jersey. I spent five lovely months in Bermuda operating a "sferics" station, followed by a month on a slow Victory Ship to Saipan. There, grateful for all I had learned at Fort Monmouth, I helped operate a station running the Pacific Sferics Net for the rest of the war. "Sferics" is the technique of locating storms by using radio direction finders to track lightning. It was not very effective in the Pacific.

[1] This section is based on an interview with Thomas Haig at the National Reconnaissance Office Headquarters on 26 September 2000.

With the war over, I reunited with my wife in California, and we then moved to Cleveland where I completed my eduction at night school. A letter then arrived from the Air Corps, offering to send me to the California Institute of Technology for a Ph.D. in physics-meteorology if I would return to active duty. My wife, my new son, and I all jumped at the chance, but when the orders arrived, I was sent to Tinker Field, Oklahoma as the Supply Officer in the 2060th Mobile Weather Squadron. The Cal Tech program had been cancelled. After a furious exchange of letters, I learned that it was a one-sided contract, and I had to report to Oklahoma.

For almost three years, I was one of only five officers in an organization of 280 men. My positive memories from this period include square dancing and the birth of my second son. Then in 1951 came the TWX that said, in effect, "Send Haig immediately to the Air Force Cambridge Research Laboratories for an urgent project."[2]

**Bombs, Balloons, Academics, and Atmospherics**

The project was to measure the "after wind" which many believed was created by the rising fireball of a nuclear explosion. My main task was to develop an anemometer that, after withstanding a nuclear blast, would be capable of measuring winds up to 200 miles per hour. In this assignment I was responsible for supervising the manufacture of a dozen of the anemometers and for testing (and calibrating) the instruments in a wind tunnel at Langley Air Force Base. I was then responsible for taking the anemometers to Eniwetok Atoll, exposing them to three blasts during the Greenhouse test series, collecting the instruments' recordings, and finally analyzing the data. We found that the "after wind" did not occur. I can remember this project very well. On the day of the second test shot, I received a letter telling me of my third son's birth.

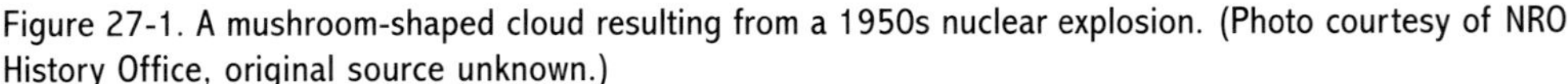

Figure 27-1. A mushroom-shaped cloud resulting from a 1950s nuclear explosion. (Photo courtesy of NRO History Office, original source unknown.)

[2] The term, TWX, refers to Teletypewriter Exchange, a pre-internet way to send and receive messages.

Immediately after the Greenhouse project, I was appointed Director of the Moby Dick program. The purpose of this program was to develop the large plastic balloons, the payload package, and the associated equipment and techniques for launching reconnaissance cameras across the Soviet Union in project Genetrix. We tested dozens of balloon designs and fabrication alternatives for the program, during which we launched some 780 units from three west coast sites. We tracked the balloons across the continent to where they terminated on the east coast. Moby Dick was publicized as a project to measure upper-air winds for meteorological research, while the real purpose was classified. Much uidentified flying object (UFO) hysteria was caused by sightings of the balloons at sundown when the earth was dark, but the balloon was still illuminated. Despite our efforts to properly identify the balloons to the public, people reported seeing fantastic colors and motions, and believed they were extraterrestrial. One unfortunate West Virginia pilot crashed after trying to intercept a balloon floating at 70,000 feet.

Figure 27-2. An example of an unmanned balloon that was used for reconnaissance missions. (Photo courtesy of NRO History Office, likely U.S. Air Force photo.)

In 1954, after the completion of Moby Dick, my request was approved to pursue my college education. I had been submitting this request annually since the Cal Tech disappointment in 1948. I was sent to the University of Illinois, where I celebrated the birth of my fourth son and received my Electrical Engineering degree.

I then was assigned, again, to the Air Force Cambridge Research Laboratories, and I was selected to direct a joint Department of Defense (DoD)-Weather Bureau-Air Navigation Development Board program. The purpose of the program was to develop a means to measure and inform pilots of atmospheric visibility during their landing approaches. In this project, we worked with members of the Airline Pilots Association.

We tracked several hundred bad weather landings by radar, and recorded the instant the pilot reported sighting the approach lights. From these data, combined with improvements in ground equipment, we developed a system that told the pilots the altitudes along the approach path at which the probabilities of seeing the approach lights were 20% and 85%. At the lower altitude, if a pilot had not sighted the lights, the pilot was to abort and make another attempt at the landing. The system was tested at Newark, New Jersey, for a year. The pilots liked the system, but it was not implemented because a new generation of automated landing equipment arrived. During this program, I designed a recorder for the rotating-beam ceilometer, standardized as RO-57, which was used at most Air Force weather stations.

**From Clouds to Meteorological Satellites**

I was reassigned in 1958 to the Air Force Missile and Space Division in California, where I served for two years as chief of the requirements office for satellite ground support. In this position, I worked on the design and development of many parts of the tracking stations and the control center for Corona, Missile Defense Alarm System (Midas), Space and Missile Observation System (Samos), and related programs.

The first successful Corona pictures in August 1960 convinced the authorities that timely and accurate knowledge of cloud cover over Asia was necessary, and could be obtained only by a satellite system. The National Aeronautics and Space Administration's (NASA) Television Infrared Observation System (Tiros) program was unable to meet the requirement, so Joseph Charyk, the Director of the National Reconnaissance Office (DNRO), authorized an interim meteorological satellite program to support U.S. photoreconnaissance satellites.[3, 4] The plan was to develop and build four satellites using Tiros technology to be launched by "Blue Scout" boosters, and operate the system for one year, which presumed that the full Tiros system would be operational by then.[5]

Figure 27-3. Example of clouds in Corona image of Kapustin Yar missile complex during the late 1960s. Kapustin Yar, located near Volgograd, U.S.S.R., was used to train and test surface-to-air (SAM) missiles and ballistic missiles of less than inter continental ballistic Missile (ICBM) range. (KH-4 Satellite photo.)

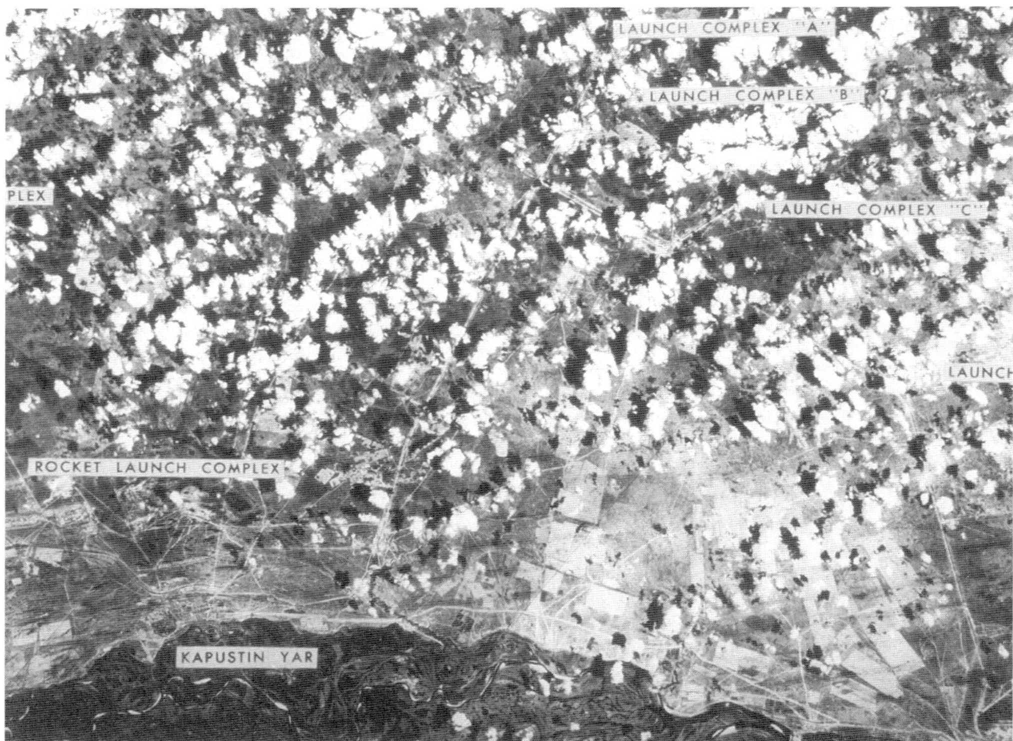


[3] Joseph V. Charyk served as the first DNRO from 1961-1963. During his tenure, the NRO operated the U-2 reconnaissance aircraft program and managed development of the A-12, and later the Air Force SR-71, program.

[4] Tiros was NASA's first experimental "wheel-mode" weather satellite, launched on 1 April 1960.

[5] Blue Scout, or just Scout, was a launch vehicle built by Chance Vought and procured under NASA direction. It was a small, four-stage, solid-propellant rocket.

I was selected to direct the interim NRO meteorological program. I was told the first launch had to be within ten months, and the total program costs were fixed, and I was to cancel all work if either the schedule or cost requirements could not be met. I accepted the job with several requirements of my own. I wanted to use fixed-price, firm-schedule contracts; I wanted to select the personnel for my program office myself; and I did not want to use a civilian Systems Engineering and Technical Direction contractor. I wanted the decisions to be made by military people in the program office. These provisions were accepted, and the program officially started on 1 August 1961, when it was called "Program II." The booster failed during the first launch on 23 May 1962. The second launch, a month later, was successful. The system fully met the NRO's requirements.

**Figure 27-4. Third launch of Program 417 on 19 February 1963. (Photo courtesy of NRO History Office, likely U.S. Air Force photo.)**

Delays on the TIROS system continued, so the DNRO authorized a second satellite procurement and changed the program name to "Program 35." We faced problems operating the meteorological satellites within the ground system designed for Corona. This spurred me to propose the building of two ground stations and a control center for Program 35, and to have them operated by Air Force personnel—no contractors. The DNRO approved, the Air Force Chief of Staff (General Curtis E. LeMay) approved, and the Commander-in-Chief of the Strategic Air Command (General Thomas S. Power) approved. Ten months later, the nation's first operational meteorological satellite program became a reality, manned entirely by AF personnel.

The "Blue Scout," a NASA project, was plagued by management and reliability problems. After looking at all available alternative boosters, I proposed refurbishing Thor Intermediate Range Ballistic Missiles (IRBMs) returning from England and Italy, and adding a second stage using parts from other boosters to create a new launch vehicle, later called "Burner I." In addition to solving the program's booster problem, this proposal provided a way to dispose of excess Thors and, therefore, received the DNRO's prompt approval. Later, a new second stage development called "Burner II" was approved.

The program was no longer "interim," and was renamed Program 417 in 1965. The program's mission was expanded to provide cloud cover information for aircraft flights during the Cuban missile crisis, the evacuation of civilians from the Congo, and air operations in Vietnam. My blue-suit crew and I were deeply involved in the engineering and development aspects of the program.[6] We designed the essential parts for the Burner boosters, invented

[6] The term, blue-suit, refers to Air Force.

Figure 27-5. Early Defense Meteorological Satellite Program (DMSP) image of Mexico and portions of Southwestern United States at 1/3 nautical mile resolution, 1971. (Photo courtesy of NRO History Office, DMSP Image.)

the satellite magnetic spin-rate control, introduced innovations that greatly reduced the cost of the ground stations, developed simplified tracking software, and produced many other improvements to the system. The approaches to program management that we generated in our program office persisted even after the program was declassified and became the Defense Meteorological Satellite Program.

**Other Challenges**

In the fall of 1965, I chose to attend the Industrial College of the Armed Forces (ICAF), the DoD's "Ph.D.-level school" founded by President Eisenhower. It is an excellent school with a worthy purpose.[7]

My last military assignment was to the NRO staff in the Pentagon, where I worked as the Assistant Director of Research and Development for Dr. Flax.[8] In 1968, finding myself under severe financial strain with four college-age sons and a new daughter, I requested permission to retire, which Dr. Flax approved. During my military career, I received a great education, my share of campaign ribbons and battle stars (three Commendation and two Legion of Merit medals), and a lasting admiration and respect for the men and women in uniform.

My search for a job was brief. I became a program manager at General Electric, but the program was cancelled a year later. Then I spent a year helping to find jobs for the 280 people in my division.

**Global Weather System: My Final Technical Achievement**

Somewhat disillusioned by my GE experience, I listened closely to an offer from Professor Verner Suomi, with whom I had worked putting infrared sensors on the Program-35 satellites. He offered me the position of Executive Director of the Space Science and Engineering Center (SSEC) at the University of Wisconsin. It meant half the pay, but an opportunity to work with young people doing useful things. I accepted the position and took my family to Wisconsin in 1970.

At SSEC, I became the lead person in the development of the Man-Computer Interactive Data Access System (McIDAS). That hardware and software system, combined with the concepts and software developed in a four-year program that I started—called Innovative

[7] The ICAF has prepared selected military officers and civilians for senior leadership and staff positions. It has been a postgraduate program that deals with national security strategy and the resource component of national power. The ICAF became a component of the National Defense University.

[8] Alexander H. Flax also served as the DNRO from 1965-1969. He presided during the period when the second generation of imaging systems became operational and began to play a major role in U.S intelligence collection during the Cold War.

Video Applications in Meteorology—changed weather systems throughout the world. The program used the geosynchronous spin-scan satellite (also conceived and developed at the SSEC) to develop the world's first global weather system. We worked through the United Nations' World Meteorological Organization to give all of the details of the McIDAS hardware and all of the software freely to anyone who wanted them.

By the year 2000, there were five satellites from four nations in geosynchronous orbit producing hourly images of the entire globe. These systems were deriving and distributing worldwide data related to cloud cover, air motion, humidity, sea surface and atmospheric temperatures, and much more. Whenever weathercasters around the world stood before a map on which clouds move and fronts appear, they were using techniques and software we first developed at SSEC. Instead of a Mercator projection, the universal basic weather map became an image of the earth as seen from geosynchronous orbit. The image could be analyzed and combined with surface observations using McIDAS hardware and software. The international impact of the work done at SSEC has been enormous.

Although my role in the development of the global satellite meteorological program did not contribute directly to U.S. national security, it certainly has made the world a better place, and ranks up at the top of my list of achievements. My NRO work did contribute to national security, and also was rewarding in many ways. I am proud of the contributions my team and I made to U.S. national reconnaissance capabilities.

*Although my role in the development of the global satellite meteorological program did not contribute directly to U.S. national security, it certainly has made the world a better place, and ranks up at the top of my list of achievements.*

—

Figure 27-6. Example of a weather image used by weathercasters in the late 1990s. The image is of Hurricane Fran from Geostationary Operational Environmental Satellite-8, 20 September 1996. (Photo courtesy of NASA..)

### Life on the Periphery—A Wife's View[9]

Perseverance was the quality that most contributed to my husband's success in his work. He never accepted failure. I imagine that trait did not always make him popular, but his pursuit of his objectives obviously brought him success. Tom never just dropped a problem because he was unable to solve it. He either found a solution or found someone who had the solution. Tom also loved his work. One indication of this is that I saw his satisfaction when his projects succeeded. There were, of course, some disappointments, when I heard him make statements like "It could have worked better," or "If only I had…," but mostly I saw satisfaction.

Secrecy was very much a part of Tom's work and career. Conversations with my husband always were centered on what the kids were doing and what they needed, not about work. And I was so busy with our children that I never really wondered about Tom's job. However, it was difficult having a husband who could not talk about his work. People would ask our boys, "What does your Dad do?" They would have to reply, "Well, he goes to the office in the morning, and goes on a lot of trips"—that sort of answer. I should add, however, that although not being able to talk about Tom's work was inconvenient, and sometimes stressful, it never really changed our relationship or affected our children.

To be honest, I suspect our children were so involved with what they were doing that it did not matter a great deal that they were unable to talk with their dad about his work. When Tom would come home with a medal, or when he would go to Washington, DC, and stand there with some General for a picture with new medals, I think they just took it for granted that he was good at his job.

There were other inconveniences. There were times for Tom when something would come up, and he would have to leave the house immediately to take care of it. I never knew what these things were. That was all right, because Tom would call and say, "Whew! It's okay now." We would just leave it at that. He was glad when his programs were declassified. Unfortunately, that was long after the fact.

I did some socializing, like big "potlucks," with the families of my husband's colleagues, but I was always more involved with our children's activities. Tom obviously did not have as much time for them, but he did what he could. He encouraged them to do well and go to good schools, and he helped them to follow through on their individual interests. My own time was pretty much consumed with family.

Figure 27-7. Barbara Haig being interviewed by Paul Burgess on Pioneer Recollection Day, NRO Headquarters, 26 September 2000. (Photo by Candi Campbell, NRO Visual Design Center.)

I never went to any of the launches or other such events, but I was very happy when *60 Minutes* showed his balloons and their launches

[9] This section is based on an interview with Barbara Haig, wife of Thomas Haig.

after the program had been declassified. Our children were thrilled too. The NRO Historian, Cargill Hall, has been a good friend to us in this regard. He wrote a history of my husband's project after it had been declassified.

Now we are years away from Tom's "secret" work. During Tom's career, we moved 23 times. Although we lived in a lot of interesting places, we always wanted to retire to the country. When we found an 80-acre farm with abandoned buildings, we foolishly bought it and have been renovating it ever since. We have been converting a barn into a house since 1972. First we fixed the farmhouse so we could live in it. It had been thoroughly vandalized. Holes had been knocked in the walls, and the walls had been filled with newspaper for insulation. When I took out one paper, the headline was, "Admiral Perry enters Japan." I gave it to the local Historical Society.

I believe the recognition of my husband's work from the 1960s has been long overdue. I am very proud of him, as are our children. This recognition and attention is a good experience for us because it allows us finally to understand the reason for all of the secrecy and worry surrounding his work. I learned more about his work during our two-day visit to the NRO in September 2000 for his Pioneer induction than I had ever known before. Overall, my relationship and interactions with my husband always have been more about the present and future than the past.

—

## Pioneer Award Presentation and Citation

Figure 27-8. Pioneer Thomas Haig (second from the left) being recognized at the 2000 Pioneer Recognition Ceremony, 27 September 2000. The pioneer award plaque was presented by DNRO Keith Hall (left) and DCI George Tenet (third from the left). Brigadier General Craig P. Weston (Director, Corporate Operations Office and Chief Information Officer, NRO) joined in the presentation. (Photo by Sara Judy, NRO Visual Design Center.)

**Thomas O. Haig, Colonel, USAF**

In 1961, Colonel Thomas Haig led a small Air Force team that developed for the NRO the operational polar-orbiting meteorological satellite, its launch vehicle, and associated command and control stations in less than a year. This system provided accurate and timely cloud cover forecasts of target sites, vastly improving the efficiency of early film-limited reconnaissance satellites.

*Career in National Reconnaissance: 1961-1965*

—

## Annex

### Selected Publications Authored or Co-Authored by Thomas Haig

1 Haig, T.O., "Technical Direction: Outmoded Management Concept?," *Perspectives in Defense Management*, May 1967.
2 Haig, T.O. and Lally, V.E., "Meteorological Sounding System," *Bulletin of the American Meteorological Society*, Vol. 39, No. 8, August 1958.
3 Haig, T.O. and Norton, W.C., "An Operational System to Measure Compute, and Present Approach Visibility Information," *AF Surveys in Geophysics*, No. 102, June 1958.
4 Hoffman, R.B. and Haig, T.O., "Space Uses of the Earth's Magnetic Field," *Transactions, 11th AF Science & Engineering Symposium*, October 1964.

# Robert J. Kohler

Robert Kohler was innovative in his approach to the challenges of developing satellite imagery reconnaissance programs. He was instrumental in the development of the National Imagery Interpretability Rating Scale for the assessment of image quality and usefulness. Kohler also was a key participant in almost all phases of the development of an imagery program, including its sensor design. He also served as the Chief of Facility for a ground support facility.

## Challenges in Developing Satellite Imagery Reconnaissance Programs[1]

Several recollections are representative of the challenges I faced during my involvement in developing satellite reconnaissance programs. These include recollections about the establishment of the criteria for image evaluation used by the Intelligence Community, some unique problem-solving methods we used for the early collection systems, the outcome of the competition that existed between the different collection programs in the early years, and the impact that this work had on my family.

### The National Imagery Interpretability Rating Scale— A Solution for Assessing Image Quality

The National Imagery Interpretability Rating Scale (NIIRS) was developed at the request of Roland Inlow, who was the chairman of the Director of Central Intelligence (DCI) Committee on Imagery Requirements and Exploitation (COMIREX). At this time I was the Deputy Director of an early satellite program using the film-recovery system One day, Inlow called me into his office and said, "We have got a real problem. We have to find a way to verify the Strategic Arms Limitation Treaty (SALT). The White House wants us to be able to say that the quality of the imagery products is good enough to verify the treaty. We need to develop a method for the image interpreters to evaluate the quality of the overhead products so that they can confirm that they have the capability to determine compliance."

*Edge Matching—An Early Tool For Evaluating Imagery Quality.* My early experience in problems associated with assessing the quality and usefulness of imagery goes back to what was called edge matching. This technique, developed by the Itek Corporation in the 1960s

[1] This section is based on written input that Robert J. Kohler submitted to the Center for the Study of National Reconnaissance.

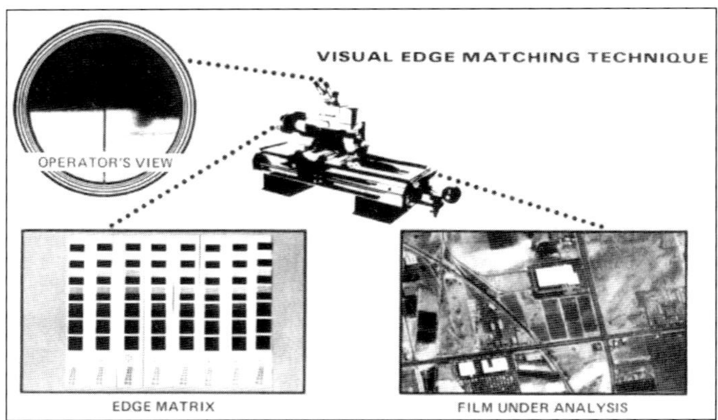


Figure 28-1. A graphic representation of the visual edge matching technique that used the visual edge-matching (VEM) machine, which is depicted in the graphic. (Photo courtesy of Lawrence Maver and Jon Leachtenauer, original source is a marketing brochure.)

and 1970s, created a process for determining the resolution of overhead operational imagery.[2] Prior to this, resolution was established by imaging objects with known size and dimensions at military installations in the desert regions of the western United States. These images enabled us to determine the resolution capabilities of the system, but they did not provide the capability to assess the resolution of collected images in foreign countries. Contractor fees were tied to image quality, so it was necessary to accurately evaluate wide swaths of imagery on a daily basis.

The technique involved taking a number of test images that had different conditions of defocus and smear. Analysts used an edge matching technique to obtain quantitative measures of system performance by scanning image edges and comparing them with reference standards, such as test images of known objects with known resolutions. The technique was performed first by ITEK Optical Systems' Visual Edge Matching machine, and later by the Automated Edge Measurement System. The technique worked very well, and we were able to evaluate a considerable amount of satellite reconnaissance imagery in this manner. The edge-matching technique was also a critical precursor to the development of the NIIRS scale, and the NIIRS process would not have been possible without the development of the edge matching technique by Itek and the National Photographic Interpretation Center (NPIC).[3]

*Developing the National Imagery Interpretability Rating Scale—Analyzing Imagery.* The first step in developing the NIIRS scale was to assemble a three-person team consisting of myself, Lee Brown from the COMIREX staff, and Dave Gifford, a statistician from NPIC. We gathered a large sample of imagery from a number of sources that included satellites and reconnaissance aircraft. The imagery also had a wide variety of image quality. We collected pictures of order of battle related to all types of air, ground and naval equipment.[4] Next we asked NPIC to develop a list of criteria against which an analyst could judge the image quality and usefulness. For example, for air order of battle, the criteria included a number of specific details associated with airplanes, airfields, support vehicles and buildings. As the image quality improved, the analyst was able to see more of the details and was better able to distinguish between the objects in the image. At the lower criteria levels, the analyst could determine if there were aircraft present at an airfield. At a middle rating, the analyst could determine what types of aircraft were present. At a high rating, the analyst might be able to observe refueling or maintenance activities. We developed similar criteria for naval and ground order of battle, as well as for the equipment and facility types that were routinely observed by the imagery analysts.

---

[2] Resolution is the quality of imagery, and relates to the ability to differentiate between objects or parts of an object in an image.

[3] NPIC was a Central Intelligence Agency organization that was responsible for analyzing reconnaissance imagery and producing intelligence reports based on that imagery.

[4] Order of Battle refers to the quantity, disposition and readiness level of military equipment and units.

*Refining the NIIRS Scale—Geometric Progression and the NIIRS 9 Example.* We decided that the next step in the development of the NIIRS was to base the scale on a geometric progression. Each NIIRS rating would be a factor of two better than the one below it. For example, the quality of a NIIRS 4 product, as measured by the ability to determine ground resolve distance would be twice as good as a NIIRS 3 image, and a NIIRS 5 product would be twice as good as a NIIRS 4 image.[5] We compared the set of test images with known resolutions, and using the edge matching technique, we assigned a ground sample distance (GSD) and hence a NIIRS rating to each of the images we collected.[6]

The resulting NIIRS scale enabled imagery interpreters to determine the resolution and quality of imagery that was related directly to the essential elements of information that intelligence analysts required to perform their assessments. The process permitted analysts to validate the quality standards of the products, and provided a metric for determining the value of imagery to customers.

The best quality rating on the scale is NIIRS 9, but we could not produce a single NIIRS 9 image to use as an example. We even examined some of the low-altitude airborne imagery, but there was nothing we could find that would meet the criteria for a NIIRS 9 product. One day the NPIC Director, John Hicks, sent me the NPIC "imagery board," a collage of imagery products for the day. On it was a picture taken from an aircraft flying in the Berlin Corridor. The picture was a low-level, oblique image into East Germany that captured an East German soldier relieving himself. The NPIC put a large label at the top "German Soldier Pissing in Snow—NIIRS 9." We finally found the example we were looking for.

**Making Hard Decisions—A Lesson in Managing Your Boss**

In the early days of one of our imaging systems, we faced a number of electrical difficulties related to controlling the attitude and scanning of the satellites. There were two sets of problems as I recall, one involved sparking and the other was a more fundamental problem with circuitry. Of the two problems, the first problem was less severe. We referred to it as "arcing and sparking." Electrical discharge led to a temporary shutdown of the system, and although this malfunction presented a significant challenge, it was not catastrophic, and we were able to work through it.

The second problem, which involved the actual circuitry, was significant and caused the entire circuit to fail. For months we were unable to determine the cause of the problem, and only discovered the cause just prior to the launch of another system. The satellite was on the launch pad, mated with the booster, and going through the final pre-launch check when the contractors determined that the cause of the problem was a component failure in the circuitry. When the component failed, it took down the entire circuit. An interesting aspect of the problem was that this particular part did not really appear to be critical. The

Figure 28-2. The Automated Edge Measurement System (AEMS) machine that was used for visual edge matching, subsequent to the VEM machine. (Photo courtesy of Jon Leachtenauer, original source is unknown.)

[5] GRD is the distance between which two objects can be separately distinguished.

[6] GSD is the distance on the ground covered by a single detector; it is the means of expressing the ground resolution of a sampling focal plane.

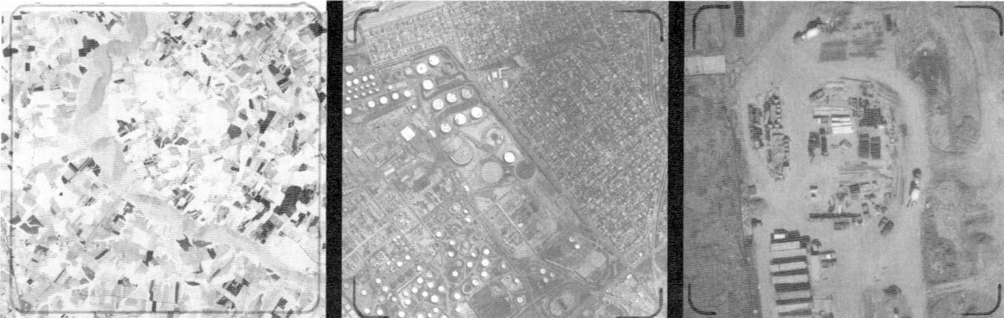

Figure 28-3. Examples of NIIRS-rated imagery from poorer to better quality. From left to right the images are rated NIIRS Level 1, Level 4, and Level 7. The NIIRS 1 image shows a rural landscape, the NIIRS 4 image shows a petroleum storage facility, and the NIIRS 7 image shows a construction terminal and storage site. (Photos Courtesy of NIMA, National Imagery Research Laboratory.)

circuits continued to function normally when the components were removed.

*Three Options to Consider.* We asked the contractors to look at the issue and make recommendations on the best way to resolve the problem. Clearly, we could not launch the next satellite without a solution. All of the players and my staff got together for a meeting one night in my conference room to discuss the contractor assessments. We came up with a set of three options to resolve the problem, each with a different approach on where to accomplish the necessary repairs.

*Option One: Back to the Factory for Repair at the Subcontractor.* The first option was the most obvious one: de-mate the satellite and send it back to the contractor's factory to take out the failing parts. The contractor would send the bad parts back to the subcontractor's factory for repair, after which time the parts would be sent back to the main contractor's factory and re-mounted in the satellite. The satellite would then be shipped back to the launch site, re-mated with the launch vehicle, and then re-do the pre-launch checkout. We projected that this process could take many months, and would cost millions of taxpayer dollars.

*Option Two: Back to the Factory for Repair at the Contractor.* The second option was similar: de-mate the satellite and send it back to the contractor's factory to remove the bad parts. However, instead of sending them to the subcontactor, the main contractor would fix the parts at its facility. This option would save time and money, and would cost approximately half of what option one would cost.

*Option Three: Remain at the Launch Pad.* The third option was to leave the satellite on the pad. We would remove the parts needing repair from the satellite without de-mating it from the launch vehicle. This would minimize the length of the delay and reduce the cost of fixing the problem. Of the three options, we estimated it would be the fastest and cost approximately one percent of what option one would cost. The down side was the risk factor. This kind of work had never been attempted on the launch pad before, and there were a number of things that could go wrong.

The proposal was that we would use a drill to take out this particular component. Because of the small size and the need for precision, we used a set of dentist tools. We had to be meticulous; we could not see the parts, and only knew where they were because of design drawings.

*The Solution and a Lesson in Management.* We decided to go with option three because it was cheaper and quicker, and because the engineers believed the process would work. However, because of the risk factor I was not sure that I could sell the plan to my boss, Les Dirks, who was the Program B Director (and who was dual-hatted as Central Intelligence Agency (CIA) Deputy Director for Science and Technology). Les loved to get involved with this type of technical issue, but we were afraid that we would end up with all kinds of people wanting to participate in the review process, and that the end result would be a decision to go with one of the other options against our best judgment. So the question became what should we tell our management?

I wrote a memo to Les about the problem, knowing full well that he very seldom read anything in his inbox. The two-page memo outlined the issue, explained the options, and stated the decision that we reached. At the bottom, I attached a note that said, "If you have any issues with this, please let me know." As I suspected, Les never read the memo.

After we completed the project and successfully replaced the failed component, I went into Les' office and told him what we had done. He was absolutely livid! He went into a rage and said, "How could you make this kind of decision without telling me?" I told him that I had submitted a memo on the project to him before we started and added, "It's probably in your inbox." He furiously shuffled through his pile of papers and sure enough, there was the memo. I said, "You'll notice Les, it says at the bottom, if you have any questions about this please give me a call. When I did not hear from you, I figured that you did not have a problem with it."

I do not tell this story to suggest that going around your management is necessarily the right thing to do. Rather, I tell it to illustrate what one of my mentors, John Crowley, another Pioneer, once told me.[7] He said, "It is important for every employee to remember that his first job is to manage his supervisor." I have never forgotten that. I think this was a case where I recognized that as the program director I would be held accountable and responsible for doing the right thing. There was no question in my mind that my job was on the line if the plan did not work correctly. I did not feel that we could afford the time necessary for an external review, and I was willing to accept the consequences for our decisions and actions. In the end everyone was very pleased with the outcome.

**Improving the Product—A Lesson in Marketing and Competition**

Another challenge we had to address was salesmanship, specifically how to convince senior management to support the concept of improving the system despite the problems we initially encountered with the program. At the time the original program was approved, there was no clear idea within the program office about what would happen after the original procurement was completed. Would the program continue or not? There were a number of reviews, and a presidential commission was set up to examine the long-term options. The program office went into protection mode, trying to mitigate any concerns about the existing system and to argue for a long-term imaging capability that would be responsive to requirements.

*After we completed the project and successfully replaced the failed component, I went into Les' office and told him what we had done. He was absolutely livid!*

*Lack of Support for Additional Funding.* At the same time that all of this was going on, Program B, in another component, was trying to take advantage of some of the new technologies of the time and develop what was being called an "advanced" version of the original system. This advanced option was briefed to the Director of the National Reconnaissance Office (DNRO), who turned the proposal down because of the problems we already were experiencing with the system. The program office saw benefits from the rapid technological advancements that were available and briefed the option to the DNRO a second time a year

[7] John Crowley is a Pioneer of National Reconnaissance (inducted into the NRO Hall of Pioneers, 27 September 2000). See chapter 26 for his recollections.

later. The program office was successful, and the option made it into the National Reconnaissance Office (NRO) budget that year. Unfortunately, the White House removed it from the budget because of the current problems we were experiencing. The fundamental White House concern was why should we fund an advanced version of a system that already was having problems?

*The Challenge to Come Up With A Plan.* About this time I was promoted to Program Director and the people who had been working on the "advanced option" had been reassigned to my program office. The head of that team, Fred Evans, became my deputy. I remember the staff and I being called into the boss's office. He told us that the improvements associated with the "advanced" option were not going to make it into the budget that year. He did not have a plan but told us to begin planning a new initiative to take the place of our system. His plea to us was, "You have got to come up with something different." Our only real guidance was to come with a plan that we could sell to the Intelligence Community, White House, and Congress.

*Our Solution.* We concluded that much of the resistance to future funding was the result of associating the future funding option with the original design and its problems. By using the adjective "advanced," it sounded like we were proposing an advanced version of a problem system. This was not the case. By this time we had solved many of the problems, and the program was in better shape than it had been, but people continued to recall the original difficulties. We needed to get the decision makers to focus more on the improved capabilities and less on the growing pains that had plagued us in the past. Our solution was to refer to our option as an "improvement," rather than to imply that it was an "advanced version" of a system with continuing problems. We put the focus on "improvement." No one could argue with that.

The second thing we did was to redesign. But rather than starting from scratch, as was proposed in the advanced option, we used as much of the original design as we could. We changed some of the details, but we kept the fundamental design. We also used as much of the original software as we possibly could, and by combining the best of the new ideas with as much of the original design as we could, we got the cost factor down to half of the original proposal.

Still, it was not clear that our "improvement" proposal would be funded. At the time defense and intelligence activities were not getting much new money. We anticipated that the NRO might be able to fund one new program. However, at the same time we were submitting our proposal, Program A was proposing an initiative of its own. Our submission for that year was not funded, while Program A's initiative received substantial startup funds.

*[T]he rivalry and competition between Programs A and B for research and development resources often reduced overall program costs, which in turn resulted in savings for the NRO.*

During this time period there was intense competition for development funds both within and across the programs. For example, one year while Program A was developing its own options, another component of Program B was tasked to develop a replacement program for a previously terminated effort. However, in the end, neither the other Program B effort nor our Program B initiative was funded because of a decision to fund a Program A initiative.

*The Lesson We Learned.* From this experience we learned that developing new overhead capabilities was only half the battle; we still needed to build support within the Agency and the customer community in order to get sufficient resources to continue with our new ideas. Consequently, we developed a marketing strategy to sell our program. What we did the next year was to put together a proposal which laid out a plan that could fund all three initiatives by doing the Program A proposal in concert with our initiative (as a Program B program). We briefed our proposal to the DNRO. Not surprisingly, this resulted in some scurrying around in Program A, which saw their program slipping away from them. The NRO leadership put together a number of panels to determine if our proposal was practical, and in the end it was decided that it could be done.

Even though this experience was a painful process, the end result was a significant savings for the nation. In the end, all programs were funded (two for Program B and one for Program A), thanks mostly to our initiative to present options. As this experience demonstrates, the rivalry and competition between Programs A and B for research and development resources often reduced overall program costs, which in turn resulted in savings for the NRO.

*Importance of The People.* Certainly the best thing about working in Program B was the people. It was not just those who were full-time Office of Development and Engineering employees. It was the security personnel assigned to us from the Office of Security, the finance personnel assigned from the Office of Finance, and many other individuals that provided outstanding support. The support we received from the CIA as a whole was outstanding. The CIA always assigned the highest quality individuals to support our projects.

*The support we received from the CIA as a whole was outstanding. The CIA always assigned the highest quality individuals to support our projects.*

We also received great support from the CIA Office of General Counsel (GC). Three signatures were required for any major procurement: the National Contracting Officer, the GC Procurement Officer, and mine as the Contracting Officer's Technical Representative. Decision making and execution was relatively easy.

The team spirit was fantastic, and the support of the National Contracting Officer was invaluable. On many occasions he said, "You have gone through the source selection process, and it is your decision. Tell us what you want to make happen, and we will do it. And if we cannot, we will go up the chain until we find someone who can." The contracting officers became part of the team, and they were more interested in getting the job done than in putting administrative obstacles in the way. We never encountered a procurement problem that GC could not solve. The CIA General Counsel, Stanley Sporken, who later became a federal judge, put it this way, "Our job is to figure out ways to make things happen, not to get in your way." It was a unique time in a unique environment. it is unfortunate that this environment does not exist today.

**Impact on the Family**

My experiences serve as an example of how mission needs impact on the quality of life for those who work in the field of national reconnaissance. For me, it all started one day when Les Dirks, then Director of Program B, called me into his office and said, "I am sending you to a ground station as the Chief of Facility. I want you to straighten that place out." I was not sure that this was possible, so I told him, "I will do it, but this is probably the end of my career."

The problem with the position was that it was a seven-day week, 24-hour-day job. I had a secure phone installed in my house, one of the old Vietnam-era systems, not one of the more modern ones that are now available. There were only seven of these phones still in use in the DC area. I needed this secure capability at home because the office was calling me all of the time—Saturdays, Sundays, two and three o'clock in the morning. Of course nighttime was when I was called most often, since that was when the satellites were operational over the Soviet Union. That phone became the most hated instrument in my house.

At the time I arrived as Chief of Facility, the program was experiencing a number of difficulties. The satellite was not working properly, and there were personnel management issues. To make things even more difficult, the Chief of Operations was dual-hatted as the Deputy Chief of Facility, which put the facility engineers in a subordinate position to Operations. Of course we had a number of contractor groups, and a number of government organizations represented, so we had a real mess on our hands.

*I had three teenage daughters...The middle one was interested in what I did, but it was because she was convinced that I was a government assassin. She was about fifteen at the time, and she could see no reason why the CIA needed engineers.*

To add to the challenges I faced at work, I had three teenage daughters. One of them did not particularly care what I did for a living. The middle one was interested in what I did, but it was because she was convinced that I was a government assassin. She was about fifteen at the time, and she could see no reason why the CIA needed engineers. Of course, I could not tell her what I was doing. For about two years, she was absolutely convinced her father was a killer, and that the reason I was being called out in the middle of the night was to 'take care of somebody' on the streets some place. It was not until her first year of college, when she was selected for a summer job in the Intelligence Community that she finally understood that I was not a killer. I think my time at the ground station was the toughest time for the family in terms of interaction with the children and impact on my wife, who could not even plan dinner with any confidence on most days.

**Conclusion**

The phenomenal thing about Program B was that we always were considered a part of the CIA. We considered ourselves intelligence officers, and we thought that part of our responsibility was to sell national reconnaissance programs inside the Intelligence Community. On more than one occasion we were directly responsible for building the support for a program at the request of the Deputy Director for Intelligence (DDI). I spent a lot of time with Bob Gates when he was the DDI, explaining and justifying CIA initiated programs for the NRO. We coordinated directly with the Collection Requirements Staff in the CIA Directorate of Intelligence, and assisted them in preparing their intelligence requirements list with the NPIC Imagery Exploitation Group to tailor the intelligence products they received.

Program B served a very unique role in the National Reconnaissance Office, notwithstanding its rivalry with Program A. I personally think that this role has been diminished in today's environment, and that it would benefit the NRO to recreate some aspects of its original organization. In those days, when the CIA wanted something, it was never turned down, and when the CIA did not support a project, it was not funded. The role of CIA Director of Development and Engineering, as the Deputy Director of Program B, was to focus on the CIA national missions. The Program B Director (the Deputy Directory of Science and Technology), who was in a direct reporting chain to the Director for Central Intelligence (DCI), provided the DCI with a conduit into the acquisition process for space assets. For this reason and others, Program B played a significant role in the history and success of the NRO.

—

**Pioneer Award Presentation and Citation**

Figure 28-4. Pioneer Robert Kohler (second from left) being recognized at the 2000 Pioneer Recognition Ceremony. The pioneer award plaque was presented by DNRO Keith Hall (left) and DCI George Tenet (third from the left). Joanne Isham (Deputy Director for Science and Technology, CIA) joined in the presentation. (Photo by Sara Judy, NRO Visual Design Center.)

**Robert J. Kohler**

A photo scientist, Mr. Robert Kohler developed for the Central Intelligence Agency Program B unique imaging techniques, and introduced photographic edge measurement and sharpening tools for evaluating and enhancing overhead imagery. He later served as the Chief of the Facility at a key imagery operations facility, and led the Central Intelligence Agency Office of Development and Engineering.

*Career in National Reconnaissance: 1967-1985*

# Ellis E. Lapin

Ellis Lapin served as the Aerospace Corporation's program director for General Systems Engineering and Technical Direction for its work on National Reconnaissance Office satellite imagery reconnaissance programs. He contributed to the on-orbit reliability of these systems by managing the design and development of the spacecraft and hardware, and assisting in the management of the on-orbit operations. By expanding on National Reconnaissance Office successes from the Corona era, he ensured the future success of the systems that followed Corona, thereby ensuring United States overhead superiority.

## Thoughts on My National Reconnaissance Career[1]

The Aerospace Corporation and the National Reconnaissance Office (NRO) had their genesis at nearly the same time. The company was established to furnish technical assistance to government agencies, principally the Air Force in those early days when the cadre of trained officers with experience in the technologies associated with rocketry and satellites was small. Our principal "product" was general systems engineering and technical direction (GSE/TD), an unfortunate nomenclature that displeased some people in the military and other contractors. An alternative designation for the offensive "TD" would have been "technical counsel," which would have been more representative of the actual situation.

The success of many of the satellite reconnaissance programs could be attributed to three principles: redundancy in equipment, exhaustive testing, and meticulous attention to detail. The Aerospace Corporation's tasks consisted of broad technical check-ups, investigation of alternative possibilities, evaluations, and recommendations for direction by the Air Force. These tasks amounted to applying these principles to the design, development, and operation process. The particular effort for which I was named a National Reconnaissance Pioneer was a straightforward by-product of the general systems engineering work.

### Early Missions and Operations

My early work on the first NRO program in which I was involved featured some exciting struggles and successes. These activities and events demonstrated the importance of prepa-

[1] This section is based on written input that Ellis E. Lapin submitted to the Center for the Study of National Reconnaissance.

ration and teamwork, and some of the factors of success and failure. A few stories about these initial years are especially illustrative.

**Sensor and Booster Development**

In the early days of national reconnaissance, we were not generally enjoying noteworthy success. Aerospace suggested that because the system's second-stage booster had a capability for stable on-orbit flight, a step-by-step approach would be an effective way to test the adequacy of the principal sub-systems of the satellite. An outcome of this approach was the establishment, with the very first flight, of the considerable potential of the principal sensor, which was then a matter of great concern. It was not until the third flight that the operation of the spacecraft's systems, completely independent of the stabilizing second booster stage, was demonstrated.

*The success of many of the satellite reconnaissance programs could be attributed to redundancy in equipment, exhaustive testing, and meticulous attention to detail.*

At the termination of operations, I had an opportunity to address the technical advisors assembled at the Satellite Test Center (STC), which was located in a rambling old structure that pre-dated the "Blue Cube." I recall only the congratulatory remarks—something to the effect that, "An airplane was a contrivance that almost did not work, but a missile or satellite was a device that almost did, so welcome to the ranks of the aeronautical designers!"

The years following the initial flights saw a great explosion in satellite capabilities, which justified the faith and perseverance of the early advocates. This success occurred to the benefit of my employer, which retains the respect and confidence of the government.

**Factors of Success and Failure**

An important aspect of those early flights was the demonstration of the qualities of perseverance and patience toward programs of national importance, which were not so discernable at the end of the Cold War. This perseverance relates, I believe, to an appreciation of the inherent importance of the program for national security. These flights, and the astonishingly realistic rehearsals that preceded them, established a vital teamwork and an esprit de corps among the military and contractor entities that contributed greatly to program success. Equally important, the occasional on-orbit mishaps led to the formation of "tiger teams," whose inspired technical detective work tracked down the culprits of the failures. These teams, organized and led by the Aerospace Corporation's technical advisor resident at the Satellite Test Center, featured subgroups engaged in technical forensics. Another prominent feature was the meeting room, papered from wall to wall with the relevant graphs and charts, and the stand-up meeting at regular intervals, day and night, summarizing the investigation process. It has been said that times of trial as well as success are needed to forge a team. I believe that.

Quality control, or rather the lack of it, was the nemesis during the early days. The achievement of reliability was a major hurdle in those days. Work and ideas at all levels were sought to improve the situation. One of those ideas was to try to improve quality control by inspiring the workforce to do a better job. In a lunch meeting with Brockway McMillan at the Pentagon, I suggested that he personally address the workers—the people behind the soldering irons or fabricators of the cable harness—as opposed to the top-level managers and vice-presidents.[2] His objective should have been essentially to exhort them for better work as a patriotic duty because of a national necessity. He demurred. Some time

[2] Brockway McMillan served as the Director of the National Reconnaissance Office from 1963-1965.

later, I stood in a contractor's large assembly hanger and listened to a general officer make that plea. It was not quite the same, and we did not get the hoped for results. We did not score every time at bat.

**Operations Demanded Quick Calculations**

Each flight was an adventure. During one mission, the tracking station's initial acquisition was startlingly early, which could only mean a highly improper orbit. Bill Sampson and I made some hurried calculations and estimated that the perigee was only a little above the airplane altitude record at the time! Further hurried calculations—this was slide rule and back-of-the-envelope stuff—and we had the numbers for orbit-adjust propulsive firing commands for the next "station pass" a few minutes later. We were lucky, as the calculations were good enough, and the commands were generated, verified, transmitted to the remote station, and executed in time. The orbit was repaired well enough to conduct an almost normal flight. Nothing quite that exciting happened for a while.

**The Reality of Security**

Awareness of security pervaded our activities. The security system was designed, I suppose, to prevent inadvertent disclosure of classified information. Although the security was strict, it obviously could not prevent all instances of adversaries' assembling a comprehensive story from bits of information from disparate sources. I stumbled upon a small example of this in a visit to the Montreal Fair of 1967. In the Union of Soviet Socialist Republics exhibition hall, lying on the floor of the USSR exhibition hall, almost below a replica of Sputnik suspended from the dome, was a largish solid glass cylinder of familiar size and shape. It was a blank of high quality glass from which mirrors and lenses for telescopes or telescopic cameras were made. On it was a small placard with the name, a familiar one, of the object's source: a European instrument glass company from the U.S. imported materials as, quite evidently, did the Soviets. The company name and the size, shape, and material were revelatory. It seemed clear that if we were ahead of the Soviets, it was not by much.

*Awareness of security pervaded our activities...[It] affected our private lives as well.*

Security affected our private lives as well. It diminished the dinner table conversation, and perceptibly narrowed the scope of common interests. While my wife had become somewhat accustomed to my taciturnity by virtue of some of my earlier classified assignments, my children evidenced little interest in my work. They more or less were aware, however, of what I had done earlier in the aircraft industry. Just recently, my younger daughter, then a pre-adolescent, volunteered that at the time she thought maybe I was a jewel thief because of my frequent absences and trips to undisclosed destinations with locked briefcases. While security was not the proximate cause of the dissolution of that marriage, it did not help.

**Relations with My Colleagues**

The military and Aerospace program offices, which were staffed at about the same time, were adjacent. This proximity made for a comparatively easy accommodation to one another. My acquaintanceship with General Greer began with a casual introduction by Colonel King (whom Greer always addressed as Willy).[3]

[3] Robert E. Greer, Major General, USAF (Ret) served as Program A Director from July 1962-June 1965. William G. King, Brigadier General, USAF (Ret) served as Program A Director from August 1969-March 1971.

During our ride out to lunch that day, the propounding of a problem by the General, to which I managed to suggest a solution during the general course of conversation, delighted King. It was an examination that I had passed. The association between the three of us was cordial during the General's tour of duty, and remained so during his civilian work after his retirement from the Air Force.

Things were a little informal in those days. We would fly down to Vandenberg with the General to hear the "stand-up" readiness reports for "booster, bird, and range." A decision to launch or "stand-down" was made quickly, and, if the decision were the former, we went back onto the plane, and flew to Moffet Field and the STC. On the evenings before the launch, General Greer favored a pizza joint known as "Big Al's," a sawdust-on-the-floor hangout along El Camino Real, with pitchers of beer on the table, bench seating, and a nice, noisy, Oktoberfest atmosphere—interspersed with phone calls to the STC to keep track of the count down. During the operations, Colonel King liked an occasional visit to another hangout called the "Rendezvous Room," which featured music, a nice bar, dancing, and regular phone calls to the STC to check on the news from the last "station pass." I liked both places, even the Holiday Inn on Highway 101, my home away from home for 5 years or so. It was a great place, even though, over all that time (the breakfast menu to the contrary), they *never* had any bananas to go with the cereal!

Relations between the Air Force "blue-suiters" and the contractors were not restricted merely to working hours. There was social interaction: an occasional formal dinner, the annual Yule-season gala, and a summertime golf tournament day. On one occasion, there were repercussions from the golf tournament. Asked to supply scorekeepers and lemonade servers, I declared a day off for the secretaries, but neglected to keep a single office phone staffed. The "just in case" event actually happened, but no one was there to respond. My boss, a man with a memory good enough to remind elephants, scored this as a mark against me years later for a crucial promotion. Nevertheless, our relationships were closer than the work relationships that usually existed between the military and contractors, and between Aerospace Corporation and the other associate contractors. For me, a few of these relationships grew into long, continuing friendships.

**My Impact on National Reconnaissance**

What stands out in my memories of being involved with national reconnaissance? There are several aspects that stand out: the feeling of being engaged with something truly important, the interaction with people in industry and at high levels in government, and the ability to engage a significant fraction of my company's technical force on problems that seemed to be intractable. It is, of course, gratifying that the NRO chose to recognize me as a Pioneer for my NRO work. A similar recognition was accorded me for a subsequent assignment in 1967, as management of the GSE/TD for the Defense Support Program, where Colonel King and I were once again able to work together.

While my ambition exceeded my accomplishments, I was fortunate to have had some exciting experiences, and to have been present at some significant moments. These include my work as a student under people at the forefront of aeronautics; as an engineer taking part in the transition from wood and fabric to supersonic flight and beyond; as a member of the cadre at the formation of RAND; as manager of the groups vying for place in the early space programs; and, as noted above, as a participant in the military satellite programs. Along the way, I was fortunate to have met and worked with some of the people who are listed as Founders of the NRO, and others who are identified as Pioneers in the NRO Hall of Pioneers.

## Conclusion

What message to leave? Benjamin Franklin, in a letter to Joseph Priestly some two hundred years ago said it best:

> "The rapid progress true science now makes, occasions my regretting sometimes that I was born too soon. It is impossible to imagine the height to which may be carried in a thousand years the power of men over matter. We may perhaps learn to deprive large masses of their gravity and give them absolute levity, for the sake of easy transport. Agriculture may diminish its labor and double its produce; all diseases may be by sure means prevented or cured, not excepting even that of age, and our lives lengthened even beyond the antediluvian standard"

But it is the remainder of the quote that is of even greater importance.

> "O that moral science were in as fair a way of improvement, that men would cease to be wolves to one another, and that human beings would at length learn what they now improperly call humanity."[4]

—

### Pioneer Award Presentation and Citation

Figure 29-1. Pioneer Ellis E. Lapin (second from left) being recognized at the 2000 Pioneer Recognition Ceremony. The pioneer recognition plaque was presented by DNRO Keith Hall (left) and DCI George Tenet (third from the left). Jon Bryson (Senior Vice President for National Systems, Aerospace) joined in the presentation. (Photo by Sara Judy, NRO Visual Design Center.)

[4] Letter from Benjamin Franklin (1706-1790) to Joseph Priestly (Unitarian theologian and natural scientist), 1780.

**Ellis E. Lapin**

Mr. Ellis Lapin managed the Aerospace Corporation's system design and engineering efforts for early Program A imaging satellites. He introduced improvements in controlled flight operations that nearly doubled the time a particular imaging satellite could function on orbit. This advance permitted greater exploitation of spacecraft imagery and thus increased its intelligence value.

*Career in National Reconnaissance: 1962-1967*

# Richard S. Leghorn

Richard S. Leghorn first articulated the concept of peacetime strategic reconnaissance in 1946. This concept recognized that in the nuclear age nations needed reliable information of force levels and technological capabilities in order to reduce temptations for surprise attacks, and to reduce the dangers of an arms race resulting from mistrust and potential misperceptions of capabilities and intentions. Leghorn made the case that high-altitude aerial reconnaissance, and later satellites, were the best way to obtain this information and achieve a level of confidence and strategic stability. His concepts of peacetime strategic reconnaissance were adopted as national policy during the Eisenhower Administration, and became a reality with the development and subsequent operation of the U-2 aerial and Corona satellite photoreconnaissance systems. In 1957, he founded and was the first president of the Itek Corporation, which produced lenses and cameras used in photoreconnaissance satellites.

## Enhancing Stability Through Strategic Reconnaissance: U-2 and Corona[1]

Upon completing the requirements for a degree in physics at the Massachusetts Institute of Technology, I went to work for Eastman Kodak, initially as a development physicist and then with responsibility for evaluating submitted inventions. In December 1940, Major George W. Goddard, Chief of the Army Air Corps Aeronautical Photographic Laboratory (APL) at Wright Field, near Dayton, OH, visited Kodak.[2] He came looking for Reserve Officer Training Corps officers willing to volunteer to be called to active duty and be assigned to his laboratory. He was building up his staff in anticipation of American involvement in World War II, which was already underway in Europe. We all knew something was going to happen in terms of American participation.

I was intrigued by the opportunity to work for Goddard provided he could arrange a transfer from Army Ordnance Reserve to the Army Air Corps. I had a private pilot's license

[1] This section is based on an interview with Richard S. Leghorn, 26 June 2001.

[2] George W. Goddard retired from the United States Air Force as a brigadier general. He made life-long contributions to long-range, high-altitude aerial photography and advocated the full range of aerial camera development. As a lieutenant in 1925 he accomplished the first successful night photography by an airplane.

and a strong interest in flying. Goddard arranged for my transfer, and I accepted an active duty assignment at APL. I arrived as a second lieutenant at Wright Field in March 1941, and my work involved research and development of aerial photography generally, with specific responsibility for improving methods of assessing and exploiting data from aerial photographs. While at the lab I worked with Dr. James G. Baker of Harvard, who was a brilliant designer of lenses for a number of aerial reconnaissance cameras, Amrom H. Katz, a physicist who tested, analyzed, and greatly improved camera shutter systems among other contributions, and Walter J. Levison, who tested and analyzed light-sensitive materials in the sensitometry unit. We would collaborate again in 1946 during the two atomic tests of Project Crossroads at Bikini Atoll.

**World War II and the Lessons of Surprise Attack**

The next thing that happened was Pearl Harbor, and like many people I will never forget it. I heard the news over the radio and immediately called my then commanding officer and senior pilot, Lieutenant Colonel Donald Hardy. Given my background in aerial photography, my interest in flying, and the fact that the "action" had now shifted from the lab to combat operations, I wanted to go to flying school so that I could get into combat reconnaissance. He fully understood, and he was very helpful in arranging a transfer despite the critical need at the time for research and development officers.

Upon completing flying school, with the rank of captain, I was assigned to organize the 30th Photoreconnaissance Squadron at Peterson Field in Colorado. We trained for overseas

Figure 30-1. Explosions in the forward magazine compartments of the *USS Shaw* at Pearl Harbor, 7 December 1941. (Photo courtesy of Naval Historical Foundation, original source unknown.)

Figure 30-2. Captain Richard S. Leghorn in England when assigned to the 30th Photographic Reconnaissance Squadron (Light), spring 1944. (Photo courtesy of Air Force History Office, original source unknown.)

duty, although we did not know where. It turned out that we were deployed to England, and as the first reconnaissance unit of the new Ninth Air Force, we were directed to focus on photography in preparation for the invasion of Normandy. After the invasion and our race across France, I became a Lieutenant Colonel and Commander of the 67th Reconnaissance Group. The Group then consisted of one squadron of P-38s (F-4s) without guns photographing from 30,000 feet and higher, two or at times three squadrons of P-51s (F-6s) flying in pairs with machine guns, and a P-61 squadron for night photography with flash bombs and Edgerton lights. To adapt these North American fighter planes for our missions, we replaced all or much of their armaments with cameras.

Following Victory-in-Europe Day, I was ordered back to the Pentagon to help prepare the reconnaissance effort for the invasion of Japan. But while I was there, the war ended in August 1945 with the dropping of atomic bombs on Hiroshima and Nagasaki. I went on terminal leave with the expectation of returning to civilian life, when I received a call from Brigadier General Paul E. Cullen, who was the senior reconnaissance officer in the Pentagon. He asked if I would return to active duty to participate in the aerial photographic and sophisticated instrumentation elements of Project Crossroads, which involved the testing of two atomic bombs at Bikini Atoll in the Pacific. I readily agreed to his request.

Before this assignment, I read a summary report issued in September 1945 titled *United States Strategic Bombing Survey (Europe)*. The report reviewed the effectiveness of the allied strategic bombing campaign against Nazi Germany and the reconnaissance that supported it. I was impressed by a few key points, namely the critical importance of aerial photography in Europe, and that the United States "needed more accurate information, *especially before* and during the early phases of the war."[3]

[3] United States War Department, 1945. See also R. Cargill Hall, "Post-War Strategic Reconnaissance and the Genesis of Project Corona," in Robert A. McDonald, ed., *Corona: Between the Sun and the Earth* (Bethesda: ASPRS, 1997), 24-34.

Figure 30-3. Atomic tests at Bikini Atoll, 1 July 1946. (Photo courtesy of Air Force History Office, likely Army Air Force photo.)

This assessment remained in my thoughts as I observed the two atomic explosions in July 1946. We were circling and observing the explosions visually and with our equipment. The first test (Able) was a surface explosion. The second test (Baker) was an undersea explosion. One effect that I'll never forget was that the explosion produced such force that a Japanese battleship was tossed well into the air. To see an actual atomic bomb, and the immediate and complete damage it did, was a fearsome experience.

As a result of that experience I was convinced that warfighting—as we had known it—was over, and it would never again play the same role in our overall security strategy. Deterrence and arms control would be the new focus of international nuclear security. I was distressed by the claim of many that the concept of war had not changed as a result of these weapons.

Additionally, the frightful possibility of a surprise nuclear attack on the United States made clear the urgent need for peacetime strategic reconnaissance, as separate and distinct from tactical wartime reconnaissance of the World War II variety. I recognized that in the nuclear age nations needed reliable information of force levels and technological capabilities if both sides were to reduce temptations for surprise attacks, and reduce dangers of an arms race resulting from mistrust and misperceptions of capabilities and intentions. I saw extremely high-altitude aerial reconnaissance and overflight of adversary territory with a small number of specially designed and fabricated aircraft capable of avoiding interception, and hopefully detection, as the best way to obtain this information and achieve a level of confidence and subsequent strategic stability.

*To see an actual atomic bomb, and the immediate and complete damage it did, was a fearsome experience.*

**Returning to Civilian Life**

Following my return to civilian life, in December 1946, I was invited to speak at the dedication of Boston University's Optical Research Laboratory (BUORL). Many luminaries, both civilian and military, attended the ceremony and spoke of the continuing importance of research and development in the field of optics and photography. Major General Curtis

LeMay, who subsequently commanded Strategic Air Command (SAC), was among those in attendance.

I used my remarks to address the requirements for, and obstacles to, peacetime strategic reconnaissance. I argued that a world in which more than one nation possessed atomic weapons was a world that required aerial reconnaissance prior to the outbreak of hostilities. I maintained that military intelligence had become the most important guardian of our national security. The nature of atomic warfare was such that once attacks are launched, it would be extremely difficult, if not impossible, to recover from them and counterattack successfully. Therefore, it obviously became essential that we have prior knowledge of the possibility of a nuclear attack, so that defensive action could be taken before it was launched. My principal point was that getting this information ahead of time from a closed society such as the Soviets maintained behind its "Iron Curtain," only could be accomplished through peacetime aerial reconnaissance.

I recognized that unauthorized overflight of a foreign state in peacetime was denied by treaty and law, and could be considered an act of military aggression sufficient to provoke military conflict. The irony is that peacetime espionage is considered a normal function of national security, while aerial reconnaissance, which is merely another method of intelligence collection, is viewed as an act of military aggression. Viewed in the context of peacetime intelligence collection, aerial reconnaissance could be viewed as similar to the efforts of military attaches, who also collect intelligence of strategic and military significance. I recognized that there were essentially two options: to change international perceptions and law on the subject (most unlikely), or to develop a capability for overflight reconnaissance with extremely low chance of detection or interception. In terms of the latter option, an aircraft with this type of capability did not exist in 1946.

*The irony is that peacetime espionage is considered a normal function of national security, while aerial reconnaissance, which is merely another method of intelligence collection, is viewed as an act of military aggression.*

In my remarks at BUORL, I argued that the accomplishment of this objective was not as technically difficult as it might first appear. Aircraft with extremely long-ranges capable of flying at very high altitudes were on the drawing boards, and in some cases prototypes had been constructed. Effective means of camouflaging them at high altitude against visual observation were well known, and countermeasures for preventing radar detection were being considered.

The Soviet detonation of an atomic bomb in August 1949, the victory of communist forces in China in October of 1949, and the surprise attack of communist North Korean forces on the Republic of South Korea in June 1950, all contributed to heightened tensions in the Cold War. The threat of a surprise atomic attack on the U.S. was taken seriously, and national security considerations now required that the political risks of reconnaissance overflights be reconsidered. Consequently, a review of U.S. strategic overflight reconnaissance capabilities, practices, and policies became imperative.

**Conceptual Beginnings of the U-2**

In 1951, I received a call from Colonel Dick Philbrick, a colleague from my days at APL and also a reconnaissance commander in Europe during the war. I was a reserve officer at the time, again working for Kodak. He asked if I would return to active duty to help get a reconnaissance branch going at Wright Field. In view of the Korean War, Kodak cooperated and provided a leave of absence for up to two years.

Figure 30-4. An RB-47E photoreconnaissance aircraft. (Photo courtesy of Air Force History Office, likely U.S. Air Force photo.)

Immediately upon arriving at Wright Field, I reviewed the requirements for a reconnaissance aircraft and compared these with available aircraft. I said that we wanted something that flies very high—no military specs, no armaments, no protective armor, no guns, just altitude. We went through the whole inventory of planes then under development. The British Canberra, a two-engine light bomber made by British Electric, seemed the best candidate. After consultations, we invited British Electric to submit a proposal to design and build a limited number of planes to meet our requirements. Part of the modifications included replacing the existing wings with longer wings that added lift and made the plane more like a glider. We also directed them to disregard military specs as if the planes would be pre-production prototypes. The plane, as proposed by British Electric, would penetrate enemy defenses at 63,000 feet, fly in 700 miles, and losing fuel it would fly out at 67,000 feet. At this time the MiG topped out at 45,000 feet, and surface-to-air missiles (SAMs) were nowhere in sight. We estimated that we could get 85% of the intelligence targets in the so-called denied territories within the 700-mile penetration.

The Air Staff in Washington didn't accept my proposals and recommendations. However, then Colonel Bernard A. Schriever, who was responsible for Air Force Development Planning, supported my ideas and arranged that I immediately report for duty on his staff.

Thus, my last assignment before again returning to civilian life in January 1953 was to prepare an Intelligence and Reconnaissance Development Planning Objective (DPO), which provided direction for research and development activities. The strategic reconnaissance part of this DPO called for high-altitude balloons and eventually Earth orbiting satellites to provide wide area surveillance of the Soviet Union, with more detailed observation to be provided by high-altitude aircraft, and later by second-generation satellites. Merton Davies of RAND made a major contribution to our DPO effort in terms of integrating RAND's work on satellites with our work.

To achieve detailed surveillance, the DPO identified a requirement for a single engine, lightweight reconnaissance airplane designed specifically for peacetime strategic reconnaissance at altitudes of 70,000 feet or higher.[4] It is interesting that an estimate prepared for

[4] Robert A. McDonald, ed., *Corona: Between the Sun and the Earth* (Bethesda: ASPRS, 1997), 37.

this project regarding vulnerability to SAMs indicated that this program would probably be ended by 1960 due to technical progress the Soviets would likely make in developing SAMs that could challenge these aircraft. The estimate noted that a satellite reconnaissance capability should be in place by that year.[5]

From 1952-1954, aerial reconnaissance overflight missions of Soviet territory were reviewed on a case-by-case basis by the President, and approved when the national security interests of the U.S. outweighed the political risks of such overflights. President Eisenhower recognized the importance and value of strategic reconnaissance, both with regard to reducing the risk of surprise attack, and to helping bring the arms race under some measure of control. Consequently, he continued periodic overflights of Sino-Soviet territory, while balancing the political risks of these missions.

In May 1954, a U.S. RB-47E photoreconnaissance aircraft was attacked and damaged by Soviet MiG-17s while on a daytime photographic overflight of the Kola Peninsula and airfields east of Leningrad. This incident provided further support for the argument that if strategic reconnaissance overflights were going to continue at reasonable risk levels, a different type of aircraft was required, specifically an aircraft that could fly at extremely high altitudes beyond the reach of Soviet air defenses. In November 1954, Eisenhower approved a program to develop a high-altitude photoreconnaissance aircraft, and this decision was followed by the government's selection of Lockheed's CL-282 aircraft, which later became known as the U-2.

I returned to Kodak after I completed my work at Wright Field in January 1953, and this return to civilian life essentially ended my participation in the planning phase for high-altitude strategic reconnaissance aircraft and satellites. My involvement with the actual U-2

Figure 30-5. The U-2 reconnaissance aircraft, unknown date. (Photo courtesy of NRO History Office, likely U.S. Air Force photo.)

[5] Consistent with that estimate, a SAM shot down a U-2 piloted by Francis Gary Powers on 1 May 1960. The Corona photoreconnaissance satellite returned its first imagery of the Soviet Union on 19 August 1960.

Figure 30-6. President Dwight D. Eisenhower with Special Assistant Harold Stassen, in the Oval Office on 22 March 1955, a few months before announcing his "Open Skies" proposal at the Big Four Summit Conference in Geneva. (Photo courtesy of Air Force History Office, original source unknown.)

program came in 1955 when I returned to the U.S. after working for Kodak in Europe for two years. Despite several inquiries I could find no evidence of any actual development of our proposed plane. In frustration, I wrote about the need and possibilities for strategic reconnaissance in an article published in *U.S. News and World Report.* James Killian, President Eisenhower's Science Advisor and a friend of mine from college days, called in alarm and asked me to speak with Richard Bissell, Special Assistant to the Director of Central Intelligence. I agreed, and Bissell briefed me at length on the U-2 program. Needless to say, I assured them both that my lips were sealed, and I expressed appreciation for being included in the program.

*Eisenhower proposed an "Open Skies" arrangement under the political framework of an arms control and disarmament proposal that he believed would address the absence of trust and the presence of "terrible weapons" among states.*

The President was preparing for the Four-Power Summit Conference scheduled for July in Geneva. As suggested by others and myself, Eisenhower proposed an "Open Skies" arrangement under the political framework of an arms control and disarmament proposal that he believed would address the absence of trust and the presence of "terrible weapons" among states. To mitigate the "fears and dangers of surprise attack," he urged that the Soviet Union and U.S. provide "facilities for aerial photography to the other country," and conduct mutually supervised reconnaissance overflights.

There was another element to the Open Skies proposal beyond the arms control feature. I told Bissell that eventually a U-2 was going to crash or get shot down, and a political action program was necessary to help protect the program and mitigate political repercussions if that occurred. The Open Skies proposal placed the U-2 program in a broader context of arms control rather than covert strategic reconnaissance. As most others including myself had anticipated, within a day the Soviets rejected Eisenhower's proposal as a U.S. attempt to gather targeting information. While Khrushchev rejected the proposal, it provided some measure of political cover for the covert program. If we could not negotiate an agreement with the Soviets, then strategic stability simply demanded that we obtain the reconnaissance information some other way.

On 4 July 1956, a camera-equipped U-2 took off from Wiesbaden, West Germany to survey the USSR's naval shipyards and submarine construction program. Contrary to American expectations, Soviet radar detected and tracked this first mission at its design altitude of 70,000 feet. This unauthorized overflight caused considerable consternation among Soviet leaders, because it was perceived as collecting targeting intelligence potentially for a preemptive nuclear first strike. Soviet vulnerability to the U-2 resulted in an order by the Kremlin for new high-altitude surface-to-air missiles, high performance fighters and interceptors, and accelerated work to develop and deploy intercontinental ballistic missiles.[6] After 1956, the U-2 program proceeded cautiously and President Eisenhower personally authorized each mission.

**Itek and Corona**

I had spoken with Killian and Bissell about the need to implement a vigorous satellite program. I had become interested in starting a business, and with the help of Rockefeller funding I left Kodak and formed a company called Itek.

Itek was incorporated on 27 September 1957, and at the time everyone said to me it was no time to start anything with defense contracts in mind because we were in a defense funk. Then, on 4 October 1957, Sputnik went up. Talk about luck!

*Itek was incorporated on 27 September 1957 and everyone said it was no time to start anything with defense contracts because we were in a defense funk. Then, on 4 October 1957, Sputnik went up. Talk about luck!*

Sputnik spurred considerable discussion and debate among President Eisenhower's military and scientific advisers about the preferred design and operational features of space-based strategic reconnaissance systems. The President gave his formal approval for a highly classified satellite photoreconnaissance project during a White House meeting in February 1958. The Air Force had a program underway, the Weapon System 117L (WS-117L), but this program was publicly cancelled and a more covert joint Air Force-CIA program was created in its place. The reconnaissance system that was selected involved the recovery of film capsules ejected from orbiting satellites, as opposed to the electronic readout system envisioned for WS-117L. Six weeks later, Bissell identified the contractor team, and selected a new classified name for this covert effort: Project Corona.

Figure 30-7. The KH-4 Corona camera. (Photo courtesy of NRO Reading Room, original source unknown.)

Itek submitted an unsolicited proposal in February 1958 to supply a camera system for Corona. The government already had decided to procure Corona's primary camera from Fairchild. However, in March 1958, a government panel reviewed the proposals of Fairchild and three other companies (General Electric, Eastman

[6] R. Cargill Hall, "Post-War Strategic Reconnaissance and the Genesis of Project Corona," in Robert A. McDonald, ed., *Corona: Between the Sun and the Earth* (Bethesda: ASPRS, 1997), 41.

Kodak, and Itek). The Itek system was chosen as the backup camera because the system offered better resolution than the Fairchild or other systems. In April 1958 the government reversed its position and decided to use the Itek-designed camera as the primary camera on Corona, thereby replacing the Fairchild system.

The timing of Corona becoming operational was fortuitous in light of the U-2 shootdown in May 1960. By that point, Corona had flown twelve test missions where one thing or another went wrong. Then on 12 August 1960, an experimental recovery bucket was returned from space with an American flag inside. This was followed six days later with the launch of the first mission to successfully acquire on-orbit imagery that was recovered from space on 19 August. These initial images were of a Soviet bomber base at Mys Shmidta, located on the northeast coast of the Soviet Far East. Corona was now operational and would remain in service until May 1972. In terms of relative geographic coverage, the second successful Corona provided coverage of more area than all of the U-2 overflight missions combined.

**Soviet Reaction**

Later that year, in December 1960, I went to Moscow to attend one of the series of Pugwash Conferences on Science and World Affairs. Although I had demurred at an earlier planning session in Paris, I was nevertheless invited as a guest of the Soviet Academy of Sciences. It was evident that they had knowledge of my role in reconnaissance overflight programs, because Academician Topchiev, deputy chairman and "political boss" of the Academy cordially requested, and I promised, not to attempt to arrive in a U-2! We met in the House of Friendship just down the street from the Kremlin. I sat opposite a scientist named Federov who was the General Secretary of the Soviet Academy of Sciences. It snowed for almost the entire two weeks we were there, and Russian women were clearing the streets with straw brooms. Federov chided me with a veiled smile saying in effect, "You are nuts. You see what

Figure 30-8. A Corona image of Red Square in Moscow taken in May 1970, ten years after Corona began regular imaging of Soviet territory in August 1960. In 1960, the General Secretary of the Soviet Academy of Sciences suggested that poor weather would prevent successful satellite imagery reconnaissance of the Soviet Union. (KH-4 photo.)

our weather is like. There's no point in putting those satellites up. You'll never be able to get pictures of us."

The Soviets knew the characteristics of our photographic systems from the U-2 cameras they obtained from the May 1960 shootdown. They discussed the quality of our photography at the conference. Everyone was very careful not to say anything classified, but there was open banter about the fact that they had tracked our U-2 flights from the beginning. We did not think they would be able to do this. And now they were tracking our satellite flights, and they had to surmise that we put cameras in them. They did not know exactly what we had in the satellites, but based on the captured U-2 equipment they could estimate the quality of the imagery and intelligence we were getting from Corona. It was a rather challenging and bizarre experience, living simultaneously in the "black world" and in the open.

Another interesting episode at that conference was that we spoke with all of the Soviet officials we came into contact with about the possibility of releasing two American prisoners. Two RB-47 crewmembers were being held at Lubyanka prison after their plane was shot down on 1 July 1960 during a reconnaissance mission over international waters in the Barents Sea. We suggested to the Soviets that if they wanted to improve the bilateral environment between the U.S. and USSR and advance a nuclear test ban treaty, then why not give President Kennedy an inauguration present and release those two crew members. Well, the Soviets released them on 24 January 1961 without bringing them to trial, and such a treaty was signed not too long after.

**Conclusion**

The concepts and systems that I helped develop, and their contribution to strategic stability during the Cold War, is a legacy in which I will always take pride and will always remember. The same can also be said for the rewarding and deep friendships developed with many talented and dedicated Americans. Peacetime strategic reconnaissance became a critical element of national security since I first proposed it in 1946. The information collected by these systems provided greater transparency and knowledge of Soviet systems, capabilities, and limitations during the Cold War. This knowledge helped American decision makers better understand the nature of the military and strategic threat, and helped reduce the risk of military surprises. Consequently, American presidents were more confident that they could enter into arms control agreements that would serve the national security and economic interests of the United States.

Effective and reliable concepts, strategies, doctrines, policies and supporting technologies and systems are required to protect and advance U.S. national interests. Similar to the triad concept that forms the basis of the U.S. strategic deterrent, there are three legs of a triad in terms of international security and stability.[7] The first leg is deterrence under arms control provided by conventional and nuclear military capabilities and international agreements. The second leg is information—information from arms control agreements and information from intelligence sources. The third leg is means for non-violent conflict resolution (e.g., diplomacy, and the World Court). The information component of this triad is critical. I am extremely grateful for the opportunities I have had over the years to contribute to acquiring and disseminating that information to decision makers, thereby advancing both U.S. national security interests and international security and stability.

[7] The underlying concept of the U.S. strategic deterrence triad is that submarines, bombers, and ICBMs provide a secure second-strike capability that should be adequate to deter an adversary from launching a surprise nuclear first strike, since a sufficient number of strategic forces would survive a first strike and would be capable of delivering a retaliatory strike that would inflict unacceptable losses.

—

**Pioneer Award Citation**

**Richard S. Leghorn, Colonel, USAF**

Colonel Richard S. Leghorn first articulated the concept of peacetime strategic reconnaissance in 1946 as a means to warn of military and strategic surprise, and saw it adopted as a national policy within ten years. In 1957, he founded the Itek Corporation that produced the lenses and cameras used in Corona and other overhead imaging satellite programs, helping that national policy become a reality.

*Career in National Reconnaissance: 1946–1962*

# Walter J. Levison

Walter Levison designed the cameras for the Genetrix balloon and Corona satellite photoreconnaissance programs. These programs were at the forefront of peacetime strategic reconnaissance and helped shape the discipline of national reconnaissance. Levison was honored posthumously as a Pioneer of National Reconnaissance. His wife, Ann, describes the impact of her husband's classified work on their family. Her recollections follow a brief description of Levison's contributions to national reconnaissance.

## A Designer of Cameras[1]

The United States Intelligence Community pursued peacetime strategic reconnaissance beginning in the late 1940s. The reconnaissance programs used heavy, long focal-length telephoto-lens cameras that produced close-up images of the earth in film formats as large as 18x36 inches. The Intelligence Community required panoramic search cameras that had sufficient resolution of objects on the earth's surface. At the same time, the cameras had to be lightweight and compact enough to be flown and operated at extreme altitudes, which included the vacuum of outer space. Walter Levison met that challenge. His cameras permitted overhead strategic reconnaissance at extreme altitudes with balloons and film-recovery satellites.

Levison's lightweight high-acuity cameras were used in the 1956 Genetrix balloon overflight program and in the follow-on WS-461L balloon overflight program in 1958. The Genetrix balloons were ballasted so they would not exceed 50,000 feet altitude, an altitude that would have prompted the Soviets to develop weapons that might threaten future operations of the U-2. The U-2 was about to begin overflights of the Soviet Union and would be operating around 70,000 feet.

Ultimately, Levison's team designed the remarkable counter-rotating optical bar panoramic camera employed in several aerial and space reconnaissance programs. In addition to his balloon-borne cameras, Levison directed the design of the swiveling (later rotating) lens panoramic search camera employed in the Corona program, which eventually achieved a resolution of six feet at the earth's surface. His team at Itek Corporation also designed the

[1] This section is based on information contained in the material submitted with the nomination of Walter Levison for selection as a National Reconnaissance Pioneer. We prepared this section in lieu of recollections.

Figure 31-1. Walter Levison at the Air Force Aeronautical Chart and Information Center in St. Louis in late 1946 evaluating imagery of the Bikini Atoll atomic tests. (Photo courtesy of NRO History Office, original source Walter Levison.)

optical bar camera used on the Apollo Lunar Orbiter.

After designing the Hyac I panoramic camera for the WS-461L program, Levison modified it as the Hyac II camera. This Hyac I balloon-borne 24-inch focal length triplet C-element lens panoramic camera kept the field narrow. Simultaneously, the camera obtained wide-angle coverage by swiveling the 24-inch lens past 70mm film, substituting a mechanical motion for an optical field. Modified as the Hyac II, it became the C camera series used in the Corona satellite photoreconnaissance program. In the camera's Corona application, the lens eventually rotated in a complete circle.

Although another firm would build the later, large-scale counter rotating optical bar cameras for the Corona program, the original camera design by the Levison team still was considered at the beginning of the 21st century to be the apex of panoramic cameras used in overhead photography.

Levison died in 1999. On 27 September 2000, the Director of the National Reconnaissance Office (Keith Hall) and the Director of Central Intelligence (George Tenet) posthumously honored Levison as a Pioneer of National Reconnaissance. Levison's wife, Ann, accepted the award on his behalf. Following Levison's induction as a Pioneer, his wife provided her thoughts and recollections of his work and its effects on his personal life.

—

## Secrecy, Success, and Satisfaction[2]

I knew almost nothing of my husband's work in reconnaissance until the 1990s. The secrecy was difficult, and his work had an impact on our family. The secrecy created an atmosphere of silence that I never would have expected.

Toward the end of the Corona program in the 1970s, I learned that Walter had designed a very innovative camera. I was told that the camera was used on a mission to the moon. By the time the entire Corona program finally was declassified, which occurred years after it had been discontinued, Walter had relaxed enough to tell me that he had worked on a secret program. He did not, however, say what it was. I later learned that many pioneers had told their wives that much from the beginning. At the time of the Corona declassification, I was overwhelmed with admiration for what Walter and his colleagues had done.

I expect that there may be more declassifications coming, and I hope I will learn of them, because it is hard not to know what my husband accomplished during those years. What I do know is that Walter invented something innovative and advanced. He was a

[2] This section is based on written input from Ann Levison, wife of Walter Levison.

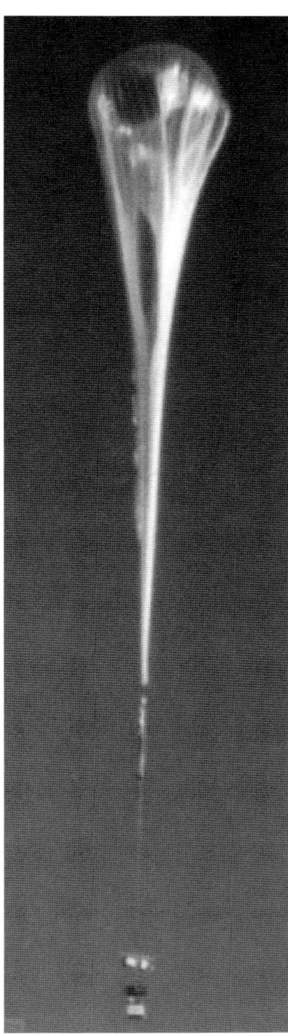

Figure 31-2. The WS 461-L unmanned balloon launch that carried the Hyac I camera system in 1958. (Photo courtesy of NRO History, original source unknown.)

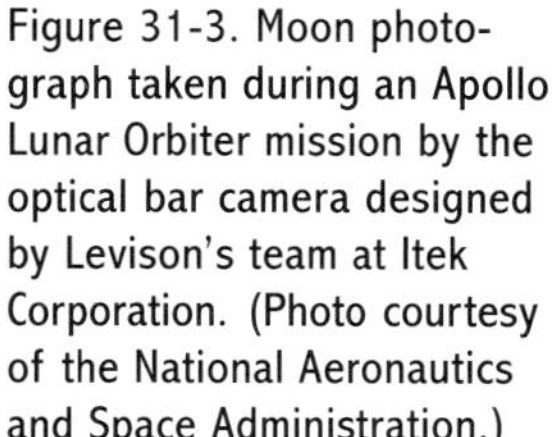

Figure 31-3. Moon photograph taken during an Apollo Lunar Orbiter mission by the optical bar camera designed by Levison's team at Itek Corporation. (Photo courtesy of the National Aeronautics and Space Administration.)

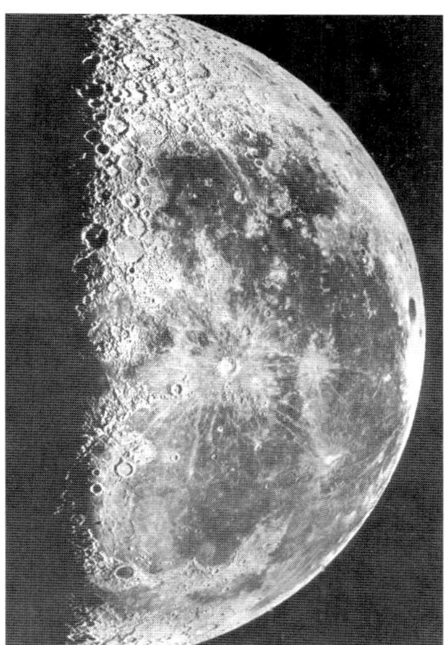

Figure 31-4. Levison's Hyac II Panoramic Camera. (Photo courtesy of NRO History Office, original source Walter Levison.)

wonderful manager, and he made innumerable contacts that helped him in his second career after he left Itek. The greatest thing for me was that my husband enjoyed his work in national reconnaissance very much. He always had the satisfaction of knowing that his work ultimately helped with the monitoring and verification of arms control treaties and agreements and, later, of learning that it had helped avert armed conflict.

—

## Pioneer Award Presentation and Citation

Figure 31-5. Ann Levison, (second from left), wife of pioneer Walter Levison. Mrs. Levison received the pioneer award plaque for her husband, who was recognized posthumously at the 2000 Pioneer Recognition Ceremony. The award was presented by DNRO Keith Hall (left) and DCI George Tenet (third from the left). Dr. Terry Heil (Senior Vice President of Intelligence Programs, Raytheon) joined in the presentation. (Photo by Sara Judy, NRO Visual Design Center.)

### Walter J. Levison

Mr. Walter Levison designed the lightweight duplex camera for the Genetrix overflight program, the high acuity camera for the WS-461L overflight program, and its rotating lens panoramic variant used in Corona satellites. He also supervised the Itek Corporation's design of the unsurpassed panoramic camera employed in other air and space reconnaissance programs.

*Career in National Reconnaissance: 1942-1975*

# Francis J. Madden

Frank Madden was an original member of the Itek Corporation, and was instrumental in the development of Itek's camera system for the Corona satellite photoreconnaissance program. In his recollections, Madden discusses Itek's contributions to national reconnaissance and various technical and organizational challenges he and his colleagues encountered.

## A Dedicated and Focused Activity and a Great Deal of Teamwork[1]

I served in the Army Air Corps as an aerial photographer in the Pacific theater during World War II. After my military service, I went to college on the "GI Bill" and graduated in 1951 at the age of 29. The GI Bill was a great thing for the country in general, and for us as individuals. I never would have attended college without the GI Bill, because I came from a poor family.

I was older than most of the other graduates and had a different perspective. I was in military service for over three and a half years, and there was a different maturity level for those of us who were coming out of the service during that period. While in the service we learned to take numerous risks. We tended to be highly motivated, and we developed relationships with people that fostered working in a cooperative way.

After college, I went directly to work for the Boston University Upper-Atmosphere Research Laboratory developing rockets and performing studies of the upper atmosphere. After some time, I moved to the Boston University Physical Research Laboratory (BUPRL), which had been doing photographic work for the Air Force for several years. Within the Air Force shop, we were building various camera systems and working on developing balloon reconnaissance cameras. The university environment allowed us to take the time to be inquisitive and to pursue innovative ideas. My first experience with national reconnaissance activities was at the BUPRL working on the development of balloon reconnaissance cameras, which were the forerunners of the satellite reconnaissance cameras.

### The Origins of Itek and Early National Reconnaissance Attempts

In 1959, Dick Leghorn and a few other people acquired the facilities and personnel of the

[1] This section is based on an interview with Francis J. Madden at the National Reconnaissance Office Headquarters on 26 September 2000.

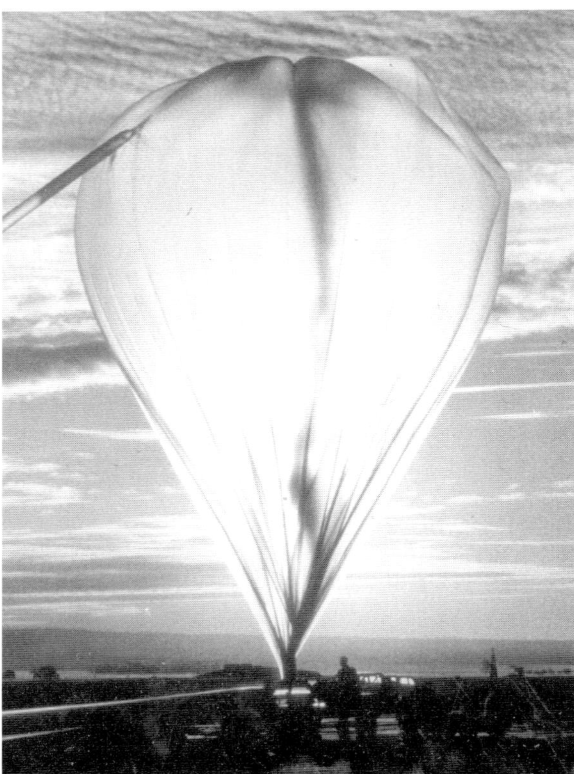

Figure 32-1. Reconnaissance Balloon. (Photo courtesy of NRO History Office, original source unknown.)

BUPRL, which became the nucleus of the Itek Corporation.[2] The Itek team was a unique group. Itek's involvement with national reconnaissance activities began when it inherited BUPRL's work with balloon reconnaissance. There were two phases of the balloon camera systems. One was Genetrix, and the other was called Broad Area Mapping System. The Broad Area Mapping System camera had a 6-inch focal length and utilized a pair of Metragon lenses. The camera produced 9 x 9-inch pictures. The concept was to integrate a compass system and clock that would indicate where the images were taken. We probably launched some 400 systems, but we only recovered 40. It was a fiasco! The balloons would go where the air currents would take them. We flew them too low, and the Soviets knocked many of them out of the air.

The U-2 program had been initiated before we started our balloon work. One problem with the U-2 was the aircraft produced vibrations and, as a result, the photography was affected. Nevertheless, the U-2 delivered very good imagery over a relatively narrow field. Naturally, the mission planners had to know precisely where to send the aircraft if they were going to acquire useful images of a particular target. A significant drawback of the U-2 was its limitation in target coverage. If there were an imaging target, like an airfield, a hundred miles to the west or east of the flight path, the camera would miss it. Obviously, if the planners were not aware of a target, they could not photograph it. Satellites mitigated this problem as a result of their dramatically increased area coverage. Satellite photoreconnaissance became critically important after a U-2 was shot down by the Soviets in May 1960. Once we had satellites, we could image a very large area without the risk of interception.

### Developing a Camera for the First Photoreconnaissance Satellite

Itek was involved in designing and building the camera for Corona—the first photoreconnaissance satellite. We developed a panoramic camera for Corona. It would have been simpler to make a different kind of camera, but we were trying to get the utmost detail onto the film. We initially used only a small portion of the camera's capabilities at any given time, but we were able to improve performance as we gained experience.

Our panoramic camera had a somewhat semi-circular focal plane. The more typical camera had a flat focal plane, with the lenses out in front of the focal plane. When the latter type of camera takes a picture, the scale across the whole photograph is the same. That consistent scale

[2] Richard Leghorn is a Pioneer of National Reconnaissance (inducted into the NRO Hall of Pioneers, 27 September 2000). See chapter 30 for his recollections.

is good in many instances, but resolution is sacrificed because of the large, wide angle that is covered with the one lens. The panoramic camera sacrifices geometric capability by scanning the lens across the film and exposing only a very small portion of the film at any instant in time.

The advantage to scanning the lens is that the highest resolution is in the center of the film, which allows for the exploitation of the lens' resolution capability by painting a picture on the film. This puts a flat surface on a curved surface, but it produces geometrical distortion. The way to fix the distortion is by inverting everything: putting the piece of film on a semi-circle section, then projecting it onto a flat surface—basically turning everything upside down. By doing this, the resolution is high and the geometry is correct. Itek built rectifiers to perform this function. It was a combination of these elements that made the Corona camera, with its high-resolution capability, different from most of the other cameras.

We faced several problems while we were developing the Corona camera. However, we were able to overcome them with innovative system design, materials, and engineering. Three problems of particular interest were temperature variation, brittle film, and a "Corona glow."

*Temperature Variation.* The issue of controlling the temperature inside the satellite in the space environment was a daunting problem. We knew there was a vacuum in space, but we did not know what kind of temperature variations to expect inside of the vehicle. If a space vehicle orbits close to the earth—and relatively speaking we were close to the earth—the vehicle would be in the sun half of the time and out of the sun half of the time. The vehicle absorbs radiation when it is in sunlight, and releases heat into space when it is not in sunlight, and these changes create a temperature variation. We turned to Lockheed to find out what we should expect in terms of the environment and asked, "Can you tell us what it is going to be like in the vehicle? We have to know so we can tell you the tolerances for the temperature control."

After Lockheed told us about the temperature environment inside the vehicle, we told them that their design had to maintain an operating environment of seventy degrees, plus or minus ten degrees, because if the temperature went beyond that, the lenses would expand or contract and their surfaces would change. In essence, it would be a different camera.

Lockheed found a solution. The company designed a paint pattern for the orientation with the sun. The pattern would heat or not heat the inside of the vehicle, and in that way the temperature was controlled.

*Brittle Film.* In the beginning, the film was a disaster because it was acetate film. If the film got cold, it would break. Also, in the vacuum of space the film would dry out, moisture would boil off, and it would crumble like an autumn leaf. If an autumn leaf is crunched, it becomes powdery, and that is what happened to the film. It became very brittle. This presented a significant problem, because when we had to move the film in the cameras, we could not move it gently. The film had to be moved rapidly from the rear of the vehicle, through the camera, and out to the front of the vehicle into the recovery section. This meant rapid twists and turns of the film, and the film just could not cope with that.

Figure 32-2. Itek's KH-3 Corona camera. (Photo courtesy of NRO History Office, original source unknown.)

Dupont and Eastman Kodak solved the problem. Eastman Kodak gets the credit for putting it to work, but Dupont developed the Mylar film and licensed Eastman Kodak to coat it and make photographic film out of it. If Dupont hadn't developed Mylar film, we still would be breaking film. Eastman Kodak did very innovative work in developing emulsions and processing methods to cope with the tremendous illumination variations experienced while on orbit.

*"Corona Glow."* One of the worst problems with the Corona camera and film was the "Corona glow." When the film was transported in a vacuum at high speeds, it generated a glow that would simply expose itself on the film, which rendered the film useless.

Figure 32-3. An example of film with an electrostatic discharge phenomenon that was caused by the sparking of static electricity from moving parts of the film transport subsystem. This phenomenon commonly was referred to as "Corona glow." (Photo courtesy of NRO Reading Room.)

My team solved the problem. The team worked out a way of designing the rubber rollers that transported the film in such a way that the rollers would inhibit the glow. While they did not eliminate the glow entirely, they were able to reduce it enough so as to avoid ruining the film entirely. There sometimes was streaking on the first few frames, but then for the bulk of the path there were no problems. The solution was to start rolling the film early before beginning to photograph.

### Itek's Dependence on Outside Suppliers

Itek could not have done all the work for Corona alone. We had to depend on many outside suppliers. For example, we needed lens cells. The acquisition of a lens cell generally is a minor issue, but the lens cells we needed were very large with extraordinary tolerances. In addition, we had to have the lens elements (five per lens) related to each other in a very exact sense. That was the key to getting the optimum resolution for the system. I give great credit to those who made our lens cells. When we went to them looking for huge lens cells they asked, "What do you want them for?" We answered, "Don't ask what we want them for, just do it!" They cooperated.

All of this external manufacturing was done in an unclassified environment. Even some internal work within Itek was done in an unclassified environment. There was only a nucleus of us who held security clearances and were authorized to know about the purpose of our work.

We had high standards, and our manufacturers could have made more money by doing business with other customers, but they stuck with us. I remember one manufacturer who delivered about fifteen or twenty cells that we rejected. We needed extreme tolerances and very thin walls because we could not afford the weight. If the lathe were not operating within those tolerances, we could not use the cells. Also, if the casting core shifted a little, the walls would be too thin on one side, too thick on the other, and we would have to reject the cells. It was tremendously difficult for the manufacturers, but they stuck with it. At that time, American manufacturers had more of a conscience.

### The Importance of Leadership and Teamwork

The Corona program experienced twelve consecutive failures. I suspect that if someone other than Eisenhower had been president at the time, he might not have stuck with the program. Eisenhower did, however, and he gave the program top-level leadership. The Corona program finally came together, and it was very successful. Eventually we were up to thirty-day missions.

When I say "we" I am referring to Lockheed, General Electric, the Air Force, and all the other people involved. I believe that cooperation was the key to getting things done in the Corona program. We were dedicated and focused on the program, and we had a great deal of teamwork. We had confidence that the fellow beside us was going to do his job, and we were going to do ours. We worked cooperatively to get the job done, and I believe that was very important.

I still maintain relationships with many of the people I was involved with during that period. We developed a sense of comradeship, which is not always easy to do. I believe part of the reason we were able to do that was that we learned about teamwork during our military service days. During wartime we learned to rely on our buddies.

**The Divorce of CIA and Itek**

From the beginning of Itek's work in satellite reconnaissance, we worked mostly with the Central Intelligence Agency (CIA). There was some interaction with the Air Force because the Air Force performed the recovery activities, but that interaction was limited. The technical control, definition of objectives, contractual monies, and other support all came from the CIA.

Some might have argued that Itek (and other contractors) controlled the CIA to some extent. That was not the case. We gave the CIA our best technical advice. It was one thing for the CIA to set technical objectives; it was another thing to be able to achieve those objectives. If the CIA told us to do something impossible, it was our responsibility to tell the boss what we could and could not do. The technical objectives (e.g., the quality of lenses, how quiet the system could be) all had to be satisfied by people who technically knew what could be done. The CIA realized we were the experts in the optics field, and we had a good relationship with the CIA in that regard.

*I believe that cooperation was the key to getting things done in the Corona program. We were dedicated and focused on the program, and we had a great deal of teamwork. We had confidence that the fellow beside us was going to do his job, and we were going to do ours.*

Unfortunately, Itek's relationship with the CIA became strained during the early days of developing the Corona replacement system. The difficulty arose when the CIA wanted to impose organizational arrangements that would, at least in their view, better control security and give the Agency more direct control of the contract. This ran counter to Itek's organizational interests and philosophy, and this disagreement resulted in a divorce between Itek and the CIA.

Even though the CIA said the issue was security, I had the impression that the Agency was going to start telling us how to run Itek. We could not accept that. When I say we, I exclude myself because I was not involved in these decisions. My role involved engineering, and making sure all the technical aspects of the system were satisfactory. However, I regularly communicated with people like Walt Levison and Frank Lindsay who were in the upper management of the company.[3]

---

[3] Walter J. Levison is a Pioneer of National Reconnaissance (inducted into the NRO Hall of Pioneers, 27 September 2000). See Chapter 31.

Itek's management and some CIA people tried to work out a compromise, but it was not possible. Itek did not want to follow the management philosophy that the CIA attempted to impose. Consequently, Itek decided to back away, and it proved to be a very traumatic experience. We had a large group of people working on CIA-related projects, and I had to break them up. The Agency shipped our work to another company, and I do not know what became of our design after that.

In my opinion, the "divorce" led to the gradual downfall of Itek. As a result of the CIA-Itek split, Itek could not afford to purchase the kind of equipment and facilities that were necessary to be competitive on classified government contracts. We did develop some sophisticated photographic equipment in support of the Apollo Lunar Orbiter and the Mars Lander for the National Aeronautics and Space Administration, but our national reconnaissance work essentially came to an end.

Itek subsequently became a small company that later combined with other companies. By the late 1990s, the residual of Itek became part of Raytheon in Lexington, Massachusetts. There was a small group that was still called Raytheon Itek. That group still did reconnaissance work, but on a much different scale of activity than when reconnaissance was a full-blown effort for the company.

**Impact of My Work on My Family**

After the declassification of the Corona program, my family finally learned what I was doing for all those years when I was unable to speak about my work. This helped them develop a better understanding and appreciation for the importance of the work and its contribution

Figure 32-4. Pioneer Francis Madden and his family in front of the Corona Imagery exhibit at NRO Headquarters, Pioneer Recognition Day, 27 September 2000. (Photo by Sara Judy, NRO Visual Design Center.) Back Row, left to right: Susan Madden, daughter-in-law; Gregory Madden, son; Elaine Schatzel, daughter; Anne Madden, wife; Frank Madden; Annette Jackson, daughter; William Madden, son; Richard Jackson, son-in-law. Front Row, left to right: Kara and Kerry Madden, and Alanna Jackson, granddaughters; Jake Schatzel and Ricky Jackson, grandsons; Susan Madden, daughter-in-law, Elizabeth Schatzel, granddaughter.

to our country. However, at the time when my children were young, the nature and demands of my work did not seem to matter much to them. All they knew was that the old man went to work and came back. They got an ice cream cone once in awhile, and everything was all right. But it was very different for my wife. It was difficult for her because at the time we had young children, and I often stayed at work for two days consecutively. I rented a couple of rooms across the street from the plant so I could go over there to lie down and take a four or five-hour nap before going back to work. It was better to stay near the plant instead of trying to get on the highway to go home, only to have to return. There also were many times when I found myself unexpectedly called away from home. I would be gone for a day or a week, and my wife could not call me. I would tell her to call my secretary if she needed me in an emergency. It was very difficult for my wife, and probably the other families as well.

People often fail to recognize how difficult this type of life and the associated strains can be for the wives. I have to give my wife a great deal of credit for having persevered. I had to disrupt vacations and other plans so often that finally we ended up buying a cottage in New Hampshire. During the summer when the children were out of school, they and my wife went there and I would show up when I could. This was one way my family coped with the fact that I was away so often.

**Sharing My Experiences After Retirement**

After leaving Itek, I worked for a company called Compugraphics, which was involved with the application of optics and film transports. I was with that company for quite a few years, and I ended up being its technical liaison between the United States and Europe. That was a nice way to end my professional career.

After I retired I started a program called Reseed at Northeastern University. This is an acronym for the long-winded name Retirees Enhancing Science Education by Experimentation and Demonstration. The program's objective is to bring together older people, generally retirees, who share their experience with students in terms of teaching physical principles. We also perform experiments to illustrate the principles we are teaching. Retirees can draw on their knowledge and experience to address questions like, "What is momentum about?" "What are force and acceleration?" "Why is it important to know something about these kinds of things?"

The Reseed program has been successful because it is good to get "old guys off their rear ends" and make them think. The students look to us as grandfather figures. We look at things from a different perspective, because many of us have had quite challenging and rewarding life experiences behind us. I believe the program is excellent for the students, as well as for us older people. The classroom teachers also are quite enthusiastic about the program and they see value in it. These new opportunities and contributions keep me going and allow me to make use of some of the experience I gained over the years.

**Conclusion**

I was kidding with my daughter recently, and I said that my wife probably would never have fond memories of Itek because it disrupted her life for so long. But the fact is that my wife and family are proud of my involvement with national reconnaissance. Most things we do in life are happenstance. I could have been doing something entirely different and perhaps made more money, and perhaps less. However, as a result of the path I chose, my grandchildren know that I was involved with something that was significant for the country. That should be impressive to my grandchildren.

—

## Pioneer Award Presentation and Citation

Figure 32-5. Pioneer Francis Madden (second from the left) being recognized at the 2000 Pioneer Recognition Ceremony. The pioneer award plaque was presented by DNRO Keith Hall (left) and DCI George Tenet (third from the left). Dr. Terry Heil (Senior Vice President of Intelligence Programs, Raytheon) joined in the presentation. (Photo by Sara Judy, NRO Visual Design Center.)

**Francis J. Madden**

As chief engineer of the Itek Corporation's camera systems development program, Mr. Francis Madden directed the design and production of the Corona cameras. He solved technical problems that threatened the camera in space, such as the Corona electric discharge associated with the film rollers, which assured success of the nation's first space imaging system.

*Career in National Reconnaissance: 1956-1975*

# James T. Mannen

James Mannen led the National Reconnaissance Office's development and operation of a major Air Force imaging reconnaissance satellite, and also contributed significantly to the success of a signals intelligence satellite system. He helped deliver a vital new reconnaissance capability to national collection assets that quickly became a demand-driven system. He directed the advancement of improved target tasking strategies, enhanced ground resolution, increased output capacity, and improved system reliability. In his recollections, Mannen shares lessons from his career and why achieving the rank of General was not as important as feeling rewarded by his work.

## Managing National Reconnaissance Programs[1]

I entered the Air Force in 1965 after working with the Navy as a co-op student. I told the Air Force my preference was to do research and development. My first assignments were doing communications with Air Defense Command in Duluth, Minnesota. I then went to the Air Force Training Command and obtained a masters degree in Electrical Engineering through an engineering graduate school program the Air Force had just instituted. After completing my education, I began my satellite work at the Vandenberg Launch Base launching Thor/Burner IIA Boosted spacecraft for what became the Defense Meteorological Satellite Program (DMSP). One year later, I became the field manager for a launch program run by Captain (later Colonel) Gary Geyer, supporting Pete Wilhelm from the Naval Research Laboratory (NRL).[2] My work on satellite reconnaissance systems continued through my Air Force career.

### Supporting an Imaging Reconnaissance Satellite Program

My work as the program manager for an imagery intelligence (IMINT) program became my greatest contribution to national reconnaissance. It was a very busy and rewarding time.

[1] This section is based on an interview with James T. Mannen at the National Reconnaissance Office Headquarters, 26 September 2000.

[2] Later, I was reassigned to Vandenberg to run additional Titan launches. I launched various vehicles and payloads from both pads between 1981-1983. Gary Geyer and Pete Wilhelm are Pioneers of National Reconnaissance (inducted into the Hall of Pioneers, 27 September 2000). See chapters 11 and 19, respectively, for their recollections.

There were few weekends off. This system required state-of-the-art developments for many of its components. I believe that our team was able to develop a system with unprecedented intelligence capabilities that contributed greatly to national security.

Years later, after I had been at Vandenberg for eighteen months of what I thought was my final tour before retirement, General Kulpa approached me.[3] He said, "I need you to come back to Los Angeles at the two-year point." Upon hearing this, my wife started crying! But it was an opportunity to work on a challenging reconnaissance satellite program, and I accepted the assignment. I later became the program's Deputy Director and Program Director.

*Budgeting and Programmatics.* I learned a lot in this assignment about programmatics and budgets. The Air Force was competing to keep this satellite program, and consequently we had to minimize our budget. We put together an architecture on a shoestring budget with bailing wire and spit. We were going to use existing communications resources, existing mission control complexes, and existing multi-mission modules that Fairchild had built for the National Aeronautics and Space Administration (NASA). On paper, that approach appeared very cost effective. In reality, it needed a lot of development. It was missing a lot of redundancy and it needed a lot of testing.

There was a serious budget problem at the time, but the program still got its additional funding. We promised Congress that we would not ask for any more money until the launch of the first vehicle. That obligation put a lot of constraints on the program, although my impression is that beginning in the 1990s the constraints on programs became much larger than those we had to deal with. Comparatively, it seems as if we had an infinite amount of money. The constraints in more recent NRO programss seem to be that the leadership lacks the authority that our program directors had, namely to spend money and make decisions.

*Finding a Market.* When I became Program Director, it was obvious that consumers in the user community were not yet sold on our system. They wanted intelligence products related to foreign technical research and development. The United States (U.S.) military was enamored by the system, but I believed that the product still needed a lot of improvement, especially on the ground. We recognized that the products in reality were not quite as good as they had been promised on paper, and we had to show the customer that the product could be useful. We eventually discovered that the products we had developed were not utilizing the total capability of the satellite, and we worked to fix that situation. I still remember our briefing to the Military Intelligence Board, when we clearly explained the advantages of our new system to the military representatives. A military crisis at the time helped to sell the system to the military.

*It is okay to be wrong in this business, but never to be uncertain, because no one will support you if it costs money.*

*Responding to the World Situation.* During a visit to a ground station, I recognized that we prioritize images and process them in that priority order for the user community. There are often very short timelines in which to get the images to military customers doing bombing missions and other tactical operations.

We decided to increase support to these military operations by increasing our system operations. Two very bright contractors figured out an innovative solution that allowed us to do this. I promised Washington that our approach was absolutely doable, even though our folks were still in the process of confirming that it was. Sometimes it is necessary to commit before the t's are crossed and the i's are dotted. It is okay to be wrong in this business, but never to be uncertain, because no one will support you if it costs money. As it turned

[3] John E. Kulpa, Jr. (retired as Major General) served as Program A Director from 1975-1983.

out, my insistence paid off, everyone was pleased with the new approach, and we launched the satellite in record time.

Figure 33-1. Cape Canaveral Air Force Station, Florida. (Photo courtesy of NASA.)

**Supporting Signals Intelligence Satellite Missions**

During my signals intelligence (SIGINT) work, I witnessed many folks taking responsibility and coming up with innovative solutions to various problems, such as improving SIGINT ground architectures and developing new vehicle control techniques. I believe it is very important always to question assumptions, and both my government and contractor colleagues displayed this same attitude with their ideas for developing and operating these systems. I also enjoyed amazing flexibility in terms of streamlined, competent and resourceful decision making, which allowed the efficient implementation of several revolutionary and innovative ideas. One event in particular stands out in my mind as a highlight of this assignment—a hurricane that brought unexpected drama.

When this hurricane came along, one of our launches from Cape Canaveral had to be delayed for some time. The mobile service tower and the rocket were designed for 100-knot winds, and they were expecting 120 knots, so we were expecting to lose the rocket, the satellite, and the launch pad. We had to evacuate 150 contractor personnel to Orlando as a part of the evacuation of all of Titusville, the Cape, and Melbourne. We set them up in a hotel, and notified their home offices. We spent three days in the shelter of a concrete building, continually putting kerosene in the generators, because the Florida power company turned off everything at 50 knots.

What made this event so special was the camaraderie and teamwork displayed after the evacuation of the pad by all of the contractors, primarily Lockheed Martin. People from Boeing and Rockwell also came over to help. They left the keys to their machine shops and the inventory lists of all their tools. They made it very clear that they were offering anything needed by the group that was stuck there with the satellite and the rocket. They said, "You are free to go in and use our facilities and supplies. Anything to support the U.S. satellite effort." It was absolutely amazing support among contractors at the working level. They made sure that nobody would run short on tools, facilities, or capabilities. It was one of the most emotional events of my satellite work.

Working together, we came up with a method of measuring the damage and proceeding with the launch. We put a monofilament line on the rocket, and used clothespins all the way up and down. There was no power to measure the deflection of the rocket with respect to the mobile service tower. With that line, we could measure the maximum deflection of the rocket with respect to the mobile service tower. When it was over, we had a record of the deflections. We used that data to clear the booster of potential overstress after the hurricane.

The booster itself suffered very little damage. The hurricane blew the doors out of the mobile service tower, but we just taped all the interfaces and put the wind-humidity-temperature gauge clocks inside the fairings. Then we took the liquid nitrogen off the pad, plumbed it up into the fairing, and let gaseous nitrogen boil off into the fairing. We had to recycle the batteries. We were concerned that they might have been off charge too long. We were concerned about the humidity and temperature, but the gauges showed that they were okay. We then went back through the complete satellite test sequence, the launch pad sequence, and launched about a month later.

After this hurricane, the Air Force committed to building a mobile service tower to last well above 100-knot winds. When another hurricane came through the area over ten years later, the Air Force still had not done anything! Neither had NASA. The Air Force almost lost another system in a more recent hurricane.

**Lessons for Managing Programs**

There are several areas of program management that will challenge the 21st Century NRO. I believe that my experiences can provide some lessons on how to conquer these challenges. These include being responsive to technological development, dealing with bureaucracy, balancing funding with performance, and maintaining acquisition authority.

*Being Responsive to Technological Development.* It is more important—and more of a challenge—to be responsive to technological development than to be on the cutting edge of technology. We built and delivered many NRO systems. They met all the requirements, but we were never happy with them. That a system met the specs and exceeded mean mission duration was good, but not good enough. The real questions were, "How do we make it last even longer? How do we make it serve the user better? How do we make some new idea work?" Look at what the systems were supposed to do. Then look at the people who, through innovation and technology, were able to maximize these systems' performance, resolution, signal designs, gain, lifetime, etc. It is amazing. My view of the NRO today is that it contains people with really good ideas on how to improve constantly the systems they build.

A lot of the new technologies are unique to the NRO and will not be developed commercially. Is the NRO as responsive to new technology as it should be? Is it as responsive to change as it used to be? The answer to both questions is probably not. The NRO needs to retain people with really good ideas on how to improve constantly the systems they build, so that the organization can be responsive to technological development.

*Dealing with Bureaucracy.* The inherent challenge in dealing with bureaucracy is moving proposals from concept through development and implementation, while remaining flexible to new and good ideas. I believe the NRO's ability to do that declined in the immediate post-Cold War era. For example, program managers used to say, "Go do this," and we would have a contract turned on to do it by the end of the day. The program directors had powerful authority and a lot of responsibility, but that authority has declined. The main cause of this change is budget cuts, although I believe it is just a bureaucratic factor.

Good ideas do not come just from one set of people, but from many different folks. The NRO needs to create a flexible environment to foster these ideas while avoiding problems such as increasing requirements without budgets. During my tenure, innovative proposals went to the NRO management who called the appropriate external authorities, and it was done. I expect that it has become more difficult to get those types of proposals through the bureaucracy. The reason for this may be that the newer generations of national reconnaissance managers did not see these processes when they worked better.

I have heard from NRO contractors that it has become more difficult to process contractual change in the NRO. Time is money. Time is people. Paperwork and decisions also require time. Creating an environment where it takes twice as long to process a change will result in more expensive and time-consuming processes. The NRO must find efficient ways to make program decisions and execute the contractual paperwork. Its people generally want to make and operate satellite systems, not perform bureaucratic work, because they believe that mission is very important.

*Balancing Funding with Performance.* "Smarter, faster, cheaper, and better." Systems development and acquisition people want to believe in these flashy words. I believe that a cheaper, better satellite is one that is designed and programmed to last two years and ends up lasting eight. Despite the NRO's continual innovation, the organization must be careful when it goes for systems that are "smarter, faster, cheaper, and better," because better is very seldom cheaper. If the NRO follows that motto blindly in developing these assets that are so critically important to our nation, it will not be innovating. The NRO must make sure that it understands the ramifications of all of the factors affecting a program.

I believe that experimenting in acquisition streamlining is a questionable decision. If the NRO is going to build these national assets successfully, can it afford to have these "experiments" in acquisition streamlining? Can it really afford to have any of its satellites fail? During my imagery intelligence (IMINT) program, we constantly were moving satellite acquisition start dates, and it saved a huge amount of resources whenever we moved a satellite acquisition start date out of the five-year defense budget. So "cheaper and better," in my mind, is not a shortcut to building a reliable satellite.

Consider a satellite that has a million and a half pieces. If its builders cut back on its testing and redundancy, there is very little hope that it is going to last seven or eight years. I believe that none of the national assets on orbit in the year 2000 would be operational if there were no redundancy. It is just that simple. Because usually we did not have any insight into why a system would fail, we had to make everything redundant.

*Exercising Acquisition Responsibility.* The NRO has unique responsibilities, and during my career had unique acquisition authorities and capabilities to accomplish its missions. However, the NRO has lost those authorities since the end of my career. I do not believe that it will ever recover those authorities, the lack of which can result in decreased system efficiency and product quality. The government's responsibility in acquisition is to ensure that it is getting the product that it contracted to buy. When there is a difference among technical advisors, the government must have the experience and knowledge base to be able to make a decision.

**Reflections on Air Force Career Management and the National Reconnaissance Office**

During the initial post-Cold War era, the mission of the NRO continued to be as important as it ever was. To accomplish its mission, the NRO must select and retain good people. What the NRO needs are professional military and civilian employees to effectively manage a professional contractor force. Building, launching, and operating satellites require experienced, dedicated people who believe that the NRO is more than just a job. I contend that the Director of the NRO (in his Air Force role) has the capability to develop a program to motivate, train, and retain the good acquisition officers, engineers, and contracting officers that the NRO needs to do its job.

I worry about career management, because of its importance to maintain a successful acquisition agency like the NRO. The Air Force must emphasize career management for its NRO folks. Through my own experience, I have identified several factors that contribute to a successful personnel management program, namely the importance of senior leadership support, job satisfaction, broad satellite experience, and assignment duration.

*Value of Senior Leadership Support.* The leadership style of senior managers is an important factor in personnel management. Knowing that someone at the two-star level cared about my career was motivating. During my Air Force career, this was an important factor in my satisfaction and success. There were three Air Force leaders who had a significant impact on my career through their vision of Air Force career management: General Kulpa, General Jacobson, and General Lindsay.[4] Their vision was basically that career officers should build experience in satellites, rockets, and operations for their entire careers, so that they would understand how things worked when they became a deputy program director or director. They needed to understand the mission in the launch base, the operations at the Satellite Control Facility, budgeting, the user community, and the work of the NRO staff, and program offices—the full system and not just the pieces.

*I was not going to make General because I did not want to go to Washington. What I really wanted to do was to build satellites successfully and on time, because that work was worthwhile.*

One example of the active interest these seniors took in my career occurred when I was a captain. General Kulpa called me into his office because I did not get the highest scores on my Officer Evaluation Report (OER), which was during the time of controlled OERs. He said to me very clearly, "Don't worry about your career—worry about the projects you are working on. You will get to be a program director if you stay on and do the kind of job you have been doing. In about fifteen years, many of your Air Force colleagues will reach twenty years for their career and retire. What you will need then is program office experience, launch-base operational experience, or satellite control facility experience, and then come to Washington." He wanted me to gain an understanding of all of the pieces of satellite systems, because the program directors at that time controlled all of the program components, including the rocket, the satellite, the acquisition, the ground stations, and the budgets.

I know that General Kulpa did care, because I went to Vandenberg in 1981 instead of joining the NRO Staff in Washington. He ranted and raved at me because I was supposed to go to Washington, and he looked after his folks and had long-term plans. The NRO generals always were fighting to make sure that their people got their share or better in Air Force assignments, educational opportunities, and promotions. They always maximized our chances for advancement.

I believe that sometimes Air Force leaders lose sight of the importance of the personal dedication of most NRO Air Force personnel when they say, "If these people do not rotate around, they would not make General." Most people want to take care of their family. But they want a job that is rewarding, and they want to know that the people in their job care about them. People will do anything for an organization that supports them and has a strong sense of its mission.

*Importance of Job Satisfaction.* Air Force officers at the NRO had personnel systems that created an environment and opportunities that did not exist anywhere else, and also helped get us promoted better than the rest of the Air Force. If I had a choice between the NRO and Space Command, I would have taken the NRO every time. At the NRO, I had assignments that gave me great job satisfaction. For example, as a captain and a project officer, I was in charge of a multi-million dollar project that was the key to the success of a major sat-

[4] These three men all served as Program A Directors. John E. Kulpa (retired as Major General) served from 1975-1983. Ralph H. Jacobson (retired as Major General) served from 1983-1987. Nathan J. Lindsay (retired as Major General) served from 1987-1993.

ellite program. Where else would a captain get that kind of opportunity? A lot of Air Force folks who worked with me as captains and majors in various Air Force Special Projects (SP) elements looked at the rest of the Air Force and said, "If I go out there, I am just going to be a captain. Here, I am a contracting officer's technical representative for a major satellite acquisition of national importance!" I could not have had the opportunities elsewhere that I was afforded in the NRO.

Because of job satisfaction, the NRO can retain an incredible number of very highly motivated people who do not care whether they make general. Frankly, if Air Force officers at the NRO have the opportunity to spend twenty years building important NRO satellites, most of them would be happy with that opportunity. Most Air Force people do not work at the NRO to make general. Making general is something that happens if it happens, so to speak. There were several things in my career that were very clear. I was not going to make general because I did not want to go to Washington. What I really wanted to do was to build satellites successfully and on time, because that work was worthwhile Someone once told me that the difference between the NRO and the Air Force is that the Air Force is practicing for war, while the NRO is at war—a technology war and an intelligence war. This is a war for which we must provide our decision makers as much information as we can in order to stay ahead.

*Need for Broad Experience in the Satellite Business.* It is important for Air Force personnel to gain broad experience in the field of national reconnaissance. In the earlier years, the Program A Director also served as the Deputy Vice Commander for Space Missile Systems Command. Therefore, SP personnel had the choice of working at the launch bases, the Satellite Control Facilities, or Space Missile Systems Command. There were many Air Force opportunities to stay in the satellite acquisition and operations business, and still be on the Personnel Control List, which was used to track a group of experienced Air Force satellite experts.[5]

In fact, I was on that list from 1972-1995. The evolution of Air Force assignment system since my retirement has made that impossible, as the Air Force started to move people around regardless of past experience. The assignment takes into account an individual's expertise and experience and their relevance to the Air Force mission. However, the system's effectiveness has declined. We had people who figured out a way to avoid losing their folks. The Air Force and NRO senior management wanted people who understood the aspects of satellite and rocket acquisition business, and consequently tried to retain them in the field.

*Criticality of Assignment Duration.* I believe that assignment duration is a critical factor for organizational success, and I heard a story that illustrated its importance. It was from a senior Central Intelligence Agency (CIA) person who said, "Statistically speaking, the Air Force Program A had been more successful in initially assigning better quality people than we were able to do at CIA. But the Air Force rotated them out after a relatively short tour. The CIA folks stayed for a long time, and after several years they were much better equipped to deal with the issues. They were more knowledgeable than the Air Force folks because of this experience, although the Air Force folks were better qualified initially because the NRO Program A had the entire Air Force from which to select its people."

If the NRO Air Force personnel continue to have short assignment durations, it will result in a lack of subsequent long-term experience and be a problem for the NRO. The NRO created multi-million dollar systems, and then assigned people with no experience to operate them. One example is that people assigned to the Heavy Lifts Special Operations Program

[5] The Personnel Control List was a mechanism to track and retain individuals with critical experience and expertise. The list was used by Program A to ensure that those assigned to the NRO could stay within the NRO or easily could be brought back to the NRO for subsequent assignments.

under Air Force Space Command spent three years in the program, but they never had the opportunity to use that experience in another assignment. This assignment system minimized their value added to that program and their later assignments. This caused Air Force personnel to feel that they were being forced out of the national reconnaissance business because of these assignment processes. They were on assignment for four years, and felt they would destroy their careers if they stayed in the NRO.

**Impact of My Work on My Family**

Despite some obvious burdens of my work, its demands did not have an excessively negative impact on my family life. The secrecy never really bothered my wife or children, and I tried to make the best of the time I got to spend at home.

*A Dedicated Wife and Mother.* Without the support and strength of my wife Lynne, my successful national reconnaissance work would not have been possible. Lynne was always willing to pick up a lot of slack from my absences. She was more interested in whether I was happy than in the details of what I was doing. If I were happy, that was good enough. She made life at home very, very pleasant. It took her no time at all to become a very independent person. The one thing I learned is that if Lynne made decisions when I was gone, I had better be very careful in criticizing when I came home!

*The one thing I learned is that if Lynne made decisions when I was gone, I had better be very careful in criticizing when I came home!*

*Security.* The only problem we had with security was early on, when we could not tell our spouses where we were going. Lynne would have the phone number of somebody in my office so that she could get a hold of that person, who could get a hold of me if there was an emergency.

On one trip to Florida, my older daughter fell through a glass window at a neighbor's house. They were running around in stocking feet at a slumber party and had two big glass panes by the front door, and she went completely through the pane. She had to have about fifty stitches. Lynne could not get hold of me, and she was not happy about that. When I got home, she told me in no uncertain terms that I was never going to go anywhere without giving her a contact number. I went back to the Program Director and security people and they said, "That's fine."

Lynne was not as concerned about not knowing what I was doing. Our kids knew I was in the Air Force, and they knew I had something to do with satellites. They were more concerned with whether I was going to be at the swim meet, the football game, or the soccer game than what I did at work. It was years before Lynne knew specifically what I did, and at my retirement the Air Force gave me what probably was the first NRO desk set.

*Time Away from Home.* The main reason that my inability to discuss my job was never a problem at home was that even if I were permitted to talk about my work with my family, I did not have the opportunity—I was traveling almost every week. Even when I was assigned permanently to the launch base—to take a break—I still worked almost every weekend. I will say, though, that when I was home, I did not watch football games. I was out somewhere with the kids, playing soccer or attending swim meets. The things I gave up as a compromise were my hobbies and my free time when I was home. I took my wife Lynne to dinner, in spite of the fact that I had eaten dinner out every night for the last four days.

**Conclusion**

I asked myself, "Why am I a Pioneer? I am a mediocre engineer of average intelligence. I liked engineering, but what separates me from the crowd?" I do not know the answer to these questions, but I do know that I would not be a Pioneer without the career opportunities I had: to be at the launch base, to work in satellite operations, to be in a program office, and to participate in both SIGINT and IMINT programs. After I had done all these things and been in these types of jobs, then I had the necessary tools to succeed. I worked with good people who created an environment for motivating their colleagues.

—

**Pioneer Award Presentation and Citation**

Figure 33-2. Pioneer James Mannen (second from left) being recognized at the 2000 Pioneer Recognition Ceremony. The pioneer award plaque was presented by DNRO Keith Hall (left) and DCI George Tenet (third from the left). Brigadier General Craig P. Weston (Director, Corporate Operations Office and Chief Information Officer) joined in the presentation. (Photo by Sara Judy, NRO Visual Design Center.)

**James T. Mannen, Colonel, USAF**

As director of a vital imagery satellite program, Colonel James Mannen introduced procedures that improved target tasking, and significantly increased ground resolution and on-orbit system reliability. During Operation Desert Storm he launched a new satellite in record time to support U.S. and Coalition Forces.

*Career in National Reconnaissance: 1971-1993*

# Mark Morton

Mark Morton directed the General Electric team that developed, tested, and built the reentry vehicle and recovery sequence for the Corona photoreconnaissance satellites. Faced with a complex challenge, it was his dedication, innovation, engineering, and management skills that contributed to America's successful recovery of pictures of the earth taken from space in the 1960s.

## The Struggle to Recover Corona Film[1]

It has taken those who worked in the early years of national reconnaissance a long time to begin to talk about what they did in the first programs, such as Corona. When the Central Intelligence Agency (CIA) recognized us for the first time in 1985, the recognition was still classified. The Corona declassification event in 1995 was the first public recognition of our efforts, but it was hard to open up to the public. Only now are people becoming comfortable with the idea of publicly revealing their work on these systems, and I am pleased to have an opportunity to discuss my own role in establishing the success of the first photoreconnaissance satellite.

I was so immersed in the challenge to develop a space-based national reconnaissance capability that I did not realize until later the dedication of all the people who were working on the project. All of the Air Force, CIA, and industry personnel displayed amazing dedication. I do not believe we ever will see an era like that again. It was really something! However, before our efforts were initiated, the requirement for a national reconnaissance capability had to be defined, and the need for satellites had to be confirmed.

### The Requirement for National Reconnaissance

During the early days of the Cold War, it was difficult for intelligence organizations to learn what weapons systems the Soviets had developed and built. The intelligence information available at the time suggested that the Soviets had acquired almost all the equipment and capabilities that the United States (U.S.) had hoped the Soviets would not be able to acquire. For the U.S. to duplicate all Soviet possibilities was neither practical nor affordable. President Eisenhower wanted a far more accurate assessment of the Soviet threat. He established a policy for urgent airborne reconnaissance, which included the U-2 project. The Soviet shoot-down

[1] This section is based on an interview with Mark Morton at the National Reconnaissance Office Headquarters, on 26 September 2000.

of Francis Gary Powers' U-2 aircraft over the Soviet Union in May 1960 confirmed for U.S. national leaders that a capability to use space-borne sensors to collect overhead intelligence on the Soviet Union was urgently needed.[2] The development of the Corona and other satellite reconnaissance systems soon followed.[3]

*To see this system work—to get the film back from orbit without having it damaged from reentry—to have it developed, and to look at the pictures—that was phenomenal!*

**The Challenge of Space-based Photoreconnaissance**

It was a long, tough program involving numerous problems and challenges. It is hard to imagine our problems back in those days when I look at the status of technology at the beginning of the 21st century. During the mid-20th century, we had no way to take relatively high quality pictures from space and electronically transmit them back to earth. As a result, we came up with the idea to take pictures in space with a traditional camera and gather the exposed film into a Reentry-Recovery Vehicle (R/V) that would eject with its own rocket system. The R/V would be maneuvered into a recovery trajectory, and the film would be protected from the heat and deceleration of reentry. After reentry, the R/V would deploy a parachute to enable a mid-air snatch by specially rigged aircraft. A challenge at any time, but back then we had never recovered a vehicle from orbit. We went through a dozen flights without successfully recovering the R/V.

**The Recovery Solution**

We had to figure out how to make this system work. At one meeting, an Air Force officer said to me that perhaps we could not get the R/V back because of an esoteric astrophysical principle that prevents something like that from returning to Earth once it is in orbit. I told him, "It's only our stupidity in finding out what we're doing wrong." We worked 24 hours

Figure 34-1. A C-119 air-catch of a Corona film capsule in the 1960s. (Photo courtesy of NRO History Office, likely U.S. Air Force photo.)

[2] Francis Gary Powers, a CIA U-2 pilot, was shot down near Sverdlovsk on 1 May 1960.

[3] The final proposal for the Corona program was approved in 1958, and the first Corona launch occurred on 28 February 1959 with the first successful R/V recovery on 19 August 1960. Much earlier in 1946, RAND had conducted the first satellite feasibility study.

a day on this, and we finally recovered the R/V on the thirteenth launch—a very difficult undertaking. I do not believe the program politically would have survived past the third failure during more recent times. To see this system work—to get the film back from orbit without having it damaged from reentry—to have it developed, and to look at the pictures—that was phenomenal!

Figure 34-2. Pioneer Mark Morton during interview session, Pioneer Recollections Day, NRO Headquarters, 26 September 2000. (Photo by Candi Campbell, NRO Visual Design Center.)

### Conclusion

When the Corona system began to operate successfully, we hardly could believe the results. We thought we were in heaven! After that first success, we were off and running. Gradually, the NRO developed technology that allowed imagery to be transmitted electronically, which meant R/Vs no longer were needed. But I still remember living through those very early days of non-stop struggles with the R/V. It was tough, but the ultimate success was certainly worth the effort and the pain.

—

## Pioneer Award Presentation and Citation

Figure 34-3. Pioneer Mark Morton (second from the left) recognized at the 2000 Pioneer Recognition Ceremony. The pioneer award plaque was presented by DNRO Keith Hall (left), and DCI George Tenet (third from the left). Dr. Vance Coffman (President and CEO, Lockheed Martin) joined in the presentation. (Photo by Sara Judy, NRO Visual Design Center.)

**Mark Morton**

Mr. Mark Morton directed General Electric's Reentry Systems Division that designed, fabricated, and tested the reentry capsules used in the Corona satellite and in the subsequent Apollo lunar landings and other satellite reconnaissance programs. His efforts produced highly reliable reentry systems that ensured the uninterrupted flow of vital overhead imagery to the nation's leaders.

*Career in National Reconnaissance: 1958-1970*

# Charles L. Murphy

Charles Murphy served as the first Field Technical Director at the Corona Advanced Projects Integration Facility. He was a crucial link between Corona operations and the strategic and tactical needs of the Intelligence Community. Colonel Murphy recalls some of the events that contributed to his success in establishing Corona as an operational satellite photoreconnaissance intelligence collection system.

## Commanding and Controlling Corona[1]

I entered the national security field around 1956. At the time, I was in the Air Force and was assigned as a crewmember on a Convair B-36 strategic bomber stationed at Carswell Air Force Base in Fort Worth, Texas. Out of the blue, I received a phone call and was told, "Come to Washington, DC." Air Force Headquarters sent me a set of orders, a plane ticket, and a telephone number to call when I arrived. I called the number as soon as I arrived in Washington. The next thing I knew, I was being interviewed by someone from the Central Intelligence Agency (CIA). He took me to one of those old Washington, DC buildings that the Agency used at that time. I do not even remember where it was.

I was hired and ended up planning missions in the CIA's U-2 program as an operations officer. I remained an Air Force officer, but I was assigned to the CIA. We worked out of an office near the White House. Few of the U-2 overflight missions we planned actually were flown, because overflight of denied territory was extremely sensitive.

Besides doing the mission planning, I also was assigned to a small Agency team of camera people and other experts. The team had its own Lockheed Constellation (known as a Connie).[2] We were the advance team, and we would fly into remote deployment areas such as Alaska or the Philippines. Our job was to plan the U-2 missions that would be launched. Some of the missions were overflights of denied territory, so it was an interesting job.

### Introduction to the Corona Program

I had been with the U-2 program for a few years when I was tapped on the shoulder and told about a sensitive photosatellite reconnaissance program called Corona. It seemed very

[1] This section is based on an interview with Charles L. Murphy at the National Reconnaissance Office Headquarters, 26 September 2000.
[2] The Lockheed Constellation was a four-engine propeller-driven aircraft.

Figure 35-1. The U-2 reconnaissance aircraft. (Photo courtesy of NRO History Office, likely U.S. Air Force photo.)

interesting, and I was shown a list of those who were working on the program. When I looked at the list of people who had access, there were fewer than fifty names. The President of the United States was on it, but the Vice President was not! I felt honored to have my name on that list.

This was an exciting and challenging time. Corona offered such possibilities for the Intelligence Community! I had to figure out how to use this new capability and integrate the Corona system with the requirements of the Intelligence Community. Basically, we needed to direct Corona to take the pictures we wanted.

For years, the original Corona mission operations center and the U-2 mission operations center were co-located in the same building in Washington, DC.[3] After making many trips out to the west coast to do planning for Corona, Richard Bissell said to me, "I think you better go out there." Around 1959 I transferred to Palo Alto, California. At least that's what my orders said. Actually, it was the Lockheed Advanced Projects Integration Facility. We called it the Skunk Works, stealing the name from Kelly Johnson's Skunk Works facility in southern California where he built the U-2. The security cover for my duty assignment was "Air Force Satellite Control Facility."

This was the facility we used to control Corona on orbit. We had to walk through the commander's office to get to the CIA communications center at the facility. One communicator and I were the only people allowed in the place. It was interesting once I got used to it. Talk about operating out of the back room—we really were! For about the first year, my total government staff included a security officer and one CIA communicator. That basically meant we had a 24-hour duty day all the time! The Agency finally assigned Vern Webb as my deputy, and he was a great deputy and a welcome addition to our team.

### Developing a Reports Control Manual

The first thing I did at the Corona Skunk Works was to write a Reports Control Manual. It provided a simplified approach for communicating through classified channels, and for providing answers to the four Ws: who, what, when, and where. The Reports Control Manual instructed imagery users how to identify their imagery requirements for us (e.g., when they had to deliver target lists of required imagery to us at Palo Alto). It also told the contractors when they had to have film or equipment ready. It was a control manual, but it did much more than that. It was an important information source for everyone in the system, from the users in Washington to the contractors and those who had to worry about getting the system in order.

Another early requirement for me was to learn how to operate the Corona satellite system. A satellite's capabilities are quite different from those of an aircraft. There are a few similarities,

[3] Mission control of Corona's photoreconnaissance activities—done in conjunction with the Committe on Overhead Reconnaissance—was different from the orbital control of the satellite that was done elsewhere.

but there is no question that there are different things to worry about. One of the first subjects I had to teach myself was orbital mechanics. That was not too difficult because I was a celestial navigator. I learned that eccentricity was not a description of a crotchety old man, but an important way of describing the shape of an orbit.

### Selecting Optimal Launch Dates and Times

For Corona operations, everything centered on the launch dates and times, which I was responsible for determining. The dates were slipping and changing constantly, because of technical problems. I also had to figure out the effects of launching at different times of the day, because the satellite obviously had to be over the target during daylight. I finally realized a very simple fact, namely that half of the earth is illuminated at any given time. Consequently, we planned each launch so the satellite's orbit would synchronize with the sun at our selected time of day. If we picked the right time of day, we would launch at times that would allow the satellite to image both sides of the globe.

We used this technique successfully against the Soviet Union. In the summer, much of the northern Soviet Union is illuminated on both sides of the globe. This provided a great opportunity because we could image the target on both ascending and descending passes. That idea alone considerably increased the amount of territory that was available for us to image. As summer approached, we were launching as late as 4:00 p.m.

I am sure not everyone understood why we adjusted the launch windows the way we did, because there were some disadvantages with our approach. For example, the satellite was at higher altitudes in parts of the orbit and sometimes the sun was very low. This orbit may not have produced the most optimal imagery, but I made it available to the analysts in case they wanted to use it. It certainly increased the amount of area available to those who made final imagery selections. In general, I spent a huge amount of time traveling back and forth to Washington to coordinate with the photointerpreters about what they wanted.

### Selecting the Orbit

In addition to selecting optimal times for launch, we also had to determine the best orbital configurations. We had to select orbits that were safe for the satellite while enabling good collection. It was safer for the systems if we injected the satellite into orbits at higher altitudes, and we launched at what was considered a safe-injection altitude for the satellites.[4] The engineers preferred higher altitudes, but the analysts always wanted lower altitudes because the resolution of the imagery would be better. This was a constant topic of discussion with the people back in Washington, particularly with the people at National Photographic Interpretation Center (NPIC), but concerns about the system's safety prevailed.[5]

*We had to select orbits that were safe for the satellite while enabling good collection... The engineers preferred higher altitudes, but the analysts always wanted lower altitudes because the resolution of the imagery would be better. This was a constant topic of discussion with the people back in Washington...*

### Controlling Imagery Operations

When we first started launching Corona, we had no control of the camera after it lifted off.

[4] In addition to altitude, other considerations included period (which determines the track across the earth's surface.

[5] The NPIC, operated and administered by the CIA, performed imagery analysis for the Intelligence Community. In 1996, NPIC was combined with the Defense Mapping Agency and Central Imagery Office to form the National Imagery and Mapping Agency.

The entire sequence, including when the cameras were turned on and off, was pre-programmed into the H-timer. It operated like a player piano, and metal "fingers" brushed against a mylar tape that had holes punched into it. When one of the "fingers" came into contact with one of the holes, it triggered an event on the satellite. It seems primitive today, but the H-timer worked very well and gave us some flexibility. Even before we obtained a ground command for the cameras, we did have a ground command to advance or retard the H-timer.

One of my early assignments in Palo Alto was to work with the program office to develop the capability for ground control of the camera H-timer. It took seven or eight launches, but we finally were successful. We wanted to image precise target areas on the earlier Corona missions, but we discovered that there was no way we were going to be over the target when we needed to be unless we were lucky.

We based our imaging of the target on how fast Corona progressed across the earth, as well as how fast the earth rotated under Corona's orbit. However, this planning method was very sensitive to small errors. Corona often operated beyond three standard deviations from the mean of the planned mission profile. Once we had the ability to control the camera from the ground during flight, we could not only choose the targets, but we also could choose targets based on near-real-time weather data. We were about a year into the program before we developed this capability. This new capability made a world of difference because we could be more selective in what we imaged. More importantly, it enabled us to save film, which was the most precious commodity we had on orbit.

*Someone suggested that the decisionmaking on imaging targets should be moved from Washington, DC, to Palo Alto. I did not believe adding this layer of bureaucracy was a good idea. All we needed to know was when to turn the camera on, and when to turn it off. Our control center consisted of a roll of dimes and a telephone booth!*

About the time Corona was modified to carry two recovery buckets, someone suggested that the decisionmaking on imaging targets should be moved from Washington, DC, to Palo Alto. I did not believe that adding this layer of bureaucracy was a good idea. All we needed to know was when to turn the camera on, and when to turn it off. Our control center consisted of a roll of dimes and a telephone booth! We sent commands to the Corona satellites through the Air Force Satellite Control Facility, which had a very elaborate command and control system. The commands were very simple: A1, A2, or something like that. These commands would tell the camera to turn on or skip a pre-program operation. It was fun working there.

**Real-Time Decisions**

The early, one-bucket Corona missions lasted only two days at a time. They were 24-hour days, and I was there for the duration.[6] Nothing was ever routine with satellites, and there were problems with almost every mission. I kept a cot in my car that I rolled into either the satellite test facility or the Skunk Works. If problems came up, and they invariably did, I had to coordinate quickly with the engineers to figure out how the situation affected the payload. We had to determine if we could still use the system or if we had to terminate the mission, and these were real-time decisions. One of the most fascinating tasks I had was working with all of the engineers trying to figure out what we could do with a few commands. We used to tell each other that we could solve any problem if only we had a 100-mile long screwdriver. Nevertheless, much can be accomplished when good people work together,

[6] The early Corona missions were seventeen orbits in duration.

and I knew how to find the people who knew how to work the problem. Whether a systems engineer or janitor, I would get them together in a room and tell everyone else to get out. I did not worry about informing the bureaucrats in Washington—we just did it!

During one mission, we launched a satellite into an orbit that was far from optimal. In fact, it was just plain bad. I thought that we could work with the H-timer to take pictures in the right place even if the orbit was unfavorable. I received a briefing from an uncleared Lockheed representative who offered some good advice. He said, "Hey, I can do this or that with the H-timer." Clearly, he was knowledgeable about that part of the system, and he offered ideas about ways to control the H-timer. I listened patiently and then said, "Hey that's great." Of course, I already knew I was going to do what he had suggested anyway. It was just amazing what we could do with those satellites to save a mission. We were able to take pictures during this mission almost exactly where we wanted to even though the altitude was not optimal.

**The Early NRO**

I was introduced to the secretive NRO right after the first successful Corona mission in 1960. We delivered the film to Eastman Kodak and then took the aircraft back to Washington. I had not seen the film yet, but I wanted to get some sleep. I was awakened by a phone call, and I was told, "Hey come on over here [to the Pentagon]—the Under Secretary of the Air Force needs to talk to you."[7] As I spoke with him I began to realize, "Hey, there is an organization within an organization here." It was the first time I was aware there was something called the NRO, but I am not sure anyone ever called it that. In fact, as far as I was concerned the NRO was a shadow organization. I did not pay attention to the rumors about what organizations people were from as long as they were cleared and briefed. If they had something we wanted, we dealt with them, and it made no difference to me if they were Army, Navy, Air Force, contractors, civilians, or from the NRO.

*If the NRO staff had tried to meet back then, I am sure we could have met in a phone booth. We were operating with ten or fifteen people [and] ...were using other people's capabilities all the time, but we never thought about drawing them into the organization.*

If the NRO staff had tried to meet back then, I am sure we could have met in a phone booth. We were operating with ten or fifteen people, but we had other people scattered all over the place. We were using other people's capabilities all the time, but we never thought about drawing them into the organization.

**Responsible for the Corona Program**

Around 1964 I transferred to El Segundo and became the Director of the Corona Program Office, basically taking Lee Battle's job.[8] I then became responsible for the entire system: the booster, the orbital stage, the camera systems, everything. Fortunately, by then we had resolved many of the initial failure modes. Things were becoming more routine, but never completely routine. Despite the redundancies, there were many single point failures on the system, and it would not take much to go wrong to have a disaster.

[7] The Under Secretary of the Air Force at the time was Joseph V. Charyk, who was serving as the Director of the NRO.

[8] Colonel Lee Battle is a Pioneer of National Reconnaissance (inducted into the NRO Hall of Pioneers, 27 September 2000). See Chapter 21 for his recollections.

Figure 35-2. Launching of a Thor-Agena from Vandenberg Air Force Base, 15 May 1962. (Photo courtesy of NRO History Office, original source unknown.)

**In Charge of Launch Operations—A Lot of Praying**

In the late 1960s, I was transferred to Vandenberg Air Force Base to be the Launch Director of Space Programs. I worked with 5,000 contractors and 1,000 military personnel. Our job was to get all the hardware delivered from the contractors, assemble it, check it out, and make sure the vehicle was ready to launch. We were not only launching Corona systems, we were also launching Titans, Thors, and other launch vehicles.

In launch operations, we were faced with real-time decisionmaking. The launches were computer controlled, and a launch could be routine if all of the procedures were followed. Even so, there were times when I had to make some tough decisions. There were two people who could stop a launch for non-technical reasons. I was one of them, and my counterpart in the program office was the other. Weather was often the problem, and we had established guideline parameters. If the conditions were outside the parameters then we did not launch. Most of the time the conditions were marginal, and I had to make the decision. There were several times we launched in spite of the weather, and I admit that I was praying a little.

On other occasions, the problem might have been related to me. During one mission we were about two hours from launch, and while we were fueling a contractor came to me holding a small cable with different connectors on it. He said, "Let me show you what I found this morning on another system. Some dunderhead mismated two plugs. The male part is shoved into the female end with no electrical contact. I don't think we could have passed a signal through if we had to." I asked if he had checked our system, or if there were any indications we had the same problem. He replied, "Nope, checked out." I recognized that he was trying to fireproof himself by placing the burden for the launch decision on my shoulders. The only information he gave me was that someone had mismated the plugs on another identical system. Though there was a possibility that the system on the pad could have the same problem, we proceeded. I admit, I was praying until the booster stage was completed.

One of the most frightening events occurred on a different mission when we were about two hours from launch. The Agena stage had been fueled, but the booster had not. Someone correctly activated "mast enable," which readies the system mast, including all of the fuel lines, to pull away from the rocket when it is launched. Somehow the launch system interpreted the activation "mast enable," as meaning we were ready for launch. The timers started to count down, and everything started running. Fortunately someone at the control panel had sense enough to keep resetting the timer to zero. Otherwise, within several minutes the whole assembly would have exploded, which would have been a quick end to the mission.

At this point we realized that we had another problem. There were two tanks in the Agena, the fuel and the oxidizer, and one was on top of the other. When the fuel and oxidizer come in contact, they explode and burn! That is how an engine starts, except we did not want to start the engine. We were concerned about a temperature buildup, so we had to plug the fuel lines back in and drain off the explosive mix.

We contemplated various ways to replace the fueling plug. We considered getting a cherry picker and lifting someone up with the plug. Of course, a cherry picker is not the safest of things. We then discussed rolling the gantry back to give the technicians access. If anything had exploded we would have blown up the whole pad, not just the missile. We did not have much choice though, so we decided to roll the gantry back into place so some brave guys could go up there and replace the fueling plug.

We rolled it back, watching the temperature all along and praying. Two or three guys climbed up and plugged the fuel lines in, and we were able to safely drain the Agena. In the subsequent investigation, we discovered the problem was a wiring error. We had a successful launch in the next day or two. I went down on the pad the next morning to thank those who risked their lives to replace that plug. They did a fine job even though they knew that if the rocket exploded, they would have been gone!

**The Pentagon and the "Mushroom Factory"**

I then experienced another career transition, which precipitated my retirement from the Air Force. John McLucas, the Director of the National Reconnaissance Office (DNRO) at the time, came out to Vandenberg and told me, "I am going to transfer you back to Washington in the military, and then you are going to retire from the military and go into the CIA. It is CIA's turn to furnish the job." That is what I did. I stayed in the same job, but as a CIA officer rather than as an Air Force officer.

I was the Director of the NRO Control Center at the Pentagon. By then the NRO was pretty well organized. The Control Center was located deep under the Pentagon's River Entrance in the basement. It was known as the "Mushroom Factory."

Most people did not know why we called it the mushroom factory. One day at a farewell ceremony, I told people to remember where we were: in the basement, sub-basement. Every time it rained we were the beneficiaries of the internal storm sewers which bubbled in and semi-flooded the place. I asked, "What is required to grow mushrooms? It takes darkness, fertilizer, and water." I said, "Nature furnishes the darkness and water, and fertilizer comes from the fourth floor," which was where the NRO front office, room 4C1000, was located.

**Conclusion**

I can sum up what I did very simply. I was "on-duty" 24 hours a day, and I made decisions. I believe some of them were good decisions, and I solved a few problems. I am honored to have been selected as a National Reconnaissance Pioneer. I was initially reluctant when I was notified

Figure 35-3. Overhead view of the Pentagon. The "Mushroom Factory" was located in the basement on the far side of the building. (Photo courtesy of the NRO History Office, original source unknown.)

of this honor. There are so many other people who deserve this honor: military, government, civilian, and contractors. I cannot over-emphasize the role of the contractors in these programs. They were just great. In fact, Lockheed's Jim Plummer should be called "Mr. Corona" as far as I am concerned. Later, he became the Under Secretary of the Air Force and the DNRO. There were many others who all worked toward a common goal. Together we got the job done.

—

### My Husband's Undercover Air Force Experience[9]

The life of an Air Force wife is one thing—the life of a woman whose husband "has gone undercover" is another. But I learned not to ask questions and just enjoy my family and the great people in "the new organization." A normal social life was difficult. Our friends were mostly "nine to fivers" who could talk of the things their husbands were involved in—there were no secrets. We maintained an Air Force cover. Frequently in visiting with friends, they would suggest riding with me when I left to pick up Charlie and I would have to make excuses as to why that would not work: such as lots of errands to run. They began to understand and would not ask.

Figure 35-4. Mary Anne Murphy being interviewed by Chanta Quillen on Pioneer Recollection Day, NRO Headquarters, 26 September 2000. (Photo by Candi Campbell, NRO Visual Design Center.)

As I look back, I guess the funniest thing was when Charlie had to depart "for the unknown" often in the middle of the night, throwing a cot in the back of his Volkswagen and returning a few days later. This became a source of curiosity to my neighbors, who must have thought he probably had another lady "on the side."

Finally, I appreciated and admired Charlie's dedication to his job and country. Our son and daughter are very proud of him and what he did for his country. I am very grateful to the NRO for honoring these "unsung heroes". It was a great tribute to some great guys.

—

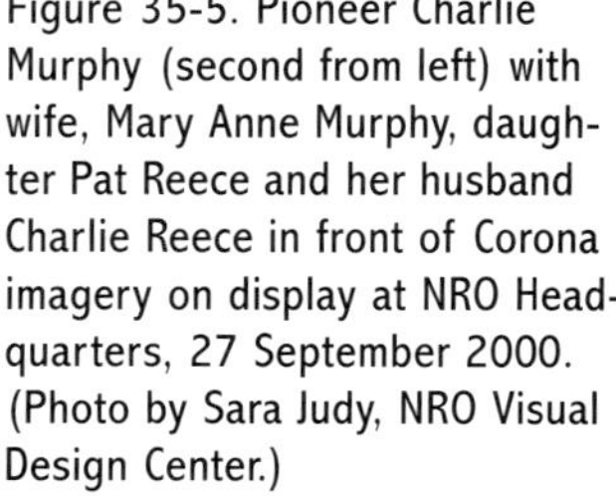

Figure 35-5. Pioneer Charlie Murphy (second from left) with wife, Mary Anne Murphy, daughter Pat Reece and her husband Charlie Reece in front of Corona imagery on display at NRO Headquarters, 27 September 2000. (Photo by Sara Judy, NRO Visual Design Center.)

[9] This section was written by Mary Anne Murphy..

## Pioneer Award Presentation and Citation

Figure 35-6. Pioneer Charles Murphy (second from left) being recognized at the 2000 Pioneer Induction Ceremony. The award was presented by DNRO Keith Hall (left) and DCI George Tenet (third from the left). Brigadier General Craig P. Weston (Director, Corporate Operations Office and Chief Information Officer) joined in the presentation. (Photo by Sara Judy, NRO Visual Design Center.)

**Charles L. Murphy, Colonel, USAF**

Colonel Charles Murphy served as the first field technical director of the Corona Advanced Projects Integration Facility, the main link to the Intelligence Community. His improvements in Corona mission planning procedures and its command system ensured Corona's on-orbit reliability and its long-term contributions to national intelligence.

*Career in National Reconnaissance: 1958-1977*

# Frederic C.E. Oder

Frederic C.E. "Fritz" Oder was the Program Manager for the Weapon System-117L (Advanced Reconnaissance System) program that made a significant contribution to early military and national reconnaissance satellites. He also contributed to many other national reconnaissance efforts, both in the military and as a civilian contractor for Lockheed and Eastman Kodak.

## A National Reconnaissance Career with the Air Force, Central Intelligence Agency and Industry[1]

I became interested in space from a very young age, long before there was a national space program. I studied and developed an expertise in geophysics and meteorology, which allowed me to get early exposure to developing payloads and putting them on the rudimentary sounding rockets of the post-World War II era.

### Laying the Foundation for My Career

I graduated from Cal Tech in 1940 having studied as much as I could in the area of geophysics. The Army Air Corps asked if I would like to continue my studies and they offered to pay my expenses as a Flying Cadet (non-flying) to get a master's degree in meteorology at Cal Tech. I visited March Field, which was the nearest Air Corps base, and I spoke with a couple of captains at the weather station to discuss whether this was the right thing for me to do.[2] The discussion left me with the impression that it was a good opportunity. I signed on to the cadet program, was commissioned on 1 July 1941, and I was called to active duty as the Base Weather Officer at a Weather Station in Pendleton, Oregon.

Not too long after that, I received orders to go overseas to a classified destination called "Plumb," which turned out to be the Philippines. I was a Weather Officer for the 7th Bomb Squadron, 7th Bomb Group. From there, I was assigned to the School of Applied Tactics in Orlando, Florida. Next, I went to Headquarters, Air Weather Service, where I was in the Operations Division. I was scheduled to go to 20th Air Force, but that never happened. Instead, I was assigned to the Command and General Staff School at Fort Leavenworth, Kansas.

---

[1] This section is based on an interview with Frederic C.E. Oder at the National Reconnaissance Office Headquarters on 26 September 2000.

[2] March Field, located in Riverside, CA, later became known as March Air Force Base.

Figure 36-1. Picture of Frederic Oder. (Photo courtesy of NRO History Office, original source unknown.)

By that time, World War II had ended, and the Army asked me if I wanted to go back to graduate school. That sounded like a good idea. When I asked where the Army wanted me to go to school, I was given several choices, one of which was the University of California, Los Angeles (UCLA). As a native of Los Angeles I liked the sound of that, and I signed up for two years at UCLA.

If I had any sort of a mentor, he was Professor Joseph Kaplan, the Head of the Physics Department at UCLA. I knew Joe during the war when he served as an operations analyst for the Army Air Corps. That was another reason I went to UCLA, because I knew Joe, and I looked forward to working with him. We shared an office most of the time I was there, and I learned a lot from our close contact. He was writing a book on quantum mechanics, and he handed me chapters to read and critique even though I knew relatively little about the subject.

After UCLA, I came back to the Air Weather Service (AWS), which at that time was based at Gravelly Point in Washington, DC. We eventually moved out to Andrews Air Force Base in Camp Springs, Maryland, and during that process AWS identified a requirement for someone to be the Director of Geophysical Research at Cambridge Field Station in Massachusetts. Cambridge Field Station was a subsection of the Electronics Subdivision of Air Research and Development Command's Wright Field.[3] The Geophysical Research Directorate had a staff of 15-20 military personnel and maybe 60 civilians, with an annual budget of approximately $4 million. The directorate was involved with things like the Aerobee sounding rockets and the V-2/Bumper sounding rockets.[4] This work provided me with knowledge and experience related to putting payloads on rockets. I also finished my doctoral dissertation during this assignment.

**Introduction to Intelligence and Space**

I finished my tour at Cambridge in 1952, and I thought I was headed for Europe. Instead I was sent to the Central Intelligence Agency (CIA). I suspect it was a computer in the Pentagon that assigned me to the CIA for a three-year tour. I became the Deputy Director of the CIA Physics and Electronics Division in the Office of Scientific Intelligence (OSI), where I got to know Richard Bissell. I developed an understanding of what we did, but did not know about parallel activities in the Soviet Union, which was our main intelligence challenge. We had some signals intelligence (SIGINT) and some imagery intelligence from overflights, but most of this information came from border region overflights that did not penetrate deep into the interior of the Soviet Union. That tour helped me to understand that there was a need for better methods of collecting intelligence on Soviet capabilities.

One of the OSI Deputy Directors told me that General Bernard A. Schriever had been

[3] The Electronics Subdivision eventually separated from Aeronautical Systems Division at Wright-Patterson Air Force Base, and became Electronic Systems Division at Hanscom Air Force Base.

[4] The Aerobee sounding rockets were a series of rockets built by Aerojet General for upper atmosphere research. The V-2/Bumper was another sounding rocket that used refurbished German V-2 rockets with an Army WAC Corporal upper stage, all under Project Bumper.

assigned as the Chief of the Western Development Division (WDD) of Air Research and Development Command (ARDC). Formed in Los Angeles, WDD was responsible for developing and fielding intercontinental ballistic missiles. The OSI Deputy Director asked if I would be interested in a job with Schriever. I already had completed two of my three years at CIA, so I thought it would be a good idea for me to join Schriever's organization.

*I suspect it was a computer in the Pentagon that assigned me to the CIA for a three-year tour.*

Upon arrival at WDD, Schriever asked me to serve for a year as the WDD liaison officer to Lieutenant General Thomas S. Power, who was the ARDC Commander. I agreed to the assignment, and during my time at ARDC I became quite familiar with a program called the Advanced Reconnaissance System (ARS). Schriever was interested in ARS because of the RAND studies that recommended the research, development, and deployment of reconnaissance satellites in space.[5] The RAND study proposed using a modified ballistic missile as a booster to place satellites into orbit.

Midway through that year the ARS program reached the point where General Schriever felt it had matured enough to be pursued more vigorously. In August 1955, Schriever presented his case to General Power that WDD should take over ARS, and Power agreed. Major General John W. Sessums, the ARDC Vice Commander at the time, was concerned because there were staff officers at ARDC who were opposed to the transfer of the satellite efforts to WDD. Power overrode them, and referred any further questions to Sessums. General Power used to have these bright lights over his desk, and when his mind was made up he would turn them off, removing any doubts. So it was a very clear decision.

**Working on Weapon System 117L—The Advanced Reconnaissance System**

Before I completed my assignment at ARDC Headquarters, Schriever asked me what I wanted to do next. I asked, "How about this project 117L?" He answered, "How would you like to run it?" I explained, "I don't know a whole lot about earth circling satellites, but I can learn." Then I added, "Put me in charge of it." He did, and I became the original manager of the Weapon System (WS)-117L family of satellites as they were integrated into the Western Development Division's Intercontinental Ballistic Missile programs. I was promoted to colonel early, and General Schriever deserves full credit for that. He was really good about taking care of his people.

I ended up with a small office of maybe six or eight people who moved out from Wright Field. We had very little in the way of funding and approved plans, but at least we had a prime contractor. The lack of organizational history helped us to be flexible, and this flexibility was beneficial in the early stages of the program.

We planned to launch our satellite on the Atlas missile, but Atlas was progressing more slowly than other programs, and this delay was going to cause a postponement in our launch. We decided to use the Thor booster as our launch vehicle, but we had to add an upper stage that later was called Agena. Lockheed built the Agena. The combination of Thor and an Agena was capable of putting a 600-pound payload into orbit (in addition to the Agena). As with all launches, the payload had to fit within specified weight and size limitations. Consequently, whatever we were going to put into orbit had to be reasonably simple in the sense that we could not afford the weight of redundancy to ensure long life.

[5] Project R and D, later the RAND Corporation, performed most of the initial Air Force investigations into the utility and design of various kinds of satellites, to include reconnaissance and surveillance. One of the most influential of these studies was Project Feedback, devoted to the feasibility of a photoreconnaissance satellite.

The WS-117L was supposed to be owned and operated (when it became operational) by Strategic Air Command, and the satellite would only be returned to us to fix technical problems. The ideas and concepts upon which the systems was designed and built had never been proven on orbit because no satellites had actually been launched prior to October 1957. Naturally, we wanted to launch something that would work and make a difference, and when we initially met with the contractor we were not even sure if we could make it fly!

When we started WS-117L, we planned to employ a read-out system called Samos as opposed to a film recovery system.[6] The rationale for this decision was that we believed the film recovery system had too many limitations. For example, the only way to get data was when a film capsule was recovered, but by that time the data were already days, weeks, or maybe even a month old. In comparison, the Samos read-out system offered the potential to receive the collected data in near-real-time. But the read-out system was limited by radio technology and power. We were limited to a six-megahertz bandwidth for the data we transmitted down to the ground, which was not very good. We could not get enough data down through that pipe. We flew the Samos E-6, which used that little six-megahertz device. We read out the film on-board the spacecraft, sent it down to the ground station, and we reconstructed all that it had seen. The concept worked, but the resolution and processing speed were not nearly as good as they needed to be.[7]

We devoted a lot of research and development resources to advancing the technology to the point where a near-real-time readout system could be employed. We knew that sooner or later satellite reconnaissance would utilize a read-out system because it offered the potential to gather more data, sooner, and at higher resolutions. Resolution was always an issue. Samos started out with very limited resolution, about 30 feet of ground resolved distance.[8] We knew that we needed to significantly improve resolution, which required advancing the technology as rapidly as possible.

The WS-117L included more than imagery satellites. The system also performed signals intelligence (SIGINT) collection. I had an interesting experience with General Osmond Ritland, who was the deputy and who later took over for Schriever at WDD.[9] Ritland asked me how it was possible to do signals collection from satellites moving so fast. I explained that while it depended on the frequency, enough signals could be captured. I told him, "As long as you get enough beeps, you'll be able to determine the structure of what's in there." He was skeptical. To show him it was possible, we put SIGINT payloads on some early Samos flights, and they performed quite impressively.

The analysts who interpreted imagery were at the National Photographic Interpretation Center, and the analysts who interpreted SIGINT were usually at the National Security Agency (NSA). We dealt with both in WS-117L, but in those days the two worlds hardly knew each other. Occasionally, NSA would get a signal from a geographic point, and they would request a picture of the location.

---

[6] The concept of a film-recovery system was separated from WS-117L and developed by the CIA as Project Corona.

[7] The read-out system was ahead of the technology of its time, and its limitations made it a poor alternative to the results already being obtained by Corona. However, after Samos technology was rejected by the National Reconnaissance Office, it became the chosen method for returning imagery from the National Aeronautics and Space Administration's Lunar Orbiter that searched for Apollo landing sites. So, the Samos E-6 system ultimately did make an important contribution.

[8] Ground resolved distance is a measure of imagery quality that reflects the ability to distinguish between two objects separated by a specified distance. For example, a resolution distance of 30 feet means that an observable object must be at least 30 feet away from another object before the image could distinguish two objects.

[9] Osmond J. Ritland is a Pioneer of National Reconnaissance (inducted into the NRO Hall of Pioneers, 27 September 2000). See chapter 39.

**The Need for Security**

Beyond the technical challenges, we also needed to protect our activities. A major flap occurred when *Aviation Week* published an article on WS-117L. We wondered how the magazine learned that there even was a program called WS-117L, Advanced Reconnaissance Systems. There was a major investigation on that one! The article reported on our effort to conceal Corona and to make it disappear from public view into the "black."

A classified program is placed at greater risk of being compromised when a program activity is visible (like launches), so we needed to develop a cover story to explain things that could be observed. For example, the U-2 program remained concealed during its development and fielding. Speculation that such a capability existed occurred only after the aircraft was already in use, and the public knew for sure only after the shootdown in May 1960. We created the Discoverer experimental satellite series as the cover story to explain Corona. Similarly, we needed to develop and implement a cover story to protect WS-117L activities.

*The analysts who interpreted imagery were at the National Photographic Interpretation Center, and the analysts who interpreted SIGINT were usually at the National Security Agency (NSA). We dealt with both in WS-117L, but in those days the two worlds hardly knew each other.*

We discussed the CIA's approach on the U-2 and Corona programs at some length with Major General Ritland. Ritland served as the deputy manager for both programs, and he took us to talk to Richard Bissell, who was the Special Assistant to the Director of Central Intelligence (DCI). Bissell managed the development and fielding of the U-2 aircraft, and was managing Corona. Bissell described for us how he approached security for the U-2. I had some exposure to the U-2 program while I was in the CIA, so I had an idea of the ramifications of a tightly compartmented program.

**Launch**

As for getting things into orbit, the environment was different on each coast. The east coast had a nice down-range chain of islands and data collectors. Launching on the east coast had the advantage of better instrumentation, but there also were strict limitations. The directions that rockets could fly away from the Florida coast were severely restricted. For a polar orbit, which was what Corona was designed for, the launch had to be nearly due south. On the east coast that meant overflying Cuba, which was a non-starter. Consequently, all of the Corona launches were out of Vandenberg Air Force Base in California.

We also had restrictions on the west coast, but they were of a different sort. For example, Southern Pacific had a train route that ran right through the base and past the launch pads, and we could not launch if a train was passing through. So we had to work around the train schedule. If a launch was delayed and a train arrived in the interim, we had to stop and wait for the train to pass. That became quite exasperating, particularly because we were launching three satellites a month.

We faced an additional challenge on the west coast. The missiles were launched out over the Pacific towards Kwajelein Atoll where we could observe and monitor their performance. However, with Corona we wanted to go into a polar orbit. The Navy controlled the range out of Point Mugu, south of Vandenberg. On one occasion the Navy did not want us to launch a particular flight, citing range safety concerns. I was a colonel at the time, and I had to take on this three-star admiral. The admiral was determined that we were not going to launch. General LeMay, who was the Air Force Chief of Staff told me, "Oh, just fire that

Figure 36-2. Corona launch from Vandenberg AFB, California. (Photo courtesy of NRO History Office, likely U.S. Air Force photo.)

damn missile!" So, we launched, and the Navy was quite unhappy. Fortunately we did not have anything break up and fall into their territory!

I never learned if the Navy was genuinely concerned about range safety or if it was a game they were playing to keep us from launching. Range safety sometimes became a real thorn in the side. In those days we never had 100 percent certainty of what was going to happen once we launched the rockets off the pad. Some of them blew up 100 feet into the air, some of them fell in the water, some of them disappeared and we never knew what happened to them.

**Leaving the Military**

The work requirements were a bit stressful on my family, but they recognized that I was involved with important work. My wife and I coped reasonably well, and fortunately her parents lived relatively close so there was some family support. One year I remember looking at my travel records, and I was away from home 48 out of 52 weeks. I was involved with all of the design decisions and analyses, so there was a real need for me to be at the Lockheed plant.

By 1960 I had served twenty years in the Air Force, but I also had pressing family responsibilities that were competing with my career. Soon, I would have three kids in college at the same time. My salary and allowances as a colonel totaled just over $10,000 a year. I was not about to be able to support three kids in college at the same time on that kind of salary! Their education was my priority, and although I did not particularly want to leave the Air Force, I figured it was about time my family got something. So, I put in for twenty-year retirement.

General Schriever held up my papers for several weeks because he did not want me to leave, but he finally submitted them and I retired on 20 September 1960. After my retirement papers were processed, I received a call from a friend who worked on the Air Staff. He asked if I would be interested in being a professor of physics at the new Air Force Academy. I explained that I had already submitted my retirement papers, and that it was too late.

**Starting My Civilian Career at Kodak**

Despite interest at Lockheed to hire me, I decided that I did not want to go to work for the company. Lockheed had been my prime contractor, and going to work for them right after leaving the Air Force did not seem appropriate to me.

Eastman Kodak was a sub-contractor with whom I had no direct dealings. They offered me a job, and I went to work there. Ironically, by that time my oldest son was at Harvard on a scholarship. I notified the university of my change in employment status, and the school responded by cutting off my son's scholarship because I was now making enough money to pay his way. Sometimes you cannot win.

I remained with Kodak for five and a half years as Director of Special Projects with a staff of about 1,000 engineers. Special Projects covered almost everything Kodak was doing related to national security and that was National Reconnaissance Office (NRO)-related. In order to maintain a reasonable cover story for our activities, we took on a program for NASA that involved the Lunar Orbiter payload. We helped create the Lunar Orbiter program, and its purpose was to photograph the Moon to identify possible landing sites for the astronauts. We utilized some of the technology that we had developed for the Samos direct read-out part of WS-117L and applied it to the lunar program.

Lunar Orbiter was a film-based system. Like Samos, the film was developed on-board and read-out with an electronic scanner, and then the data was transmitted back to Earth. So the project had two significant benefits: we took good lunar pictures, and the project provided a nice cover story for our highly classified work. It also provided a place where I could assign new employees while they were awaiting their clearances before I could move them into the classified programs.

My team at Kodak also worked on the payload for another space-based research program. Human factors were an important consideration for this project. For example, we performed a series of tests that measured human eye perception. While the human eye can integrate scene information very well, there is a problem of burnout. If we were going to have astronauts staring at the Earth looking for spots of interest to aim a sensor, we needed to determine how long they could effectively perform that activity.

**Moving on to Lockheed**

This human factors project provided me with an opportunity to switch jobs and go to work for Lockheed. My wife had problems with the cold weather in Rochester, New York, so we moved out to California and I went to work for Lockheed. I initially continued to work on the human factors project.

I became the first Lockheed Program Director for the first near-real-time reconnaissance satellite, and this was an exciting program and opportunity. There were various complex and difficult requirements, such as the satellite's long lifetime, on-board computers that were fairly programmable, high-resolution optical sensors, and several other new technologies that had to be created to meet the government's requirements. The program received the support and resources it needed, and there was a lot of good work being done in all aspects of the system: guidance, command and control, optics, structures, etc. Funding for the technology risk reduction was done very well, and the CIA made a major investment in technology before the program began.

Les Dirks and the CIA did a first class job. Les was the head of Program B, and he deserves much of the credit for the program's success. Les also was a very fine engineer. This was the only program that I was associated with where the government did so much homework, and there were many resources that Lockheed could draw on from the beginning of the development phase. This reservoir of knowledge, experience, and equipment gave us a great deal of confidence that we could build the spacecraft and that it would work the first time out.

*On one occasion the Navy did not want us to launch a particular flight, citing range safety concerns. I was a colonel at the time, and I had to take on this three-star admiral. General LeMay, who was the Air Force Chief of Staff told me, "Oh, just fire that damn missile!"*

There also was a solid relationship between the government and contractors built upon mutual trust that I have not seen elsewhere. There was a proven formula for success: a simple statement of work, a clear understanding of what the government wanted, a handshake, and then we worked together to get the job done. An important element in this success was my relationship with Charlie Roth. For example, when Charlie would say, "I want you to change the following…," I would say, "Yeah we can do that." We would shake hands and sometime later the contracts people would work it out. There was no fighting over anything,

Figure 36-3. Pioneer Frederic Oder (right) being interviewed by Lieutenant Colonel Rory Maynard on Pioneer Recollections Day, NRO Headquarters, 26 September 2000. (Photo by Candi Campbell, NRO Visual Design Center.)

we just went ahead and got the job done. It was a "your word is your bond" kind of relationship—I trusted him, he trusted me. And I never once let him down, and he never once let me down. I regret that Charlie didn't live long enough to attend the NRO Pioneer Recognition Ceremony. He was a great program director for the government.

I had spent about 19 years with Lockheed by the time I reached age 65. Lockheed had a policy, a very firm policy, that anyone who was 65 and in a management position had to retire. And that year the Chairman of the Board was turning 65 and facing retirement, so I also retired.

**Consulting After Retirement**

Jimmie Hill, the Deputy Director of the NRO, contacted me about writing histories of NRO programs in which I was involved. I spent about a third of my time working on these histories, in addition to some other consulting projects. We started out with Corona. It became apparent that we needed better logistics support, so the NRO arranged for Lockheed to support us. They had some secure offices on the west coast where we could be based, and we were provided a considerable amount of documentation to help us in our effort. In fact, we wound up with a whole wall of safes!

In those days the Air Force was not very good at keeping data of historical value. The bosses would say, "Well, we used to have all that stuff in files here, but we had to clean out the files to make room for a new office." And so the records went to the shredder! Very seldom would anyone ask, "Does this information have historical value?" Some of the early information was hard to recover. That was why Jimmie Hill asked us to capture what we could at that time.

In comparison, CIA officers are collectors of information, and they were great at collecting and storing records. They had a location where they had a permanent collection of records. The archive was computerized, and we would use that system to search for documents. We would locate the record we wanted, submit a request, and the information would appear either by fax or hardcopy.

**Conclusion**

The NRO and its leadership are confronted by a world that probably is more dangerous in the 21st Century than it was when we were working on our projects during the Cold War. We mainly had the Soviet Union to deal with, whereas in the 21st Century there are more numerous threats. The United States continues to need accurate, timely information on a range of potential adversaries, yet national reconnaissance alone cannot provide all of the answers. There also is a continuing need for high quality human intelligence in addition to the unique systems and capabilities provided by the NRO.

—

## Pioneer Award Presentation and Citation

Figure 36-4. Pioneer Frederic C.E. Oder (second from left) being recognized at the 2000 Pioneer Recognition Ceremony. The pioneer award plaque was presented by DNRO Keith Hall (left) and DCI George Tenet (third from the left). Brigadier General Craig Weston (Director, Corporate Operations Office and Chief Information Officer) joined in the presentation. (Photo by Sara Judy, NRO Visual Design Center)

**Frederic C. E. "Fritz" Oder, Colonel, USAF**

In the late 1950s, Colonel Frederic Oder directed the first United States satellite reconnaissance enterprise, the WS-117L (later Samos) Program. This effort produced the launch complexes and vehicles, spacecraft, Satellite Control Center, and tracking stations used in all early satellite reconnaissance programs, including Corona. He continued his career with Lockheed and Eastman Kodak.

*Career in National Reconnaissance: 1956-1984*

# Robert M. Powell

The late 1950s was a time of trial and error, with the latter predominating. Early attempts to operationalize the emerging concepts for space reconnaissance and theories of orbital control were often frustrating, sometimes humorous, but always educational.

## A Time of Trial and Error for Space Reconnaissance[1]

My introduction to space reconnaissance programs was as Lockheed's manager of a satellite command and tracking station at Kaena Point (known as "Hula") on the Hawaiian island of Oahu in July 1958. In early 1959, we had our first experience at tracking and commanding a satellite. We blew it!

We practiced for the big day by using a transponder and a simple telemetry unit mounted in an aircraft, and Colonel Teuvo (Gus) Ahola arranged for the aircraft to fly toward us to simulate a satellite pass.[2] Our radar operator practiced counting the commands sent, and our telemetry operator practiced counting the commands received by the equipment in the plane. As an added backup, I gave the plot board operator a stopwatch, and his task was to keep track of azimuth and error information provided by the TLM-18 telemetry antenna. He started the watch when we began sending commands, and called out when they should end. Commands were sent at a rate of one per second.

The big day arrived. The station at Kodiak, Alaska did not get commands to the satellite as planned, so it became our turn. Unfortunately, the paper tape recorder behind the radar operator jammed, and he turned to clear it. This caused him to lose count of the commands sent. Meanwhile, the telemetry operator had trouble synchronizing his equipment, causing him to also lose count. Colonel Moose Matheson came on line from the Sunnyvale control center in California and asked us to time the satellite's passing, which the plot board operators did very well, but then lost track of command sending time.[3]

---

[1] This section is based on written input that Robert M. Powell submitted to the Center for the Study of National Reconnaissance.

[2] Colonel Teuvo A. (Gus) Ahola commanded the 6594th Recovery Group, and was responsible for the Corona recovery effort.

[3] Air Force Lieutenant Colonel Charles "Moose" Matheson served as Vice Commander of the 6594th Test Wing in Sunnyvale, California, which operated the control facilities for Air Force satellite programs.

Figure 37-1. Agena-D spacecraft. (Photo by Lockheed Missiles and Space, courtesy of NRO History Office.)

Realizing that we were out of control, I halted the sending of commands. When we played back the recorded telemetry, we found that we had sent (and the satellite had received) about twice the number of commands than what was originally intended. I called Colonel Matheson in Sunnyvale and told him what had transpired. I advised him that, near as I could predict, the recovery bucket would come down somewhere around Moscow. That stirred things up a bit, particularly with Colonel Matheson.

Later on, Colonel Frederic "Fritz" Oder visited us to learn what went wrong, and what had been done to fix the problem.[4] We learned that the Kodiak station had been required to count the number of commands sent by the radar rather than relying on telemetry feedback, and we proceeded to establish the same practice at "Hula."

**Building a Reconnaissance Satellite**

During the 1960s, I was a member of a Lockheed team for one of the satellite programs in which we developed the reconnaissance payload. We used the Agena as the spacecraft to deliver the payload into orbit. As the Chief Systems Engineer, I had to sign off on all engineering drawings and interface documents.

To further complicate matters, our work as a subcontractor and as a systems integrator was classified. This work included developing new subsystems for the secondary propulsion system, and other subsystems related to tracking, command, and telemetry. All of the classified work had to be performed in areas separate from the work on the Agena spacecraft. We also were developing a new booster/Agena interface, and a new interface with the launch complex at Vandenberg Air Force Base.

**Innovation on the Launch Pad**

We also had to find innovative solutions to operational problems that surfaced. For example, we discovered problems with one of the spinning components during qualification tests for one system. Specifically, the component would heat up excessively, which caused the part to expand and seize. Prior to one of the launches, a colleague and I went to the pad armed with stethoscopes to listen for signs of friction as the system was operated. Neither of us

---

[4] Frederic "Fritz" Oder is a Pioneer of National Reconnaissance (inducted into the NRO Hall of Pioneers, 27 September 2000). See chapter 36 for his recollections.

heard any rubbing sounds, so we launched the vehicle. The mission had normal orbit operations, but near the end of the planned flight I requested permission to progressively increase demands on the system. I received permission, and we gradually increased demands. After several increases in the command rate, the part seized. This outcome confirmed the problem we suspected, but for other reasons we had already decided to change that particular component.

> *Not since the Manhattan Project had the gap been so narrow between the conjurer and the engineer.*

**Redundancy as Insurance**

We decided early in our development activities that redundancy would be an important element of our design and operations. We held regular meetings to try to anticipate and solve problems. Shortly after one of our successful missions, my boss, Hal Huntly, and I attended one of these regular meetings at the Air Force Special Projects Office in Inglewood, California. Colonel Bill King from the Air Force and Ed Clark from the Aerospace Corporation were also present. Ed suggested that we add a detonating fuse as a redundant measure to increase the reliability of releasing the payload. We implemented this idea on the next flight, and guess what happened. The primary separation system failed to operate completely, so the redundant fuse system finished the job. Ed's fixation on redundancy had borne fruit, and this justified his nickname, which was "Ed, Ed, Clark, Clark."

**New Program Management Methodology**

Beyond developing innovative engineering for spacecraft, we also developed novel program management methods. For example, I believe one or our early programs was the first to employ the "factory-to-pad" management concept. When put into practice, this approach entailed shipping the space vehicle in as flight-ready a condition as possible, with all the pyrotechnics installed. Colonel Bill King insisted on using this approach.

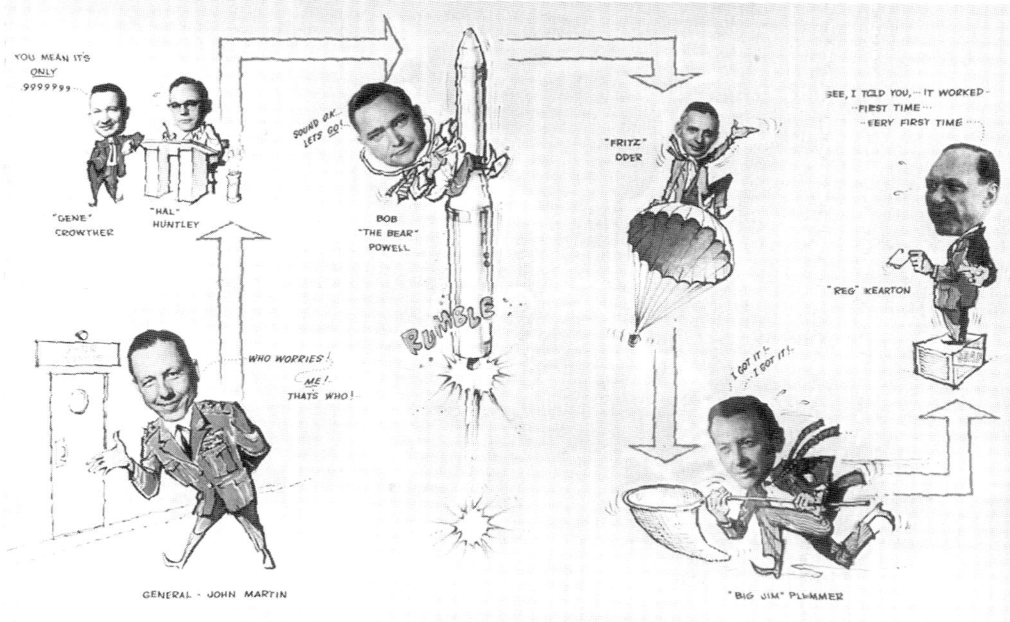


Figure 37-2. A commemorative illustration that satirizes some of the personalities involved in one of Powell's programs. Notice Powell portrayed using a stethoscope while clinging to the launcher (Illustration courtesy of Robert Powell.)

Figure 37-3. Robert Powell's Lockheed team. Powell is in the front standing row, sixth from the left. (Photo courtesy Robert Powell.)

We also developed procedures for close oversight of our program activities. Many of our programs were immensely complex, and we frequently pushed the state-of-the-practice, if not the state-of-the-art. Fortunately, the customer would often sponsor key preliminary development activities at several potential subcontractors. For one program, I asked one of my employees to relocate himself and his wife in order to be close to one of the key subcontractor facilities. I wanted him to be on location to provide oversight of the subcontractor's work. The work at this subcontractor was being performed in a research type of environment, and my representative was instructed by me to change the environment to be more schedule- and product-oriented. Consequently, he became actively involved in daily operations. In one instance, he made note of some key people's time of arrival in the parking lot, and brought that up at a program review and status meeting. This unconventional method had a salutary effect on program productivity.

## Conclusion

My experiences during the developmental years of national reconnaissance are generally consistent with those of the other pioneers. Our tasks were not so much the development of technologies and methods, as the invention of them. Not since the Manhattan Project had the gap been so narrow between the conjurer and the engineer. In this respect, the relatively large number of unexpected failures of the earliest programs is not nearly as remarkable as the speed and decisiveness with which we identified solutions.

—

## Pioneer Award Presentation and Citation

Figure 37-4. Pioneer Robert Powell (second from the left) being recognized at the 2000 Pioneer Recognition Ceremony. The pioneer award plaque was presented by DNRO Keith Hall (left) and DCI George Tenet (third from the left). Dr. Vance Coffman (President and CEO Lockheed Martin) joined in the presentation. (Photo by Sara Judy, NRO Visual Design Center.)

**Robert M. Powell**

Mr. Robert Powell, Lockheed's program manager for a key high-resolution satellite reconnaissance program, devised a novel payload-pointing mechanism that greatly extended the lifetime of satellites in orbit. Moreover, in that same program, he contributed to the design of a mechanism that dramatically increased and improved imagery coverage of the Earth's surface.

*Career in National Reconnaissance: 1959-1975*

# Edward H. Reese

Edward Reese developed the original ground system architecture for a National Reconnaissance Office satellite imagery reconnaissance program. This technical achievement allowed the success of the nation's first near-real-time satellite imaging system, which revolutionized imagery intelligence collection methods and capabilities.

## Development Techniques That Stand The Test Of Time[1]

My most significant technical achievement was the development of the Ground System Data Architecture to support an imaging satellite program. The architecture of the Operations Facility Segment (O/F) included the computers, databases, applications software, and special purpose hardware to implement command and control for the entire mission.[2] I believe my approach to the engineering discipline that was required to develop a large software system was my main contribution to national reconnaissance. This most exciting accomplishment of my career occurred when the satellite first launched and started to deliver intelligence products.

At the time of the inception of the program, I viewed it as the most challenging technical undertaking of my era. This technical challenge seemed like my calling because it had everything I wanted to do, and I felt patriotic to boot. It was the first imaging system of its kind, and the amount of leading-edge technology and the immense size of the software system were awesome. While the growth of today's technology has made the then state-of-the-art seem outdated, I find it amazing that the architecture, methodologies, and disciplines that I helped develop are still at the forefront of early 21st century intelligence collection methods.

### Impact of Our Data System Architecture

The primary value of our architecture to the government, and the country as a whole, is that it has passed the test of time. The basic architecture we developed in the years 1972-1975 has been sufficiently flexible to evolve with technology over the following 25 years without becoming obsolete. This is a very rare occurrence in the intelligence business, and one in which I am very proud to have been involved.

[1] This section is based on written input that Edward H. Reese submitted to the Center for the Study of National Reconnaissance.

[2] The O/F controlled both the space and the ground assets of the system.

My company, then General Electric (GE), also benefited as a result of these successes. The techniques for the system engineering and software development of this system, as well as other algorithms and products designed for the program, became the core business for GE's Management and Data Systems Operation.[3] The company definitely gained the government's confidence as a result of this project.

*My colleagues and I believe very strongly that the program contributed tremendously to ending the Cold War. When the Berlin Wall came down, I felt my work had contributed to its fall.*

The aerospace industry, itself, also was impacted. The methodologies for the development of the satellite system and software engineering requirements, databases, and operations concepts established standards for disciplined development practices. Unfortunately, these techniques are taken for granted today. Nevertheless, the program design positively impacted the industry as a whole.

**Colleagues and Working Environment**

I led a team of highly motivated system engineers. These were extremely talented people, who gave unselfishly to what was more of a cause than a work assignment. The team's principle motivation was patriotism, which caused my colleagues to be self-driven beyond belief. The contractors we interfaced with to develop and design the system were the best in the country, probably the best in the world. It always was challenging and often very hard,

Figure 38-1. A portion of the Berlin Wall on display outside of the New Headquarters Building at CIA. This side of the wall, with its pro-freedom graffiti art, faced West. The reverse side of the wall, which is blank gray, faced East. (Photo courtesy of CIA Public Affairs, CIA photo.)

[3] When Lockheed Martin acquired GE, Reese's division was absorbed within Lockheed Martin Management and Data Systems.

Figure 38-2. Edward Reese preparing input for his recollections at NRO Headquarters, Pioneer Recollections Day, 26 September 2000. (Photo by Candi Campbell, NRO Visual Design Center.)

but the team always believed in the program and the other players. At this writing, I am still very close friends with many of my colleagues from this program. God bless Les Dirks, whom I idolized as the technical visionary who made the program a reality![4]

Security demanded that our working environment be restricted to interaction only within our team. The closed nature of our internal society caused some problems, but at the same time it created the feeling of an elite society (in the good sense of elite). The travel restrictions imposed by security rules also led to some humorous incidents. For example the first time I went from Philadelphia to the Central Intelligence Agency (CIA) Headquarters in Langley, Virginia, I traveled by train. My CIA contact told me to get a cab with Virginia license plates at Union Station (in Washington, DC). After looking for a half hour, I was forced to call the contact and tell him I could not find one. He said to take any cab—Virginia cabs were just cheaper. I, of course, thought that the CIA must have had cab drivers with security clearances!

**Conclusion**

The camaraderie and the opportunity to work with the most talented people in government and industry is an experience that I will cherish forever. Additionally, my colleagues and I believe very strongly that the program contributed tremendously to ending the Cold War. When the Berlin Wall came down, I felt my work had contributed to its fall.

Since the end of the Cold War, it seems to me that the majority of Americans have begun to forget about the challenges of the Cold War, and the dedication and sacrifice of many "Cold Warriors." My teammates and I, even though our work was classified, confronted these challenges, and contributed to national reconnaissance and technological development. Would I do it again? I certainly would. It was important work.

*Security demanded that our working environment be restricted...it created the feeling of an elite society.*

The NRO's Pioneer Program, in recognizing those of us in national reconnaissance who were instrumental in winning the Cold War, is reminding the nation of the importance of these contributions. My time developing national reconnaissance systems—during a period when the Cold War created world tension and constant challenges—was the most rewarding work experience I ever had.

[4] Leslie C. Dirks served as NRO Program B Director from June 1976-July 1982. Dirks also served as the first director of the CIA's Office of Development and Engineering.

## Pioneer Award Presentation and Citation

Figure 38-3. Pioneer Edward Reese (second from the left) being recognized at the 2000 Pioneer Recognition Ceremony. The pioneer award plaque was presented by DNRO Keith Hall (left) and DCI George Tenet (third from the left). Dr. Vance Coffman (President CEO Lockheed-Martin) joined in the presentation. (Photo by Sara Judy, NRO Visual Design Center.)

**Edward H. Reese**

Mr. Edward Reese, General Electric's system engineer and program technical director, directed development of the architecture and the ground data system that integrated hardware and software to process digital imagery from electro-optical imaging satellites. He also directed subsequent improvements in this system that continue to meet this nation's overhead imagery requirements.

*Career in National Reconnaissance: 1965-2000*

# Osmond J. Ritland

Osmond J. "Ozzie" Ritland served for 27 years in the United States Air Force, retiring as a Major General. He was the Air Force project manager for the U-2 program in the mid-1950s, and later made invaluable contributions to the development and operation of the Corona satellite photoreconnaissance system. Ritland was honored posthumously as a Pioneer of National Reconnaissance. Martha Ritland, his wife, briefly recalls her husband's commitment to the Air Force, following a brief summary of Ritland's contributions to national reconnaissance.

## Putting the Air Force First[1]

General Osmond J. "Ozzie" Ritland was an Air Force officer who put the mission and the Air Force first in his life. In doing so, he made major contributions to the development of America's national reconnaissance capability, and he was a pioneer in that field. Ritland made significant contributions to early U.S. aerial and satellite reconnaissance programs.

From December 1954 to April 1956, Ritland served as Special Assistant to the Air Force Deputy Chief of Staff for Development, and had responsibility for the U-2 program. Ritland served as the Vice Commander of the Western Development Division and in the Air Force Ballistic Missile Division from April 1956 to April 1959. He then served as the Commander of the Ballistic Missile Division and the Space Systems Division from April 1959 to May 1962. In these positions, he guided and oversaw the early successes of the Missile Defense Alarm System and Space and Missile Observation System satellites, as well as the development and system acquisition of Air Force ballistic missile weapon systems. His persistent leadership contributed to the successful development and launch of the Corona satellite in August 1960.

Ritland died in 1991. On 27 September 2000, the Director of the National Reconnaissance Office, Keith Hall, and the Director of Central Intelligence, George Tenet, posthumously honored Ritland as a Pioneer of National Reconnaissance. Ritland's granddaughter, Elizabeth Kosich, accepted the award on his behalf. The citation on the Pioneer award plaque captures the essence of his contribution to national reconnaissance.

—

[1] This section is based on information contained in the material submitted with the nomination of Ozzie Ritland for selection as a National Reconnaissance Pioneer. We prepared this section in lieu of recollections.

## A Brief Memory of My Husband's Dedication to the Air Force[2]

My memories, and those of our two daughters—even from an early age—are clear and consistent. Our lives revolved around Ozzie's work and his dedication to it. For instance, in the course of testing a British Mosquito aircraft in 1944, the plane caught fire. He instructed his flight engineer to eject, and then he ejected himself. His parachute caught on the small escape hatch and was damaged, causing him to descend too rapidly. He broke his back when he landed. This happened on a Sunday, which was a routine workday in the years during World War II.

Ozzie was a kind and loving husband and father, but his family understood that his job had to come first in those tense years. That priority continued throughout his Air Force career, and my daughters and I are proud and thankful for all he did for his country and for us.

Figure 39-1. Osmond J. "Ozzie" Ritland, circa 1970s. (Photo courtesy of NRO Reading Room, original source unknown.)

—

## Pioneer Award Presentation and Citation

Figure 39-2. Elizabeth Kosich, Ozzie Ritland's granddaughter (second from left) received the pioneer award plaque for her grandfather who was posthumously recognized at the 2000 Pioneer Recognition Ceremony. The award was presented by DNRO Keith Hall (left), and DCI George Tenet (third from the left). Brigadier General Craig P. Weston (Director, Corporate Operations Office and Chief Information Officer, NRO) joined in the presentation. (Photo by Sara Judy, NRO Visual Design Center.)

[2] This section is based on a letter Martha Ritland sent to the NRO after her husband's induction as a Pioneer.

**Osmond J. "Ozzie" Ritland, Major General, USAF**

As the Air Force manager of the U-2 Program from 1954-1956, General Osmond Ritland developed the service infrastructure that made early overflights of the USSR possible. Subsequently, as Vice Commander of the Western Development Division, he duplicated this feat for the nation's early reconnaissance satellite programs, including Corona.

*Career in National Reconnaissance: 1954-1965*

# Robert W. Roy

Robert "Rob" Roy oversaw National Reconnaissance Office launches at Vandenberg Air Force Base from 1958 to 1964, and participated in over 175 launches. He managed the construction of launch pads, and established launch requirements and procedures that became standards for reference. These launch procedures and techniques contributed to significant growth and mission success for National Reconnaissance Office launches at Vandenberg.

## Ensuring Satellite Delivery[1]

My early Air Force experiences occurred in Florida at Patrick Air Force Base (AFB) and Cape Canaveral, two of the primary locations for space-related activities. These assignments provided me with the knowledge and experience necessary to prepare me for my subsequent assignments at Vandenberg AFB, for which I was named a National Reconnaissance Pioneer. During the period before my transfer to Vandenberg, I solidified my interest in the nation's space endeavors, and I was able to use my positions to prepare for the opportunities that I later received.

### My Early Involvement and Experiences in the Space Age

My participation in early National Reconnaissance Office (NRO) satellite activities was influenced by my interest and desire to be a participant in the nation's space program. I was interested in missile programs from the time I graduated from the Naval Academy and received my commission in the Air Force. I studied electronics and guided missiles because I thought that they were a ticket to the new missile era. Fortunately, I was assigned to Patrick AFB after school.

*Introduction to Missiles and Space at Patrick Air Force Base.* My first experience in the missile age was as the launch control officer for the Matador pilotless aircraft program. This involved the checkout and launch preparation for test flights of hardware being developed by industry contractors. I accumulated my early missile training by observing the contractors' conceptual, design, and engineering efforts in this pre-space environment. What eventually proved important about these early experiences was the modification of my thinking from civilian to military. This transition occurred when I had to transform the "contractor

[1] This section is based on written input that Rob Roy submitted to the Center for the Study of National Reconnaissance.

way" into the "military way," as testing activities changed from research and development to military deployment and operations.

I was at Patrick AFB during the period we called the "Ballistic Missile Age." These years saw the introduction into the military arsenal of intermediate range and intercontinental ballistic missiles (ICBMs). I was assigned as the field officer for the X-17 program, testing the scale models of heat sink reentry vehicles that became the warheads for the Thor and Atlas ballistic missiles.[2]

The trajectory of the X-17 carried it to over one hundred miles in altitude. The results of the reentry conditions were recorded through radio telemetry as the test warhead came back into the earth's atmosphere. The methodology of meticulously testing and retesting, evaluating, conferencing, and fixing problems was the contractor standard for all launches, with this program as well as others at Patrick AFB.

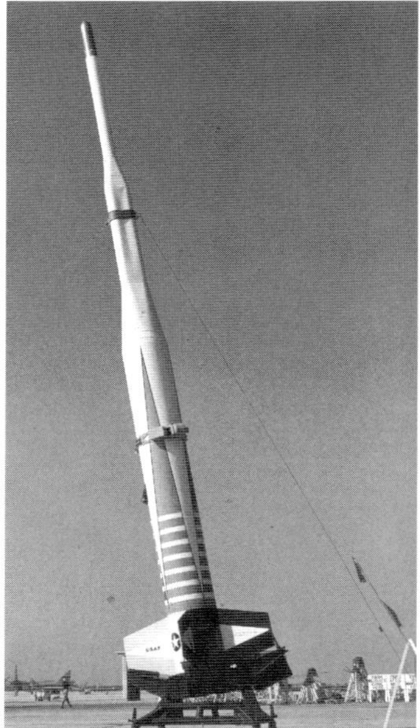

Figure 40-1. The X-17 test vehicle at Patrick Air Force Base, early to mid-1950s. (Photo courtesy of Rob Roy, likely U.S. Air Force photo.)

*Launching Ballistic Missiles at Cape Canaveral.* I was fortunate to have been at Cape Canaveral in the late 1950s when the concepts for space vehicles were first evaluated. The Army undertook the launch of the first space satellite, and demonstrated that communication from space was possible. Shortly thereafter, the concept of tracking ballistic missile launches from an orbiting space platform was first discussed, only to disappear from planning. It was a time when ballistic missile launch preparation and test launches were the big items at Cape Canaveral. I remember being excited that my assignments related to the infrared tracking of ballistic missile launches, as well as with a program called Sentry.

**Launch Responsibilities at Vandenberg AFB**

The experience I gained at the Eastern Test Range as a launch control officer was probably the primary reason that I was transferred to Vandenberg AFB in California in 1958.[3] While at Vandenberg, I accumulated more missile launch experience than anyone in an Air Force uniform, and by virtue of this experience I became an expert in launch pad operations after more than 150 launches.

*A New Concept for Launch Activities.* When I initially examined the launch activities at Vandenberg AFB, it became apparent that a new concept for launch preparation and countdown was required. The individual contractors would have to prepare, checkout, and launch only their portion. Personnel performing the checkout of individual subsystems within a single

---

[2] The X-17 test vehicle was designed for the Air Force by Lockheed. This missile was assigned the role of obtaining data on intercontinental ballistic missile warhead characteristics, while reentering the atmosphere at hypersonic speeds. Of the 26 X-17s fired by the Air Force, 20 were considered successful in obtaining aerodynamic data, and one reached over 9,000 mph. Later the Navy used the vehicle as part of the Polaris test program. The X-17 also was employed in the Argus tests in "out-of the-atmosphere" nuclear explosions. (See Frederick I. Ordway III and Ronald C. Wakeford, *International Missile and Spacecraft Guide*, New York: McGraw-Hill, 1960, 54.)

[3] The Eastern Test Range is an area that begins at Cape Canaveral and extends southeast over the Caribbean and Atlantic Ocean. The Test Range provides static and flight-testing support to military (DoD), civilian (NASA), and commercial space and missile launch customers. The range hosts telemetry receiving stations, communications stations, radar tracking stations and other type of instrumentation over an area that eventually expanded to a 5,000 nautical mile range.

contractor's vehicle stage would not necessarily know how their subsystem interacted outside of their own stage. The combination of multiple and independent stages or systems meant that the preparation and countdown of the individual subsystems would be performed in parallel, isolated from the other components, with only a single controller knowing the inter-relationships.

Communications would not be as open as in the past. Where interaction between stages did occur, the results would not necessarily be discernable to all. Countdown manuals needed to become checklists of meter readings, responses, and acceptances. With parallel actions going on, communications information and reactions required that the main controller have binaural, tertiary, or greater monitoring capability. The launch preparation and countdowns themselves became parallel functions needing integration, further compounding the communication and control problems.

*My Role as a Military Officer in Launch Operations.* For the first time a military officer (rather than a contractor) could start, stop, or interrupt a contractor's technical sequence. A military officer acted as the total system controller, and would interpret the results of a test about which the contractor had only limited information, especially with regard to its relationship with other systems. This situation presented a new and different concept for checkout and launch. Parallel countdown activity had to be represented in the manuals, and in the range and operational aspects of a launch activity.

Until that time, the idea of a military officer developing launch checkout procedures and writing the countdown manuals was a new approach. Anytime a realignment of management responsibilities must be made, it requires deft and resourceful maneuvering by a skillful arbitrator. It is rare for a person or group to relinquish its authority and prestige, and achieving this was my challenge!

At Patrick AFB, I was basically a government launch director who coordinated between the Eastern Test Range and the system contractors. In contrast, at Vandenberg AFB I became an integral part of the total operation. I coordinated the separate preparations and tests of the individual contractors, integrated the test activities, and interpreted the readiness for the subsequent launch. I felt an intense pressure not to overlook even the most minute of details in order to be able to assure the proper authorities that all was tested and the launch had a high probability of success.

Others helped me in this effort, but for security reasons many were limited in their level of information about the various systems. It was like being alone to work my way out of a very critical and intense situation. Launch operations are the collection of a large amount of information about the systems being prepared, the integration of the various parts into a composite status, and the understanding of both the technical aspects of the launch operation and the mission communications and readiness.

**My Experiences with the Discoverer/Corona Program**

My first job at Vandenberg was to build the launch pads for the multistage orbital program called Discoverer, which was the scientific cover program for the classified Corona photo-reconnaissance satellite program. There certainly were others who had more experience than I did in building cement platforms, propellant-loading compounds, and steel structure towers, but the new space vehicle launch complexes proved to be different. The launch vehicle was now comprised of several stages: the booster, the satellite vehicle, and the payload (with its associated communications link to the tracking station).

*The Challenges of Integrating and Preparing the Vehicles.* The various builders of the launch areas were knowledgeable about their individual requirements, but they only had limited information about how the different stages fit together functionally during the checkout and launch preparation, or how they inter-related operationally during the countdown and launch.

Because of this lack of comprehensive understanding of the bigger picture, for the first time the civilian contractors did not have the unlimited control as they were accustomed to having at Patrick AFB. At Patrick AFB all of the participants in the launch preparation and countdown had knowledge about almost everything concerned with the system being launched.

Initially, at the Eastern Test Range there was usually a launch vehicle contractor who had primary responsibility and who served as the controlling agent for launch operations. At Vandenberg AFB this no longer was the standard operating procedure. The launch preparation and subsequent countdown for a system such as an intercontinental ballistic missile, with additional stages and functional components, is a very complex undertaking. Consequently, the multistage vehicle had different contractors, with each contractor responsible only for their own vehicle stage or system.

The ability to assure post-launch success depends upon precise attention to the many facets of each missile stage and system during preparation and launch. This was accomplished at Patrick AFB by having a prime contractor integrate the many facets of the missile system. At Vandenberg AFB, however, there was no single prime contractor with overall responsibility.

*First Successful Discoverer/Corona Recovery.* The Corona Program employed a Thor intermediate range ballistic missile as a booster, and an Agena second stage that also performed as the orbiting satellite. I remember the first successful recovery of a Discoverer capsule after its trip through space. The capsule contained an American flag that was given to President Eisenhower. It was a big event in the news. I still have the picture that was taken with the Vandenberg launch crew surrounding the capsule. After this successful test mission under the cover of the Discoverer program, the Corona program went on to satisfy the many intelligence requirements levied on the NRO.

*Corona Configuration and Mice Cargo.* One of our publicly announced cover payloads was a capsule of mice. The cover experiment was that we were sending the mice into space in order to monitor their performance while in orbit. We experienced one point in the countdown where we did not receive the proper responses from the "mouse capsule." To determine if the mice were awake, we shook the whole Thor/Agena vehicle. Everyone agreed that this was a test that really used the modern technology of the time!

*Disposing of an Unusable Engine.* I remember one of the early Thor/Agena launches when there was a problem that caused the range safety officer to terminate the flight. The Thor main engine fell back to earth near the blockhouse. The main engine no longer was usable, and I asked

Figure 40-2. Vandenberg launch crew with a returned Discoverer capsule. From left to right: William Diener, Will Heisler, General Joseph Cody, Joe Foss, Rob Roy, and Roy Lefstad, August 1960. (Photo courtesy of Rob Roy, likely U.S. Air Force photo.)

if I could have it. This request was approved. I had the engine cut in half so the technology of rocket motors could be easily observed. I sent half to the University of California at San Luis Obispo, and the other half to my sister who was a teacher at my high school in Somerset, Pennsylvania.

Figure 40-3. A Thor/Agena launch between 1959 and 1960 (Photo courtesy of Rob Roy, likely U.S. Air Force photo.)

**Assigned to the Atlas/Agena Program**

Following my work with the Corona effort, I was assigned to the Atlas/Agena program. Here again, the first order of activity was to build the launch area including the gantry tower, fueling complexes, and blockhouse. When the boss (General William King) first visited the launch area, I asked him to join me in walking up the open staircase of the gantry tower.[4] This experience provided a pleasant view of the Pacific Ocean, but it also gave a rather precarious feeling of being hung far above the ground, similar to the steelworkers constructing skyscrapers. General King has never let me forget this introduction that I gave him to the pad crew.

Communications became an even greater problem, and I spent a considerable amount of time developing a binaural system that automatically switched checkout channels. Effective launch procedures and communications both prior to and following a launch were critical to mission success, and I was deeply involved in these efforts.

*Pre-Launch Work.* My recollections of launch operations are of a kindred spirit with many people performing complex and detailed tasks. The work began with the arrival of the booster stage, which followed the complete refurbishment of the launch area. The Atlas booster was a majestic sight. It towered far above the landscape, and yet was little more than a metal balloon that could not support its own weight without being pressurized. Shortly thereafter, the Agena satellite stage arrived for installation on top of the Atlas. The checkout that followed was a beehive of activity. Test procedures were conducted day and night in seemingly planned confusion. This frenzy was a result of the urgency to be ready on schedule, which overlapped with the work being performed by the many subsystem contractors.

Figure 40-4. Pioneer Rob Roy at the Thor/Agena Launch Complex, 1959. (Photo courtesy of Rob Roy, likely U.S. Air Force photo.)

In order to keep the various subcontractors aware that they were all working to integrate one system, I initiated a process of an end-to-end simulated flight demonstration. The process was oriented toward a consolidated checkout that went beyond the normal preparation of the subsystems. After separate subsystem testing, we performed a simu-

[4] William G. King (retired as Brigadier General) served as NRO Program A Director from August 1969 to March 1971.

Figure 40-5. An Atlas launch, circa early 1960s. (Photo courtesy of Rob Roy, likely U.S. Air Force photo.)

lation of the orbital flight. This process provided information that increased our confidence in the likelihood of mission success. If necessary, subsystems were repaired and testing and simulation were conducted again. We worked around-the-clock to meet the schedule.

*Post-Launch Work.* There were several post-launch tasks that had to be completed in order to prepare for the next launch. The launch area had to be refurbished, the equipment had to be re-calibrated, the propellants had to be replenished, and any problems had to be corrected. Union members from a local labor pool performed much of this work. The union, understandably, did not wish to hurry themselves and run out of work. On the other hand, the program managers wanted to be ready for the next launch order. In order to maintain harmony between the two parties and to keep on schedule, I frequently found myself performing additional functions. For example, I helped union crews clean up and prepare for maintenance to the damaged pad area, I performed cable wrapping with the missile contractor crews, I joked with the union stewards, and I told stories to union and contractor crews at three in the morning in order to keep them working around the clock.

### Dealing with Security

Security requirements created significant constraints on how we operated. This situation meant patience was necessary in terms of handling misunderstandings. Most of the participants in these secure programs did not hold the clearances necessary to access all levels of the program. An example was when a program representative came to the launch pad to inspect a specific portion of the payload for which he was cleared. This person was sent to the launch pad without properly alerting those involved with payload security. Because the person only had the proper clearances for a portion of the specific mission, he was denied access to the gantry tower. He became indignant, and was a very unhappy individual since he felt that he had not carried out his function for that operation. These were the kinds of security complications that we had to deal with while trying to remain focused on operational matters.

Security also had an impact on the operational schedule. We met our schedules, but sometimes with a little difficulty. A nearby railroad presented one such security challenge. A public railroad ran through Vandenberg AFB, and because of the classified nature of our launches, it was a common occurrence for us to suspend a launch countdown while a train passed through. Soviet Premiere Nikita Khrushchev was once a passenger on one of the trains that passed within a short distance of the launch area.

### Some Lessons

The challenges we faced gave us opportunities to learn lessons and to change how we did business. Four examples illustrate this point: the importance of end-to-end simulation, the value of diplomacy; the importance of details; and the value of alternative communication channels.

*The Importance of End-to-End Simulation.* The first launch of the Corona program ended in disaster, and turned out to be a lesson in why details are important. Hundreds of hours of preparation, checkout, test, problem fixing, and retesting took place. The Agena satellite had every aspect of its mission performance checked and rechecked. However, there was one aspect that had been overlooked under the dictates of safety. The testing was done without the retro-rockets installed. The first time the retro-rockets were on the combined Thor/Agena payload was the day of the planned launch. One of the last checks before liftoff was to "gimbal" the Agena engine.[5] When power was applied to the gimbal motors, a circuit routed power to the retro-rockets, and they fired into the nitric acid and hydrazine fuel tanks. This resulted in the tanks rupturing, personnel running from the launch pad in a frenzy, and the release of toxic acid fumes which frightened all who witnessed the mishap. Fortunately, no one was hurt, except for their pride. This event was one of the primary reasons for the initiation of an end-to-end flight simulation that became standard for all operations during my tour at Vandenberg.

*The Value of Diplomacy.* Corona and other classified photographic satellite launch activities were given a home at Vandenberg AFB, which was a Strategic Air Command (SAC) base. Vandenberg was new to the test range concept at that time. This arrangement differed from Patrick AFB, which was a contractor-oriented test range. At the time, SAC considered itself (probably correctly) the primary command in the Air Force, and SAC officers were not in the habit of taking orders from anyone other than SAC commanders. Consequently, SAC personnel were adamant about asserting their primary role at Vandenberg AFB, which operated in a culture oriented towards military operations.

There were many new situations and problems that arose at Vandenberg that resulted from the strained working relationship between SAC and non-SAC personnel. For example, SAC personnel believed they should develop the safety rules and range operating procedures, and control all supporting activities. They were intolerant toward the Test Wing, which launched satellite vehicles from their range. The SAC officers felt that the most important mission was the Atlas and Titan intercontinental ballistic missile force. This environment provided another motivation for me to learn to become a military diplomat.

The tensions came to a head when, during an Atlas/Agena launch, the Atlas booster lost pressurization after the camera payload had been installed. It was only a matter of time until the pad pressurization system would be depleted and the Atlas would collapse. The Agena, with the payload, fell to the pad surface as the Atlas collapsed, but fortunately there was no explosion. The SAC safety crews wanted to go to the pad immediately and stop access until the rocket propellant had been cleared from the launch area. However, it was necessary for us to first cover the payload and make arrangements to have it removed before safety restrictions could be imposed. This was my finest hour of diplomatic maneuvering between a bastion of SAC officers and a solo field officer from the Test Wing.

*The first launch of the Corona program ended in disaster, and turned out to be a lesson in why details are important.*

*The Importance of Details.* An orbiting mission failed because the camera doors on the satellite failed to open. Telemetry indicated that the explosive bolts that should have opened the doors had fired, but the doors failed to respond. The ensuing investigation was extensive and complex. The cause of the problem finally came to light when a clever engineer completely dissected the explosive bolt. Until then we did not believe the bolt was a problem because of the telemetry readings, but we were wrong. The bolt manufacturer had, without notification, modified the manufacturing method of the bolt, which resulted in a lower thrust of the piston when it was fired. Consequently, there was not enough thrust to open the door. To ensure this problem would not recur, we installed two redundant explosive bolts so that the doors would open even if only one bolt fired. The bolt was a small piston type, in which

[5] To gimbal is to move the engine nozzle right and left, fore and aft.

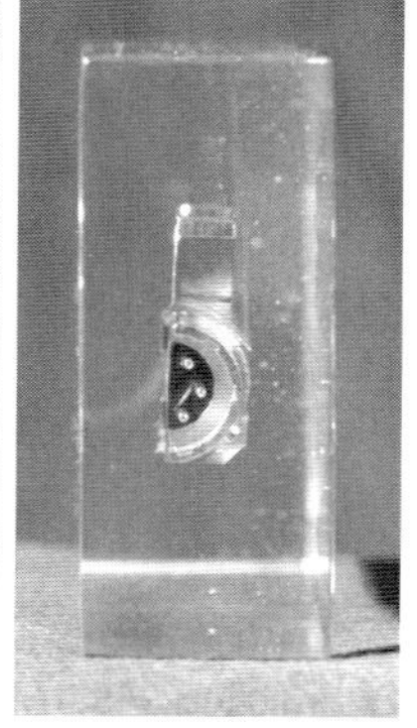

Figure 40-6. Four views of a defective explosive bolt encapsulated in plastic. Notice the inscription "Lest We Forget." (Photo by Sara Judy, NRO Visual Design Center, encapsulated bolt provided by Rob Roy.)

the explosive acted in much the same manner as an automobile piston when the gasoline is ignited. I still have a plastic encapsulation of the explosive bolt, cut in half to show the modification to the fabrication method. The inscription on this plastic block reads "Lest We Forget." This is an example of the need to pay attention to even the smallest detail.

*There is a Value in Alternative Communication Channels.* Open channels of communication between the appropriate authorities are a necessity. I cannot emphasize this aspect too strongly. This requirement may mean that at times rank or hierarchical structure gets bypassed. For example, Defense Contract Administration Services (DCAS) was assigned the responsibility for approving overtime under the government contracts. The desired launch windows and schedules came from the Program Office. More often than not these schedules required overtime work by the contractors who, for security reasons, could not adequately explain this to DCAS. The DCAS officials did not have the necessary security clearance. I discussed the problem with Brigadier General Robert Greer who informed DCAS that I could authorize overtime if necessary.[6] The DCAS complied, but very unhappily. The point of this example is that communication without going through hierarchical channels resulted in solving a problem that was causing schedule delays.

**Conclusion**

During my involvement with national reconnaissance the programs were cloaked in an aura of secrecy and technical advancement. For example, the launch pad operations were new and unique, and the safety procedures had to correspond to both the security requirements and the complexities of the new systems. The possible consequences of accidents that threatened exposure of our classified work were ever present, and procedures were required that could be followed automatically. Unique solutions and unbounded creativity were essential, and we spent many early morning hours solving problems and overcoming challenges. That hard work paid off when we launched payloads successfully.

I am very proud that my contribution to NRO successes was to put the wonderful inventions and engineering achievements of others into orbit where they could perform their important missions. In my opinion, I had the best job in the Air Force.

---

[6] Robert E. Greer (retired as Major General) served as Program A Director from July 1962 to June 1965.

**We Were a Tight-Knit Group of Test Crew Wives**[7]

Figure 40-7. Rob Roy preparing input for his recollections at NRO Headquarters, Pioneer Recollections Day, 26 September 2000. (Photo by Candi Campbell, NRO Visual Design Center.)

Back in the "old days" when Rob and I were dating, he was attending the Naval Academy, and I was at Seton Hall College in Greensburg, PA. He would talk incessantly about space and tell my friends that man would be going to the moon in the not too distant future. My friends thought that he was some "far-out guy" with a big dream. I did not see him that way, and I was not surprised when he joined the Air Force to study electronics and missiles.

After we married and started our Air Force life at Patrick AFB in Florida, it seemed natural that missiles would become a large part of his life. From the beginning —the years at Patrick and Vandenberg— I felt that the work that these men were doing was a series of baby steps to ultimately sending men into space. While we were stationed at Patrick, the space age took on a new dimension when the Russians succeeded with their launch of Sputnik. The race had begun.

By the time we went to Vandenberg AFB in California, we had two small daughters. Our son was born later. When we were at Vandenberg initially, there were only twelve officers attached to Rob's Test Wing. They called themselves the "dirty dozen." They worked constantly, and they also became a very tight-knit group, as did the wives. Of course, the test wing grew, and the work seemed to grow accordingly. The long hours presented the greatest challenge for the wives, all of whom lived in quarters on base. We were unaware of the underlying reasons for the urgency of the programs, but we could tell the work was very important, and was given top priority, because we were frozen at Vandenberg. As the Cold War escalated it became apparent that these programs, and the success of the missions, were closely connected to national security.

Figure 40-8. President John F. Kennedy. (Photo courtesy of NRO History Office, National Archives photo.)

We could view the launches of the space vehicles merely by stepping out from our front doors, and it never ceased to be exciting. I believe all the wives were proud of their husbands for their dedication to the programs. We often gathered together to have a meal and to unite our friendships. The husbands often would be late for dinner, and sometimes we had to eat without them. We even had a flair for the abstract, as evidenced by one of the family's dogs named "Payload."

Several historical events occurred while we were at Vandenberg, and they stand out in my mind as though they occurred yesterday. One of these events was the Cuban Missile Crisis. I joined some neighbors and watched President

[7] This section is based on an interview with Patricia Roy at National Reconnaissance Office Headquarters, 26 September 2000.

Kennedy address the nation. Our husbands were at some unknown place because they had been put on alert. It was a nervous time, and it was a huge relief when Khrushchev backed down and the threat of World War III was averted. Sadly, the other event I vividly recall was the assassination of President John F. Kennedy. The unthinkable had happened and the nation mourned.

In spite of the long hours of work, I believe our years at Vandenberg were by far the most wonderful years of our Air Force career. We formed close and long-lasting friendships. We were all held together by a common bond at that time, and that bond has held us all close in the spirit of recognizing the Pioneers.

—

**Pioneer Award Presentation and Citation**

Figure 40-9. Pioneer Rob Roy (second from left) being recognized at the 2000 Pioneer Recognition Ceremony. The pioneer award plaque was presented by DNRO Keith Hall (left) and DCI George Tenet (third from the left). Brigadier General Craig P. Weston (Director, Corporate Operations Office and Chief Information Officer, NRO) joined in the presentation. (Photo by Sara Judy, NRO Visual Design Center.)

**Robert W. "Rob" Roy, Colonel, USAF**

Colonel Robert Roy directed National Reconnaissance Office launch operations at Vandenberg Air Force Base at a time when these activities increased dramatically. He supervised the construction of launch pads that served more than one program, and established range safety requirements, ground safety procedures, and launch preparation procedures that became standards for reference.

*Career in National Reconnaissance: 1958–1964*

# Charles P. Spoelhof

Charles Spoelhof, an optical engineer at Eastman Kodak, developed camera systems that were widely employed in United States national reconnaissance capabilities. In his recollections, Spoelhof speaks of the challenges of developing a space-based camera system, the contributions and challenges faced by the Eastman Kodak team, and the impact of his work and related secrecy on his personal life and family.

## An Attitude of Country First, Company Second, and Self Last[1]

When people look back at the advancements in national reconnaissance it is easy for them to say, "Satellites were the new stuff and film was 'old hat'. You could have put any available film in the camera to take pictures." That assertion is simply not true. The research and development done by Eastman Kodak with its own funds, combined with the development paid for by the government, resulted in the creation of new films, special processing, and photographic technology that made Corona and other programs possible.[2]

Other companies, such as Itek and Perkin-Elmer, were doing work similar to Kodak, but Kodak had the knowledge of the film, and this was an essential ingredient for the success of the entire imagery reconnaissance mission.

### Samos

Eastman Kodak's introduction to national reconnaissance occurred with the Samos program. This program was a major step in terms of getting Kodak into space. Lockheed was the prime contractor for Samos, with Kodak and several other firms as sub-contractors. Prior to that, Kodak had an enormous amount of photoreconnaissance experience during World War II and the Korean War, as well as with the U-2 program. However, entirely new technology had to be developed for space reconnaissance.

Samos and Corona were competing technologies: Samos radioed images back to Earth, whereas Corona returned film from space. The government developed both and then determined which could better meet the mission requirements. The Corona system had the advantage of recovering the film; therefore, it was not limited by bandwidth constraints

[1] This section is based on an interview with Charles P. Spoelhof at the National Reconnaissance Office Headquarters, 26 September 2000.

[2] Corona was the first photoreconnaissance satellite program.

Figure 41-1. Amrom H. Katz. (Photo courtesy of the NRO History Office.)

and could collect more imagery. The Samos system had the advantage of continuous coverage over long satellite life.

Although the Corona program was the one to survive, I believe the Samos program, with its huge demand for high-resolution photography, was a breakthrough for all later systems.

What Kodak built for the Samos program worked well; however, only one of three Samos launches succeeded in placing a payload in orbit, and that payload was compromised by the failure to eject an antenna that partially obstructed the camera. System capability was later demonstrated when Kodak provided the National Aeronautics and Space Administration with the photographic payload for the Lunar Orbiter Program using a modified Samos design. There were five launches and all five were successful, mapping the moon for the Apollo landings.

Our team worked well together with a competitive attitude of, "Hey, this is tough, but we can do it!" The competition was not within the team; rather the competition was, in a sense, with the technology itself. We always wanted to do more, but it had to be sound engineering, so we were a bit on the conservative side to be sure our objectives could be accomplished. There was also a bit of competition between Kodak's Samos team and Itek's Corona team as to "Who could get there first?" In 1960, the Corona program successfully returned the first images from space, and Samos was second to succeed when it returned images early in 1961. However, the Samos program was then cancelled in favor of the Corona approach. I am sure there were more thoughts behind this at high-level decision-making within the government than I was aware of at the time.

**Corona and the Challenge of Film in Space**

Kodak's role in the Corona program was to provide the photographic film. The special film requirements for use in the Corona camera were so excessive that there was no film at the time that would work. We had to come up with a new and stronger film base. During the life of the program we kept making the base thinner so the satellite could carry more film and, therefore, could acquire more imagery from each mission. In addition to developing a stronger film base, Kodak developed improved film emulsions and new processing and duplicating technology.

Photoreconnaissance during World War II and the Korean War involved aircraft that required fast film because the platform vibrated. Aerial photographers had to take pictures at a very short exposure time or the images would smear. In addition, an airplane could carry a fair amount of weight, but when we went to space, payload weight was greatly limited. Early satellites could carry only a very small payload and cameras had to be miniaturized. As a result, we needed extremely high-resolution film that could push right to the limit of what could be done within the laws of physics.

Our solution was to develop a very fine-grained film that required a microscope to see the detailed image. Processing of this new film had to be done very carefully. The environment had to be extremely clean because a speck of dust on the film would look as big as a battleship when viewed under a microscope.

Our success depended upon a competent team of people at Kodak's research lab and film manufacturing. Another group of dedicated people developed new technology for processing and duplicating the film that was returned from space. These elements of the process were just as important as building the satellite. What at first seemed impossible became possible. Without this new film and technology the United States (U.S.) would never have accomplished what it did with Corona and other early space imaging systems. Photo interpreters would never have been able to see adequate detail.

**Designing More Advanced Cameras for Space**

As we worked to develop the technology to advance satellite photoreconnaissance beyond Corona, we had to consider several factors that are fundamental to camera systems: focal length, stability, image motion, aperture, and film.

*Focal Length.* Amrom Katz was involved in lens and camera design for reconnaissance systems, and he performed the first experimental simulation of electro-optical imaging by a reconnaissance satellite.[3] He was extremely bright and one of the great leaders in aerial photography.

A Katz rule of thumb was that there is no substitute for focal length, the longer the focal length the better the picture. That became the design philosophy of aerial photography. However, for satellite photography, weight and space were very limited.

*Stability.* A major factor in aerial and space reconnaissance is platform stability. An unstable platform requires a short exposure time or the image will smear. Fortunately, space is a pristine environment because there is no aircraft engine vibration shaking the camera. As a result, a longer exposure time can be used along with fine-grain, high-resolution film.

*Image Motion Compensation.* Although space is a quiet environment, the camera platform moves at orbital speed causing the image to move rapidly. Unless the exposure time is extremely short, the image will smear. The way around this problem is image motion compensation. One method we often used was to move the film, at the same speed as the image, behind a narrow slit. This type of camera is called a strip camera. The system automatically advanced the film, and the slit width controlled the exposure time.

*Aperture.* The aperture, or the diameter of the lens, is an additional constraint we had to consider. Amrom Katz believed there was no substitute for focal length, but I concluded that if a lens were constrained only by the physical limitation of diffraction, then longer focal length does not necessarily help. The longer the focal length, the fuzzier the image, but the image becomes larger and the two compensate for each other. However, if the lens aperture is enlarged, then the image gets sharper and more detail can be seen. That is why astronomers make big telescopes; it is the size of the aperture that counts. However, a larger aperture consumes more weight and space, and these factors are at a premium in a satellite payload.

*Film.* If the aperture is made as large as possible and the platform stability and quality of image compensation are known, then there is one more consideration—the film. A long focal length (slow lens speed) requires a fast film with course-grain, whereas a short focal length can use a slower, fine-grained film of higher resolution, but at a smaller image scale. It turns out that these two almost exactly compensate for each other, producing the same

---

[3] Amrom Katz was a physicist with RAND Corporation involved in lens and camera design for aerial systems. He co-directed a project on peacetime overflight reconnaissance, and co-proposed film-recovery satellites as an immediate alternative to the near-real-time readout satellite then under development as the Samos program. Amrom Katz was honored by the National Reconnaissance Office as a Founder of National Reconnaissance, 27 September 2000.

ground resolution. However, the film load will be considerably reduced for the shorter focal length and the smaller image scale. Therefore, the challenge became to make as sensitive and fine-grained a film as possible.

*Putting It All Together.* Based on fundamental work that had been done at the Kodak Research Labs, I did a study of these major factors to be considered when designing an optimum reconnaissance system for space. It resulted in a minimum weight and volume of camera and film to achieve the specified ground resolution and coverage. It was an efficient, compact system. In the end I had modified Katz's rule to be, "In an optimized system there is no substitute for aperture." Kodak's goal then became to produce larger diameter lenses that were perfectly corrected, and to make film of adequate speed with finer and finer grain.

*That chart would be almost a document in itself, although it was destroyed years ago.... We called it the dynamite chart because when it was rolled up for security purposes it looked like a stick of dynamite.*

This was new information at the time, so I put my ideas into a briefing and took it around the national reconnaissance community to anyone who had the appropriate clearance. Art Simmons told me, "I want you to give your briefing at the Central Intelligence Agency (CIA)."[4] I was instructed to go to Dulles Airport and somebody would meet me there. Sure enough, somebody met me there and secretly slipped me into a big, black car and brought me over to the CIA Headquarters building to give my briefing. When I finished my briefing, they put me back in the car and back on the airplane home. The security around this developing technology was so tight that, in many cases, nobody but my bosses knew where I was going when I took these trips.

At one point when I was briefing Edwin "Din" Land, he looked at some simulations we made—pictures of what the satellite possibly could do.[5] I remember Land looking at the pictures carefully and saying, "I do not think this is good enough. Can you do better?" I started to tell him my idea for a new system and Land said, "I've got to go to another meeting. You stay here and dictate to my secretary what you were just telling me." Of course the first thing I did was call my boss, Art Simmons. "Art, if I propose this, I am committing the company. Are you willing to back me up on this?" He said, "Go ahead."

Not long thereafter, Art Simmons got a phone call from a general at the Special Projects office on the west coast. The general said, "Art, I am told I have to give you a contract, but I do not know what it is for. Can you come out tomorrow and tell me?" Art said, "But tomorrow is Sunday." The general replied, "Oh well, come out Monday." So Art flew out and told him about the system. We put together a proposal in two weeks. Our proposal was a concept, but it sold after a more thorough engineering investigation. That contract amounted to several years of business for Eastman Kodak. Perhaps another individual or company could have conceived this idea as well, but I happened to be in the right place at the right time with the right person in order to initiate that system.

**Lessons We Learned**

I believe much of our success at Eastman Kodak was the result of our operating principles. These included: integrated management oversight, protection of secret projects, making responsible estimates of cost when bidding for contracts, and maintaining attitudes for success.

[4] Arthur Simmons was the director of Eastman Kodak's Research and Engineering Department, which included the Special Projects Group

[5] Edwin Land was the founder and CEO of the Polaroid Corporation, and served as an advisor to Presidents Eisenhower, Kennedy, Johnson, and Nixon. The National Reconnaissance Office honored Land as a Founder of National Reconnaissance, 27 September 2000. See chapter 4.

*Integrated Management Oversight—My Dynamite Chart.* Integrated management oversight is a concept that probably is best exemplified by something we called the "dynamite chart." The dynamite chart helped us to integrate oversight, and was a management tool I put together under Art Simmons' direction. The chart was a short-system summary of all the things that were going on in the programs. I would put cryptic numbers on the chart and Art and I would review it asking, "Where are the holes in the system? Where can we do better?" That chart would be almost a document in itself, although it was destroyed years ago. It represented the whole community effort of the NRO. With that insight we could see where it was necessary to invent things that could improve the programs. That is a privilege I had, and it gave me a great deal of background that I could use to develop better systems. We called it the dynamite chart because when it was rolled up for security purposes it looked like a stick of dynamite.

*Protection of Secret Programs.* At Kodak, we took great care about security. I remember some other companies advertising what they had done in their government contracts. They needed that, perhaps, in order to compete with each other, but there was no one other than Kodak supplying film, so we could be quiet and operate from a secure standpoint. We took security very, very seriously.

Our whole management philosophy incorporated security. The company stressed security over taking corporate credit. I cannot remember who made this statement, but it is a rather famous quote: "It is remarkable how much can be achieved if you do not worry about who gets the credit." That was the attitude of Kodak. At times some of us groused about it saying, "Others can say something of what they are doing, and we cannot!" Nevertheless, our management insisted our work remain totally secret.

*Making Responsible Estimates of Cost.* Kodak took great care to ensure that it not only provided sound engineering for its products, but it also did accurate estimating for contract costs. Our management refused to buy into a contract. As a result, we had the reputation for being expensive, but our estimates could be accepted with confidence.

*Maintaining Attitudes for Success.* If I were to summarize the attitude at Kodak, and probably a key attitude of the entire community, it could be captured in this bold statement: "commit to the difficult, if not the impossible." We were stretching technology to see how far it could go. Another fundamental attitude that contributed to our success was very important to me, and I believe everyone who worked on this program felt this way: Country first, and company second. Yes, we had to make a profit or our management would say, "I'm sorry, we are not going to pursue this effort." But the government was very good about covering all of the expenses and ensuring that companies made a profit.

**My Colleagues**

I am very proud of the Kodak organization and my group at the company. We were not crawling over the top of each other to see who was going to get ahead, and there was very little office politics. We did our jobs, and we respected our leaders. It was one of the finest groups with which I have worked. We all appreciated each other.

The atmosphere was intense and at times we needed relief. I recall an incident when those I supervised organized a party. They roasted me, and that was great fun. I have maintained friendly relationships with many of my colleagues long after the project ended.

Art Simmons and Ed Green are two people I consider greater pioneers than myself. I worked for both of them. My job was technical, and I did a fair amount of the inventing for the systems, but they put together organizations that made our accomplishments possible. I gladly would step aside and say that they ought to be honored as pioneers instead of me. They are both deceased. In Annex 1, I tell a little more about Art and Ed, as well as other colleagues.

### A Dedicated Team

Eastman Kodak provided many opportunities for my co-workers and me. Although part of a large corporation, our organization was an independent core comprised of some of the finest people within the company. As we expanded, we could hire the best and we put together a team second to none. It was this team effort that brought about revolutionary advances and spectacular success.

I had the privilege of representing Kodak to the government community, and used that opportunity to further the understanding of the science and technology of high-performance reconnaissance systems. I was supported by a very dedicated group of people. Some of them trained me, many encouraged me, and all were dedicated to making our projects succeed. They provided the engineering, manufacturing and analytical support that brought success to the programs for which we were responsible. While I was away talking with people like the Director, National Reconnaissance Office Joe Charyk, members of the President's Advisory Board, and Din Land, others back home were doing the hard engineering work.[6] When I would return, their concern was, "What did Chuck commit us to now?"

*As we expanded, we could hire the best and we put together a team second to none. It was this team effort that brought about revolutionary advances and spectacular success.*

The team we assembled was an extremely dedicated group of individuals, but we also had fun. By getting together with colleagues' families we made sure that our spouses also enjoyed in the fun.

### Conclusion

I was 24 years old and just out of graduate school when I first became involved in national reconnaissance, initially with the Genetrix program and then with the Samos program soon thereafter. I spent most of my Kodak career working in the national reconnaissance field.

I married when I was 23 years of age, and I am fortunate that I have a very dedicated wife. Kay recognized that I was involved in work that was very demanding, even though I could not tell her what it was. It turns out that she guessed quite accurately. She tells that story later in this chapter.

When I learned that I had been selected as a Pioneer of National Reconnaissance, I wrote a letter in which I explained to my children why there had been secrecy and absences—why I was gone so much. It was hard on the family. I got a letter back from my daughter recounting some of her thoughts on this period in our lives. Everyone in this business faced the same problems. I imagine it even rocked some marriages, but fortunately not mine since Kay was very supportive of my work. My efforts contributed to a great deal of success at Kodak, and I am very pleased with that. I remember working many long hours and often in the evenings at home. However, I tried to dedicate every weekend to my family. I was home at least on Sunday and many of the Saturdays.

When I first was involved with national reconnaissance we had two young children, and two more children were to follow. The children were always asking, "Where's Daddy? When is Daddy coming home? Why isn't Daddy home more often? Why isn't Daddy home evenings?" I often stayed late at the office because we either had an emergency to deal with or we were on overtime because it was a crash program. One day when I was home in the evening I said, "Well, I'm home tonight." My daughter, who was very young at the time, began to dance and sing, "Daddy's home! Daddy's home!" Then I began to realize how much I had taken away from them.

[6] Dr. Joseph Charyk was the first Director of the National Reconnaissance Office (DNRO), from 1961-1963.

—

### My Husband has a Curious Mind and a Commitment to Work[7]

Chuck was brought up in a minister's family where if the children knew any secrets about people they were not supposed to tell. If someone came to the door upset the children were told, "Don't talk about this to anybody else." As the daughter of a church elder, I was raised the same way.

When I became aware of Chuck's contributions to national reconnaissance it considerably broadened my understanding of his career. I did not think much about his work at the time. I knew what he was doing was important, but our children and I had no knowledge about the details. We were secure in knowing, even though we did not always know where he was, that he was doing a job for the country.

**Keeping Secrets**

Chuck and I did not make a big thing of secrecy with the children. I would tell our children, "If your father does not tell you anything, it means you will never be tempted to talk about things that would get you into trouble for saying something wrong." Our most curious son asked, "Just what does Daddy do at Kodak?" He was a little boy, five-years old, and his mind was going all the time. I explained, "That is the way he makes the money that he brings home so he can pay for the house, food, and all the things that we need." His mind kept going, and he finally said, "I sure would like to see that machine." I am sure he thought his dad was printing counterfeit money!

*Security Checks.* When Chuck first began working for Kodak I asked him, "How's your work going?" He replied, "Well, it is boring because I'm not really doing anything. I really cannot get into the work until I get my clearance." I was then introduced to "security checks."

Chuck came home from work with forms that asked for information about my family. My parents had emigrated to the United States from the Netherlands, and the government wanted to know if I had any more relatives there. They wanted to make sure that I would not be a security risk—that someone would not attempt to exploit my relationship with Chuck to get information. Our background is pretty straightforward, so that was not a problem.

Later on when we had our own little house, people would come around the neighborhood with dark suits and white shirts. My husband would say, "Well, don't be surprised if you see them go from door-to-door, because they'll probably be checking for more advanced clearance for me." I would notice that they would go to everybody's house and not stop at our house, and that is how I knew that they were not from a religious organization but from the FBI.

When Chuck was first being cleared for government work, I read in the papers about some developments in the field of national reconnaissance. And so I said, "I'm going to make a guess at what you're doing." He said, "Don't tell me. I will not tell you yes or no." So we never discussed it, but I had the "spy-in-the-sky" idea. I thought that was probably what he was doing. I did not ask details. I just accepted

*When Chuck was first being cleared for government work, I read in the papers about some developments in the field of national reconnaissance. And so I said, "I'm going to make a guess at what you're doing." He said, "Don't tell me. I will not tell you yes or no."*

---

[7] This section is based on an interview with Kay Spoelhof at the National Reconnaissance Office Headquarters, 26 September 2000.

Figure 41-2. Pioneer Charles Spoelhof (right) with his wife, Kay (left) looking at his medallion in NRO Pioneer Hall on Pioneer Recollections Day, 26 September 2000. (Photo by Candi Campbell, NRO Visual Design Center.)

the fact that Chuck was in the kind of work that he really could not talk about.

*Dreaming About Security.* Occasionally I would dream about someone approaching me for information about what my husband did. One time I had a very real dream that there were big guys, with Russian accents, and they were asking me "What does your husband do?" Nothing like that ever really happened, but I knew that if it did, I could look them right in the eye and tell the truth, "I don't know what he is doing."

Later on in his career, Chuck got involved in work that had absolutely nothing to do with national reconnaissance, but there were always things in his life that were secret. On the other hand, there were a lot of other things that were not, so I did not feel that I was part of an exclusive group. Most of the time, when we did have free time together, we did not talk about his work.

*Too Classified and too Technical.* Chuck often brought his work home, but I was never tempted to look in his briefcase. We would not have understood anything anyway. Our children were small, and I am not a mathematical whiz or scientific person. We did not hear classified information, and what we did hear about was too technical for us to understand.

My mother was visiting us when Chuck received a telephone call in the evening from someone at work. Later, after overhearing Chuck's side of the conversation, my mother said, "He was speaking English but I didn't understand anything he said." I answered, "You weren't supposed to, and I can't understand it either."

There was a company open house and I hoped to see Chuck's office. Ordinarily I never went there because it always was blocked off. I thought I really was going to see something in his office on this occasion. When we got off the elevator he said, "You see that window over there, the one that's painted over?" I said, "Yes." He said, "That's where I work, and we cannot go there." So that was a bit of a let down.

**Everything was an Adventure to Him**

Chuck always had a curious mind and an interest in science, particularly optics. He really loved his work at Kodak. If he had not had four children at home, he probably would have been very content to just remain absorbed in his work.

Everything about science intrigued Chuck. It did not matter whether it was physics, cameras, or living things, just about everything was an adventure to Chuck. He wanted to know how things worked. When we went on camping trips he would study the flora and fauna and even pick up snakes if he was sure they were not venomous.

Astronomy and space were other fascinations for Chuck. When we would see a satellite pass overhead, he would point it out and say, "Now do you see that? It looks like it is zigzagging across the sky, but your eyes are playing tricks on you." We would watch satellites, especially when we were camping where the sky was dark, darker than it was around Rochester, New York.

**For the Family, a Matter of Pride**

I was very much aware of how valuable I was in Chuck's life. With him being away on travel a lot, plus the long workdays, and sometimes eating dinner with people from out of town, the children would ask, "Where is Daddy today? Is he out of town?" When Chuck was in town I would say, "Don't worry, you go to sleep and when he gets home he'll take a look at you. I promise you." And when Chuck came home, he would. He would go around to each child and say goodnight in his own way.

My children and I watched some launches on television. When the astronauts set foot on the moon, I had an idea Chuck might have had something to do with that.

I really believe that the children did not feel that they were missing out on much. We managed to maintain a rather full life in spite of his work. It was never an obstacle. I was lonely at times, but it was not a stumbling block for anybody in our family. Recently, Chuck wrote a letter to our daughter and three sons. In it he wrote:

> "This letter primarily refers to past events when the four of you were still children at home. I did a great deal of traveling, usually for a few days at a time. For the first ten weeks after we moved into our new house, I was gone one or two days a week at least."

Even when he had a day off, sometimes he was so busy the night before that he was just wiped out, and that took a lot out of our week.

> "So I was leaving Mom with the task of watering the new lawn, keeping mud out of the house, and a myriad of duties demanded of a new home."

At the time I believed that was important. For Chuck, he missed time with the family. This is the way he closed his letter.

> "My job was often to tell government people what was possible and then to develop the new technology required to get better and better pictures. Therefore my face was one most often seen in high government offices while others back home slaved away doing the things we promised. They were exciting times, busy times, and challenging times. So I share the award not only with my co-workers, but also with Mom and the four of you who put up with my being away so much of the time."

When the National Reconnaissance Office (NRO) notified us that Chuck was to be honored as a Pioneer of National Reconnaissance, our daughter sent us a note (even in this electronic world we still send handwritten letters to each other). In the note she told us, "I remember when Dad was gone and once in a while he would be home for a night, and he would come home on time and he would not be exhausted, and we could play until we had to go to bed." She wrote, "I gave a really exaggerated jump for joy and made a big thing of it." Poor Chuck would feel guilty, but she was elated and they would play on the floor.

*When Chuck was in town I would say, "Don't worry, you go to sleep and when he gets home he'll take a look at you. I promise you." And when Chuck came home, he would. He would go around to each child and say goodnight in his own way.*

**Conclusion**

If Chuck were in any other occupation, he probably would still put a lot of time into his work. Chuck's work in the secret world has not been an obstacle. No, in fact it has been a matter of pride for us. Several years ago we were in Washington, DC, and we went to the Smithsonian National Air and Space Museum. The museum opened a section about reconnaissance, which was very exciting for Chuck. We toured the whole section, and Chuck felt he actually was showing me what he and his colleagues had done, and he was proud. I am very happy for him and very proud of him. Being with Chuck at the NRO 40th Anniversary celebration was my rich reward.

**Pioneer Award Presentation and Citation**

Figure 41-3. Pioneer Charles Spoelhof (second from left) being recognized at the 2000 Pioneer Recognition Ceremony. The pioneer award plaque was presented by DNRO Keith Hall (left) and DCI George Tenet (third from the left). Mr. Carl Marchetto (President of Commercial and Government Systems Division, Eastman Kodak) joined in the presentation. (Photo by Sara Judy, NRO Visual Design Center.)

**Charles P. Spoelhof**

Mr. Charles Spoelhof, an Eastman Kodak official, collaborated on the design of the U-2, A-12, and Samos cameras, and directed company efforts that led to the application of thin-based Mylar film in National Reconnaissance Office satellites. The latter contribution, which more than doubled the amount of film that could be carried into orbit, revolutionized film recovery systems.

*Career in National Reconnaissance: 1954-1985*

—

## Annex 1

### Notable Colleagues

There were many talented and dedicated people whom I had the benefit of working with throughout my career at Kodak. Art Simmons, Ed Green, Ken Macleish, and Bill Feldman were a few senior individuals who guided our programs, built our organization, and mentored many of my co-workers and me. I also had the distinct privilege of interacting on several occasions with Edwin Land of Polaroid. These men were key contributors to achieving what we accomplished at Kodak for the National Reconnaissance Office (NRO).

**Art Simmons and Ed Green**

Both Art and Ed were not only outstanding managers, but they were also excellent engineers. Art had a background in optics and Ed in photography, an ideal combination for the work we were called upon to do. Ed was a giant in the reconnaissance community. He strongly influenced policy as well as technology. Both were men of vision, highly respected for their judgment, and very influential within the NRO community. They drew support for our programs from all corners of Kodak and built a dedicated team of competent people. They insisted on sound engineering and accurate estimating. Above all others, I attribute our continued success to their leadership. I consider these two senior managers to be true pioneers of national reconnaissance.

A few stories may illustrate these claims and also demonstrate the human side of our group. We were asked to make a unique camera to be carried by the first astronauts to the moon. This Lunar Close-up Camera was to take detailed photographs of lunar soil in color and stereo. The problem was that it was only a few months before the Apollo launch. We asked for instructions, but nothing came for many weeks until a large package arrived with the usual boilerplate of procurement documents. Since it would have been impossible to develop and deliver the camera under normal procurement practices in so short a time, Art Simmons immediately returned a message to the National Aeronautics and Space Administration (NASA), "No Bid." I was very disappointed because I thought it would be a very worthwhile instrument for lunar exploration. But Art, in his wisdom, got NASA to concede and allow us to build the camera on a crash effort with minimum government oversight. A small team housed in an old warehouse delivered the first of several cameras within six months. It performed successfully.

Ed was a tough manager and that's what Kodak needed at the time. Faced with the urgent demands of the U-2 program, he assembled a group of photographic experts from within Kodak and confiscated processing equipment from company production operations. In a rundown building of the Naval Ordinance Division, a place where Kodak made ordinance fuses during World War II, they customized this equipment for production and installed innovative photo technology. This was the only group that the government considered reliable enough to process this precious film coming from the U-2 and later from the Corona missions. One challenge was the ultra-clean requirement in duplicating the fine detail on the film and making quality copies for dissemination.

Serious problems sometimes led to humor. On one mission the film stopped winding on the take-up spool within the recovery bucket before it should have been full. When the film came back for Ed's group to process, Ed telephoned Art and said that they found the problem and one of our people was responsible. Art was very concerned, but when Ed arrived a few hours later he handed Art a screwdriver that had been accidentally left in the bucket and got wound into the film. Ed then pointed out the initials on the screwdriver: "GE" (for General Electric, the company that built the bucket). Ed said jokingly, the screwdriver must have belonged to George Eastman.[8] I'm sure that the General Electric people made sure nothing like that ever happened again.

**Dr. Kenneth Macleish**

Ken was an able scientist, engineer and leader. He headed our Samos program and was responsible for its innovative approach and technical success. I was one of his young engineers who started the program at Kodak. From him we learned careful engineering, accurate analysis, and disciplined reporting. As one of the first managers of Kodak's work for the NRO, he had a profound influence on what we accomplished in the years to come.

---

[8] George Eastman is the founder of Eastman Kodak Company. In 1881 he established the precursor to Eastman Kodak: The Eastman Dry Plate Company.

**Dr. William Feldman**

Bill also deserves a great deal of credit. He, as well as Ken Macleish, came from Tennessee Eastman where they worked on the atomic bomb project during World War II. Bill was a careful scientist and a man of the very highest integrity. He became my boss and mentor during my early years at Kodak. He headed Kodak's Lunar Orbiter Program in which we were responsible for the imaging payload and reconstruction of the returned image. The National Aeronautics and Space Administration considered this one of its most successful programs.

**Edwin Land**

Din Land was key to the success of the NRO and our government work at Kodak. Din was head of Polaroid and was a member of the President's Scientific Advisory Committee. I was giving a briefing about our proposal for a new system and was showing how a succession of payloads would produce improvements as we went along. I referred to this as the "learning curve." Din listened very carefully and when someone else entered the room he quipped, "Chuck is telling us about his yearning curve." In other words, that was what I wanted to do.

I recall an amusing story about Din. Art Simmons, Ed Green and I went to Boston to see him on government business. When we landed in the Kodak plane there was a limousine waiting with Din in the back seat. At that time he was developing his new instant color camera. Din was one of the hardest working people I ever met. He seemed jumpy and spoke in halting sentences. He told us about the work he was doing and as he shook and wiggled he said, "I'm working on my new camera system! And I'm putting in 16-hour days! I only have to sleep four hours a day! But it's not affecting me at all!" I remember turning my face and grinning to myself thinking, "Maybe I'd be shaking like that if I put in that many hours." We were working long hours but no one could have kept up with Din.

## Annex 2

**Selected Publications and Patents**

Publications

"Exploring the Solar System with Photography," Invited paper, symposium, Calvin College, 1970.

*The Hubble Space Telescope Optical Systems Failure Report,* 1990, Report to NASA by the HST Optical Systems Board of Investigation.

"Photography of the Planets from Orbit," *Proceedings of the SPIE,* September 25, 1969, by Keene, Frome, Spoelhof.

"Photography of the Planet Mars." Invited paper, JPL Symposium, 1970.

Patent

Wide Angel Optical System—3,455,266 (1969).

# Albert D. Wheelon

Albert "Bud" Wheelon served as the first Central Intelligence Agency Deputy Director of Science and Technology from 1963-1966, after serving as the Director of the Office of Scientific Intelligence. He was responsible for U-2 overflights and the development of the Oxcart A-12 and three major satellite reconnaissance systems. Wheelon recalls his experiences at the Central Intelligence Agency during the early years of national reconnaissance.

## Early Struggles in Strategic Reconnaissance[1]

These are my recollections regarding my work in national reconnaissance, from the early airborne and film-return satellite programs that were jointly conducted by the Central Intelligence Agency (CIA) and the Air Force, to the later near-real-time satellite reconnaissance system developed by the CIA with National Reconnaissance Office (NRO) financial support. My window of direct observation began when I joined CIA in June 1962 and ended when I left in October 1966.

During my first year at CIA, I was responsible for the Agency's Office of Scientific Intelligence (OSI). Dr. Herbert Scoville, Jr. was the Director of the NRO Program B, which was responsible for CIA's reconnaissance activities and interface with the emerging NRO organization.[2] Scoville and I were good friends, and he was one of the main reasons I joined the CIA. He involved me directly in a few of his activities, and it was in these assignments that I first met Joe Charyk and his NRO staff.[3] Most of what I came to know during my first year about the larger NRO issues came from discussions with Scoville and his colleagues.

Less than a year after my arrival at CIA, I took over primary responsibility for CIA's reconnaissance activities as the Deputy Director for Science and Technology (DDS&T). My mandate was to move forward in satellite reconnaissance. Three systems came out of the mandate, each in a slightly different way. At first, I studied the considerable correspondence

[1] This section is based on written input that Albert D. Wheelon submitted to the Center for the Study of National Reconnaissance. This input includes copies of two 1999 letters from Wheelon to Richard Helms, former Director of Central Intelligence. Material in this chapter is also quoted or paraphrased from: Wheelon, Albert D. "Strategic Reconnaissance," *Physics and Society*. Vol 28 (No 4), October 1999. Copyright (1999) by the American Physical Society.

[2] Herbert Scoville, Jr. served as the first Program B Director from July 1962-June 1963.

[3] Joseph V. Charyk served as the first Director of the NRO from 1961-1963.

between Scoville and Richard Bissell.[4] Bissell left the CIA a few months before I arrived to head the Institute for Defense Analyses (IDA). We only were in occasional contact while he was at IDA, and even less when he returned to Connecticut. I had many conversations, both before and during my time at CIA, with some of the early leaders in the field of national reconnaissance. These individuals all served in various positions in national reconnaissance and in presidential advisory roles and included Edwin "Din" Land, Jim Killian, Bill Baker, George Kistiakowsky, Jimmy Doolittle, and Jerry Wiesner.[5]

**Improving Imagery Capabilities**

My first effort as DDS&T was to identify the requirements for an improved Corona or a new system to provide improved imagery capabilities.[6] We worked closely with National Photographic Interpretation Center (NPIC) photointerpreters to find out what they needed. I tasked the Drell group to see if Corona could be upgraded to provide that performance.[7] Its view confirmed my own opinion that it could not be done.

We then began a vigorous effort to design a new system that could provide what the photointerpreters needed. We hired a different contractor to develop the camera for this system, which provided the technical breakthrough that we needed. Thus, the Corona follow-on was born.

From the outset, Brockway McMillan, Director of the NRO (DNRO), did everything in his power to stop this program from proceeding.[8] The need we had identified for primary intelligence and our technical solution were irrelevant. My view was that the NRO and the Department of Defense (DoD) were opposed to any new satellite program at the CIA. Later, he publicly admitted that, "This had been the biggest mistake," he had ever made. Because McMillan refused to provide NRO funds for this work, Director of Central Intelligence John McCone agreed to let me use CIA money to begin this work. He later negotiated with Deputy Secretary of Defense Cyrus Vance to fund our activities through the NRO, and we got funding to do serious development.[9]

The program proceeded, and was enormously successful on its first flight. The program was later transferred to the Air Force when the CIA needed to unburden its best people for the challenge of a new system. It is ironic that the Air Force managed this system during its declining years after trying to strangle it in the crib.

---

[4] Richard M. Bissell, Jr. served as the CIA's Corona Program Manager. He previously served as the U-2 Project Manager and a Special Assistant to the DCI for Planning and Development.

[5] Edwin H. Land (CEO of Polaroid, chairman of the Intelligence Subcommittee of the Technology Capabilities Panel, and chairman of the President's Science Advisory Committee's Intelligence Panel), James R. Killian (President of the Massachusetts Institute of Technology, and chairman of the Technological Capabilities Panel, the President's Foreign Intelligence Advisory Board, and the President's Science Advisory Committee), and William O. Baker (AT&T Bell Laboratories, President's Foreign Intelligence Advisory Board, and President's Science Advisory Committee) are Founders of National Reconnaissance. George Kistiakowsky served as President Eisenhower's Science Advisor. Jerome Wiesner served as President Kennedy's Science Advisor. Jimmy Doolittle, who won the Medal of Honor for leading a raid of B-25 bombers against Japan in 1942, served on the National Advisory Committee for Aeronautics, the Air Force Advisory Board, and the President's Science Advisory Committee.

[6] Corona and its follow-on program were joint NRO Program A and B efforts. Program B (CIA) provided the Corona and follow-on imaging program security systems, payloads, and mission planning/requirements. Program A (Air Force) was responsible for the Thor, Agena, and Corona vehicles and their health, launch vehicle and payload integration at Vandenberg Air Force Base, on-orbit operations (launch, tracking, and controlling) of the Corona satellites, and recovery vehicle and film recovery and delivery.

[7] Sidney D. Drell is a Founder of National Reconnaissance. See Chapter 2 for his recollections.

[8] Brockway McMillan served as DNRO from 1963-1965.

[9] Cyrus Vance, the Deputy Secretary of Defense, later served as Secretary of State for President Jimmy Carter.

**Working on a Signals Intelligence Program**

The signals intelligence (SIGINT) program I worked on came about in a different way. In July 1963, I recognized that we could use U.S. SIGINT satellites to intercept data on a continuous basis. We moved rapidly to develop this concept, first with CIA money and later with reluctant support from the NRO.

Gene Fubini, the DoD's Director of Defense Research and Engineering, took the lead in trying to stop this program, while McMillan fought the battles for Corona and its follow-on program. Fubini first tried to convince outside advisors and decisionmakers that it was not technically feasible. When that claim was proven incorrect, he urged McCone and Vance to rely on an experimental synchronous satellite being developed by NASA. When that attempt failed and we kept moving forward, he started a similar program in the Air Force. This duplication continued for three decades. The CIA SIGINT program was successful on its first launching, while the Air Force program failed four times before working properly.

*In the fall of 1963, I was at home watching a football game being played in San Francisco. It suddenly struck me that if I could do that, the technology was available to view the earth's surface from orbit and to observe that scene as it was being received in the spacecraft...*

**Moving Toward a Near-Real-Time Imaging Capability**

In the fall of 1963, I was at home watching a football game being played in San Francisco. It suddenly struck me that if I could do that, the technology was available to view the earth's surface from orbit and to observe that scene as it was being received in the spacecraft, in other words to develop a near-real-time imagery system. The following day, I tasked Les Dirks with the assignment of helping me make it happen.[10] This turned out to be one of the best decisions of my career. We ended up creating a new class of satellite imaging systems that revolutionized intelligence collection.

We realized that the key technical element was the electronic retina in the camera, which was a mosaic of devices that convert light waves to electrical signals. This was the essence of a television camera, and is the basis for today's camcorders.

I visited Bell Telephone Laboratories to explore their latest technology. The company was working on the development of charge-couple devices (CCDs), which are small semiconductors. Dirks and I saw the devices, and realized that they were quite close to what we needed. We saw a way for us to accomplish the most difficult task. Almost all of our efforts during my tenure as DDS&T were focused on developing the CCD arrays.

Another question related to the technique for sending the data back to earth from low-orbit satellites. This was a major issue because the optical data were so vast in each frame that we could not afford to store it on board the satellite. The data simply would pile up and overflow the limited available storage devices. We could not downlink the data as it was being collected, because it would have to be received in denied territory. The right way to do this was to uplink the data to a relay satellite in much higher orbit, which could then pass it on to the ground station. After I left, Al Flax identified a better way to do this.[11]

---

[10] Leslie C. Dirks served as Program B Director from June 1976-July 1982. Dirks also served as the first Director of the Office of Development and Engineering. Dirks had studied physics as an undergraduate at MIT, and went on to Oxford as a Rhodes scholar before his successful career in U.S. intelligence.

[11] Alexander H. Flax served as the DNRO from 1965-1969.

### The Role of Physicists in National Reconnaissance

Physicists contributed significantly to the development of our strategic reconnaissance capability. "Physicists were influential advisors in the beginning: recognizing technical opportunities, catalyzing technical decisions and moving presidents to act. Their advice carried great weight with Presidents Eisenhower and Kennedy because they had no departmental loyalties and because they were intellectually committed to objectivity."[12]

"Physicists are optimistic and adventuresome by nature. Because they try to understand nature at a fundamental level, they have the knowledge and confidence to take large steps. This was abundantly demonstrated during the Second World War in the radar and atomic bomb projects. It was shown again in the development of overhead reconnaissance systems."[13]

"Bissell and others who managed these programs within government were economists and lawyers. That changed when President John Kennedy moved into the White House. Physicists like Harold Brown and John Foster took influential positions in the Defense Department. Bissell's place was taken by Herbert Scoville with a doctorate in physical chemistry, and I succeeded Scoville. Ten years after receiving a Ph.D. in theoretical physics at MIT, I found myself in charge of the U-2 program, Oxcart A-12 development, and the Corona photoreconnaissance program. I brought other physicists and scientists into those programs—both as government employees and as advisors. I persuaded Sidney Drell to take a sabbatical leave from Stanford to help us with Corona, and he has been an influential advisor in such programs ever since. Land recruited Richard Garwin who has been extraordinarily effective for thirty years."[14]

"When I left government service, I was replaced by Carl Duckett, who, in turn, was replaced by Les Dirks—a physicist from MIT and Oxford. An unusual Air Force officer with a Ph.D. in physics—Lew Allen—played an important role in these programs, and later became Chief of Staff of the Air Force. When the primary responsibility for such programs passed to the National Reconnaissance Office, it was led during important periods by two physicists: John McLucas and Hans Mark."[15]

### Conclusion

It is important to recognize the enormous contributions of all the individuals who worked on these programs. The engineers, craftsmen, assembly workers, Air Force officers and enlisted men, and CIA communications and security people all contributed to the success of the mission. What is abundantly clear, however, is that the National Reconnaissance Program as we know it would not have been created without the decisive contributions of a dozen physicists. These scientists should take considerable pride in the contributions that they made to this historic effort.[16]

The scientific team I gathered at CIA went on to build three new satellite systems, each more capable than Corona. When the government eventually opens the windows on the reconnaissance programs that followed Corona, the American people will learn of technical achievements more impressive than the Apollo moon landings. The National Reconnaissance Program and the Apollo moon program proceeded in parallel—one in utmost secrecy, the

[12] Quoted from Wheelon (1999).
[13] Wheelon (1999).
[14] Wheelon (1999). Richard L. Garwin is a Founder of National Reconnaissance. See Chapter 3 for his recollections.
[15] Wheelon (1999). John L. McLucas served as DNRO from 1969-1973. He later served as Secretary of the Air Force, the first DNRO to serve in that role. Hans Mark served as DNRO from 1977-1979. While director, he also served as Acting Secretary of the Air Force and later as Secretary of the Air Force.
[16] Paraphrased from Wheelon (1999).

other on national television. Both served to steady our course during the dangerous years of the Cold War. Apollo generated public confidence. Corona, and its follow-on programs, generated presidential confidence.[17]

—

**Pioneer Award Presentation and Citation**

Figure 42-1. Pioneer Albert Wheelon (second from left) being recognized at the 2000 Pioneer Recognition Ceremony. The pioneer award plaque was presented by DNRO Keith Hall (left) and DCI George Tenet (third from the left). Joanne Isham (Deputy Director for Science and Technology, CIA) joined in the presentation. (Photo by Sara Judy, NRO Visual Design Center.)

**Albert D. "Bud" Wheelon, Ph.D.**

The first director of the Central Intelligence Agency's Directorate of Science and Technology, Dr. Albert Wheelon was responsible for U-2 overflight operations and development of the Oxcart Mach 3 reconnaissance aircraft. He initiated the development of three imagery and signals satellite systems that changed the form and content of overhead reconnaissance.

*Career in National Reconnaissance: 1962-1966*

[17] Paraphrased from Wheelon (1999).

# Appendices **1–3**

Appendix 1 — Gloassary and Acronyms

Appendix 2 — Biographical Abstracts of the Pioneers and Founders

Appendix 3 — The Pioneer Recognition Program

# Appendix 1 Glossary and Acronyms[1]

**A-12** Lockheed, later CIA, designation for the Mach-3 aircraft developed under Project Oxcart, a follow-on to the U-2 program. The Air Force later employed the SR-71 reconnaissance aircraft, which was an evolution of the A-12.

**ABM** Anti-ballistic missile. An interceptor missile designed to counter strategic ballistic missiles or their elements in flight trajectory.

**ABRE** advanced ballistic reentry system

**ABS** Atlas boosted satellite dispenser

**Advent** U.S. Army program in the early 1960s to build a synchronous communication satellite.

**AE** aeronautical engineering

**AEC** Atomic Energy Commission

**Aerobee** A sounding rocket that carried payloads into the upper atmosphere where astronomical and physiological missions and experiments were performed. The Aerobee family of rockets was in use from 1947-1985.

**AF** Air Force

**AFB** Air Force Base

**AFBMD** Air Force Ballistic Missile Division

**AFIT** Air Force Institute of Technology

**AFSC** Air Force Space Command

**Agena** An upper stage of a launch vehicle used for Corona and other missions.

**AIAA** American Institute of Aeronautics and Astronautics

**American Society of Photogrammetry (ASP)** Forerunner to the American Society for Photogrammetry and Remote Sensing.

**American Society for Photogrammetry and Remote Sensing (ASPRS)** The United States professional society for imaging and geospatial information. The ASPRS mission is to advance knowledge and improve the understanding of the mapping sciences, as well as to promote the informed application of photogrammetry, remote sensing, and geographic information systems. The Society was founded in 1934 as the American Society of Photogrammetry (ASP) and has advanced its mission through publications, continuing education, and professional certification programs. It maintains a certification program for photogrammetrists, mapping scientists for remote sensing, and mapping scientists for geographic/land information systems.

**ANDB** Air Navigation Development Board

**APL** Aeronautical Photographic Laboratory (Army Air Corps)

**ARDC** Air Force Research and Development Command

**Argon** The program name for a 1960s U.S. satellite reconnaissance system designed to collect imagery for mapping purposes. It was developed in parallel with Corona and flew 12 missions between 17 February 1961 and 21 August 1964.

**ARPA** Advanced Research Projects Agency (later DARPA) whose mission is to develop imaginative, innovative and often high-risk research ideas offering a significant technological impact that will go well beyond the normal evolutionary developmental approaches; and to pursue these ideas from the demonstration of technical feasibility through the development of prototype systems.

**ARS** advanced reconnaissance satellite

**ASAT** anti-satellite

**ASICS** application specific integrated circuit system

---

[1] This appendix lists selected words, acronyms, and abbreviations that are used in this book. Their respective definitions and explanations are for the terms as used within the context of the recollections. In many cases the same term might have additional definitions that are beyond the usage of this book.

**ASP** *See* American Society of Photogrammetry
**ASPRS** *See* American Society for Photogrammetry and Remote Sensing
**Atlas** A U.S. space launch vehicle used for military, commercial, and scientific payloads.
**AWS** Air Weather Service
**azimuth** The angle of the direction between an object and a fixed point.
**B-36** A post-World War II-era strategic bomber called the Peacemaker, which served as a test bed for an early reconnaissance camera, and later considered as a platform for launching a small reconnaissance aircraft. A number of RB-36 reconnaissance aircraft also were built. They often were referred to as the "Featherweights," as they were lighter and could fly higher than the bomber version.
**Bomarc** Boeing and Michigan Aeronautical Research Center. This term also refers to the name of a surface to air missile built by the Boeing and Michigan Aeronautical Research Center.
**BSU** beta storage unit
**BUORL** Boston University Optical Research Laboratory
**BUPRL** Boston University Physical Research Laboratory
**ccd** charge coupled device. A detector that converts light to electricity.
**CIA** Central Intelligence Agency
**CINC** Commander-in-Chief
**COMINT** Communications intelligence is technical and intelligence information derived from foreign communications by other than the intended recipients. COMINT does not include information from the monitoring of plain text public broadcasts.
**COMIREX** The DCI Committee on Imagery Requirements and Exploitation. Established in 1967, COMIREX was a DCI committee to advise and assist the U.S. Intelligence Board on matters involving overhead reconnaissance and imagery exploitation.
**COMM** communications
**Corona** The program or project name assigned to development of the first U.S. photographic reconnaissance satellite, which operated from August 1960 to May 1972, and was jointly managed by the CIA and the Air Force. Its cameras were designated KH-1 through KH-4, and it was a film-return system. The term is often used to refer to the satellite itself.
**CRTF** Classification Review Task Force
**customer** The U.S. Government and its representatives in relationship to contractors.
**cw** continuous wave
**DCAS** Defense Contract Administration Services
**DCI** Director of Central Intelligence
**DDCI** Deputy Director of Central Intelligence
**DDI** Deputy Director of Intelligence
**DDR** Deputy Director of Research
**DDR&E** Deputy Director for Research and Engineering
**DDS&T** Deputy Director for Science and Technology
**Discoverer** Scientific cover program for the formerly classified Corona satellite photoreconnaissance program (*see* Corona).
**DMSP** Defense Meteorological Satellite Program
**DNRO** Director, National Reconnaissance Office
**DoD (DOD)** Department of Defense
**DPO** Development Planning Objective
**DS&T** Directorate of Science and Technology (CIA)
**EE** electrical engineering
**Electronic Systems Laboratory (ESL)** A division of Sylvania GTE based in Mountain View, California (now a part of TRW).
**ELINT** Electronic intelligence is technical and intelligence information derived from foreign electromagnetic non-communications transmissions by other than intended recipients, and

foreign non-communications electromagnetic radiation emanations from other than atomic detonation or radioactive sources.

**ExComm** Executive Committee (of the NRO). A 1960s committee comprised of the Deputy Secretary of Defense, the DCI, the President's Scientific Advisor, and the Director NRO.

**Fan Song** NATO designation for the Soviet SA-2 fire control radar.

**FBI** Federal Bureau of Investigation

**FIA** Future Imagery Architecture

**firmware** Software that is embedded in a hardware device that allows reading and executing the software, but does not allow modification (e.g., writing or deleting data by an end user).[2]

**FMSAC** Foreign Missile and Space Analysis Center

**Fort Meade** U.S. Army post in Maryland that serves as the location for the Headquarters of the NSA.

**Fortran** Formula Translator. First programming language for numerical, scientific, and engineering applications.[3]

**FTD** Foreign Technology Division

**Galactic Radiation and Background (Grab)** Operational name of the first ELINT satellite system, initially proposed as Project Tattletale and approved as Project Canes. The Grab system was developed by the U.S. Naval Research Laboratory from 1958-1960, operated from 1960-1962, and transferred to the NRO in July 1962.

**GCHQ** Government Communications Headquarters. The SIGINT organization in the United Kingdom.

**GDIN** Global Disaster Information Network

**GE** General Electric Company

**Genetrix** The project name for a formerly highly-classified U.S. Air Force balloon reconnaissance program in the 1950s. The balloons carried cameras to altitudes of up to 90,000 feet and were designed to drift over Eastern Europe, the Soviet Union, and China. The balloons employed a camera with a 6-inch focal lens length, designed by BUORL and built by Fairchild Camera Company, among others. Some consider the camera a predecessor to Corona. Genetrix was supported by the Moby Dick Program (*see* Moby Dick).

**GMAIC** Guided Missile and Astronautics Intelligence Committee

**GPS** Global Positioning System

**Grab** *See* Galactic Radiation And Background

**GRD** *See* ground resolved distance

**Greenhouse** U.S. nuclear bomb test program that took place at Eniwetok Atoll in 1951-1952.

**ground resolved distance (GRD)** A measure of image quality. The measure indicates the ability to distinguish between two objects separated by a specified distance. For example, a GRD measurement of twenty feet means that two objects cannot be distinguished as separate from one another unless they are at least twenty feet apart.

**GSE** General Systems Engineering

**HCL** hydrogen chloride

**H-timer** Horizon timer. Device used to control camera sequences, including on and off, in early Corona missions. It operated like a player piano. It had metal "fingers" brushed against a mylar tape that had holes punched into it.

**ICAF** Industrial College of the Armed Forces

**ICBM** intercontinental ballistic missile

**IDA** Institute for Defense Analyses

**Idealist** Name used to describe the overall U-2 program after the 1 May 1960 shootdown over the Soviet Union.

---

[2] Source: Federal Standard 1037c—Telecommunications Terms.

[3] Source: The Free On-line Dictionary of Computing (FOLDOC).

**IEEE** Institute of Electrical and Electronic Engineers

**IMINT** Imagery intelligence is the result of imagery analysis processed for intelligence use.

**IR** infrared

**IRBM** Intermediate-range ballistic missile (as defined by the Intermediate Range Nuclear Forces (INF) Treaty). A ground-launched ballistic missile or a ground-launched cruise missile having a range capability in excess of 1,000 kilometers but not in excess of 5,500 kilometers.

**Itek** Company that was involved with the design of space-borne cameras. Itek designed and built the Corona cameras. It was co-founded in 1957 by pioneer Richard Leghorn.

**JASON** Independent, senior-level scientific and technical think tank composed primarily of scientists and academics representing prestigious universities and research institutions in the United States. The principal function of the JASON group is to provide senior-level government managers, primarily in the Department of Defense and Intelligence Community, with scientific and technical expertise. The term, JASON, is not an acronym, nor does it have any particular meaning. Some have considered the term to stand for "July, August, September, October, November"—the time of year when the JASON group often performs studies.

**JOT** junior officer training

**Land Panel** A presidential advisory panel led by Edwin Land in the 1960s. Its formal name was the National Reconnaissance Panel of the President's Science Advisory Committee (PSAC). It was composed of distinguished scientists and engineers from academia and industry. The panel was responsible for maintaining an overview of the NRP. It paid particular attention to the technical characteristics of intelligence requirements, the status of existing projects, and the adequacy of existing intelligence programs.

**Lanyard** The program name for a 1960s U.S. satellite photoreconnaissance system designed to solve the information gap of high-resolution photographs. It was expected to provide 2-foot-resolution imagery, but only collected a best resolution of 6 feet. It flew only one successful mission, which was during the summer of 1963.

**LEO** low-earth orbit

**LMSC** Lockheed Missile and Space Corporation

**MAD** Mutually Assured Destruction. A concept that was the foundation of bipolar strategic nuclear deterrence during the Cold War. Specifically, the concept was based on the premise that neither the United States nor Soviet Union would have an incentive to launch a first nuclear strike because both sides had adequate strategic forces that could survive a first strike and deliver a retaliatory counterstrike that would inflict unacceptable losses. Consequently, both sides recognized that if one side launched a nuclear first strike the most likely outcome was mutually assured destruction.

**Manhattan Project** U.S. project that developed the atomic bomb during World War II.

**McIDAS** Man-computer interactive data access system. Developed by the Space Science and Engineering Center at the University of Wisconsin, McIDAS uses the geosynchronous spin-scan satellite, and was the world's first global weather system.

**MIB** Military Intelligence Board

**Midas** Missile Defense Alarm System. A U.S. infrared sensor that was designed to detect the launch of Soviet intercontinental ballistic missiles. MIDAS is the precursor to the Defense Support Program.

**MIT** Massachusetts Institute of Technology

**MMS** multi-mission module satellites

**Moby Dick** A joint Air Force/CIA 1950s program that developed high-altitude balloons, payload packages, and associated equipment and techniques for the classified Genetrix reconnaissance program. The Moby Dick program also included overt balloon meteorological experiments (*see* Genetrix).

**MRI** magnetic resonance imaging

**$N_2$** nitrogen molecule

**NASA** National Aeronautics and Space Administration

**National Imagery and Mapping Agency (NIMA)** The Intelligence Community agency responsible for providing imagery, imagery intelligence, and geospatial information including mapping data of the Earth's surface. NIMA supports Intelligence Community requirements and is a DoD combat support agency that contributes to the operational readiness of U.S. military forces.

**National Imagery Interpretability Rating Scale (NIIRS)** A perceptual measure of image quality or interpretability developed by the Intelligence Community in the 1970s. It consists of a series of imaging quality ratings based on a sequence of interpretation tasks. The tasks associated with each rating increase in difficulty and therefore require improved image quality.

**National Photographic Interpretation Center (NPIC)** NPIC was the multi-agency, CIA-managed component responsible for analyzing national-level imagery and disseminating derived intelligence products to the Intelligence Community. It was disestablished in 1996 and became part of NIMA.

**National Reconnaissance Office (NRO)** The Intelligence Community agency responsible for designing, developing and operating U.S. reconnaissance satellites.

**National Reconnaissance Program (NRP)** The NRP is the national program to meet U.S. national security requirements through space-borne and assigned airborne reconnaissance.

**National Security Agency (NSA)** The Intelligence Community agency responsible for planning, coordinating, directing, and performing foreign SIGINT and information security functions.

**NATO** North Atlantic Treaty Organization

**Naval Research Laboratory (NRL)** The Navy's corporate laboratory, established on 2 July 1923, headquartered in Washington, DC.

**NIE** National Intelligence Estimate

**NIIRS** *See* National Imagery Interpretability Rating Scale

**NIMA** *See* National Imagery and Mapping Agency

**NMR** nuclear magnetic resonance

**NO** nitrogen oxide

**NORAD** North American Aerospace Defense Command

**NPIC** *See* National Photographic Interpretation Center

**NRL** *See* Naval Research Laboratory

**NRO** *See* National Reconnaissance Office

**NRP** *See* National Reconnaissance Program

**NRT** near real time. A time relationship between the collection of specific intelligence and its availability to users.

**NSA** *See* National Security Agency

**$O_2$** oxygen molecule

**OER** Officer evaluation report. Performance evaluation for Air Force officers.

**OF** operations facility segment

**Office of Scientific Intelligence (OSI)** The CIA office created in 1948 to produce national estimates on scientific topics.

**OMB** Office of Management and Budget

**OSI** *See* Office of Scientific Intelligence

**OSP** Office of Special Projects. Created on 1 October 1965 within the CIA DS&T, OSP consolidated all of the CIA's satellite activities and later became Office of Development and Engineering.

**Oxcart** Name given to the CIA A-12 project to develop a follow-on reconnaissance aircraft to the U-2.

**PCM** pulse-code modulation

**PCS** permanent change of station
**PEM** program element manager
**PET** performance evaluation team
**PFIAB** President's Foreign Intelligence Advisory Board. The PFIAB provides advice to the President concerning the quality and adequacy of intelligence collection, analysis and estimates, counterintelligence, and other intelligence activities. The PFIAB, through its Intelligence Oversight Board, also advises the President on the legality of foreign intelligence activities.
**prime** prime contractor (as opposed to subcontractor)
**Program A** The U.S. Air Force element of the NRO from 23 July 1962 to 31 December 1992. It was located on the West Coast and consisted of the Secretary of Air Force Special Programs (SAFSP). When Program A was disestablished, it was integrated into functional NRO directorates.
**Program B** The CIA element of the NRO from 23 July 1962 to 31 December 1992. It consisted of the CIA's Office of Development and Engineering in the DS&T. When Program B was disestablished, it was integrated into functional NRO directorates.
**Program C** The U.S. Navy element of the NRO from 23 July 1962 to 31 December 1992. It consisted of elements of the NRL and the Naval Security Group. When Program C was disestablished, it was integrated into functional NRO directorates.
**PSAC** President's Science Advisory Committee (*see* Land Panel)
**PSQ** personnel security questionnaire
**R&D** research and development
**radiosonde** Miniature radio transmitter that is carried aloft with instruments for broadcasting humidity, temperature, wind velocity, and pressure.
**RAF** Royal Air Force (United Kingdom)
**RAND** A nonprofit institution that performs research and analysis to help improve policy and decision-making. Its name, originally Project R and D, is a contraction of the terms "research and development." The then-U.S. Army Air Forces created RAND in 1946, and RAND performed most of the initial Air Force studies into the utility and design of various kinds of satellites, including reconnaissance and surveillance.[4]
**recovery bucket** Bucket-shaped container that held the film acquired by a film-return satellite photoreconnaissance system. The exposed film was spooled into the bucket and returned to earth inside a protective capsule. The capsule was retrieved in mid-air by an aircraft during the capsule's descent to earth. If the aircrew were unsuccessful in snatching the capsule in mid-air, a naval vessel, or aircraft using frogmen, would recover the capsule from the ocean.
**RF** radio frequency
**RFP** request for proposal
**ROTC** Reserve Officer Training Corps
**RV** reentry vehicle or recovery vehicle
**SAC** Strategic Air Command
**SAFSP** Secretary of the Air Force for Special Projects
**SALT** Strategic Arms Limitation Talks. These were negotiations on limiting the strategic nuclear forces of the U.S. and Soviet Union. A number of treaties and agreements resulted from these talks. The 1972 SALT I agreements included the Treaty on the Limitation of Anti-Ballistic Missile Systems and the Interim Agreement on the Limitation of Strategic Offensive Arms. The second series of negotiations resulted in the 1979 SALT II Treaty.
**SAM** surface-to-air missile
**Samos** Space and Missile Observation System. An early U.S. satellite program that focused on "read-out" photographic systems, either by television transmission or video recording tape. One Samos photo-satellite project also attempted to develop a high-resolution film-return

[4] Based on "About RAND," http://www.rand.org (10 April 2002).

system. Other Samos projects involved ELINT satellites. The Samos program was preceded by the Sentry project (*see* also WS-117L).

**SAMSO** USAF Space and Missile Systems Organization. A subordinate element of the Air Force Systems Command. SAMSO operated between 1967 and 1979.

**SCF** satellite control facility

**SEI** specific emitter identification

**Sentry** Name to which WS-117L was changed in 1959. In 1960 Sentry included Discoverer; Samos Projects 101A, 101B, and 201; and Midas.

**SETA** systems engineering and technical assistance

**Sferics** Technique of locating storms by using radio direction finders to track lightning.

**SIGINT** Signals intelligence is intelligence information that includes any communications intelligence, electronics intelligence, and foreign instrumentation signals intelligence, however transmitted.

**SLC** space launch complex

**SLGS** space-ground link subsystem

**SoLRad** solar radiation

**SP** Special projects. A term applied to classified "black" programs.

**SPO** Special Program Office or System Project Office

**SPS** Special Project Staff. Created in 1964 to focus CIA management attention on its reconnaissance satellite programs.

**SR-71** The Air Force designator for the Mach-3 production aircraft developed by the Lockheed Corporation from the original CIA A-12 reconnaissance platform.

**SSCI** Senate Select Committee on Intelligence

**SSEC** Space Science and Engineering Center at the University of Wisconsin

**STC** Satellite Test Center

**TCP** Technological Capabilities Panel. A panel formed at the request of President Dwight D. Eisenhower in 1954 to perform a comprehensive scientific assessment of the nation's offensive, defensive, intelligence, and communications capabilities.

**TD** technical direction

**TDRSS** tracking and data relay satellite system

**telemetry** The use of telecommunications for automatically indicating or recording measurements at a distance from the measuring instrument.

**teletypewriter exchange (TWX)** A public switched teletypewriter service in which teletypewriter stations are provided with lines to a central office for access to other stations.

**Television infrared observation satellite (Tiros)** NASA's first experimental "wheel-mode" weather satellite, launched on 1 April 1960.

**Thor** An early U.S. space launch vehicle.

**Tiros** *See* Television infrared observation satellite.

**Titan** A U.S. space launch vehicle that has been a principal part of the U.S. Air Force's launch system since 1977. Versions of the Titan also have been used to fly scientific missions for NASA, as well as to support commercial launches into space.

**TK** Talent-Keyhole. The name of the security control system established to protect the products of satellite reconnaissance activities.

**TT&C** telemetry, tracking, and command

**TWX** *See* teletypewriter exchange service

**U-2** The high-altitude reconnaissance aircraft designed in the mid-1950s by the Lockheed Corporation. President Eisenhower initiated the U-2 Program with the initial purpose of monitoring strategic capabilities in the Soviet Union. The aircraft became the best source of national reconnaissance information until the first successful Corona satellite was launched in August 1960. The U-2 continues to be an operational reconnaissance system into the 21st century.

**UCLA** University of California, Los Angeles

**UFO** unidentified flying object
**UK** United Kingdom of Great Britain and Northern Ireland
**United States Army Air Corps** Forerunner of the United States Air Force
**USAF** United States Air Force
**USIB** United States Intelligence Board
**USSR** Union of Soviet Socialist Republics
**VAFB** Vandenberg Air Force Base
**VE Day** Victory-in-Europe Day
**WADC** Wright Air Development Center
**WDD** Western Development Division
**WS-117L** Weapon System 117L. The 1950s Air Force effort to develop reconnaissance satellites. It evolved from RAND's Project FEED BACK and was the precursor to all U.S. satellite reconnaissance programs (*see* also Sentry, Samos, and Corona).
**X-17** Program designation for the Air Force reentry test-vehicle program to determine intercontinental ballistic missile data.

# Appendix 2 Biographical Abstracts of the Pioneers and Founders

**C. Lee Battle, Jr.**
Colonel Lee Battle, USAF (Ret) was born on 25 May 1915 in Alpine, Texas . He attended the University of Oklahoma, where he received a bachelor's degree in chemical engineering in 1941. In 1946 Battle received a master's degree from the Massachusetts Institute of Technology (MIT). Battle also spent a year at MIT as a Visiting Fellow in 1947.

Battle was commissioned in the Air Force in 1942. His first assignment in national reconnaissance was as the director of the Discoverer and Corona programs. Before his retirement from the Air Force in 1968, Battle held the position of Military Assistant to the Deputy Director of Research and Development in the Central Intelligence Agency (CIA). After leaving the Air Force, Battle joined the Lockheed Missile and Space Corporation. He retired from Lockheed in 1980.

After retirement Battle moved to Hawaii, where his hobbies included music and physical fitness. Battle married the former Velma Norton; they have two children, Barbara and James. Battle died on 2 August 2002.

**John T. Bennett**
Mr. John Bennett was born on 8 December 1920 in Linton, Indiana. He attended Indiana University, and received an associate of arts degree in aeronautical engineering from Spartan School of Aeronautics in Tulsa, Oklahoma. In addition, he completed all curricula for Aircraft and Power Plant Mechanics and took courses in advanced structural analysis, aerodynamics, dynamics, rocket propulsion, and orbit mechanics.

From 1941-1960 Bennett worked at Northrop Aircraft on aircraft and missile design. His first national reconnaissance work was the concept and prototype of the photoreconnaissance nose section for the Snark missile. Bennett moved to TRW in 1960 where his career focused on the classified world of national reconnaissance and military satellite systems.

At the time of his recognition as a national reconnaissance pioneer, Bennett was retired and living in Sedona, Arizona, pursuing several hobbies, including hunting, fishing, and exploring the Arizona back country.

**John W. Browning**
Colonel John Browning, USAF (Ret) was born on 18 May 1924 in Wheeling, West Virginia. He earned a bachelor's degree in engineering from West Virginia University and a master's degree in management from Purdue University.

In 1943 Browning joined the Army Air Corps and served as a combat pilot during World War II and the Korean War. Browning began his association with national reconnaissance in the 1960s as the team leader for the design, development, and deployment of an NRO signals intelligence (SIGINT) system. Browning later became the deputy commander of satellite operations at a ground station.

Browning retired from the Air Force in 1982. He married the former Myrnie Priest in 1946; they have one daughter, Marilyn. Browning died on 3 January 1996.

**Jon H. Bryson**
Colonel Jon Bryson, USAF (Ret) was born on 4 March 1942 in Asheville, NC. Bryson received a bachelor's degree in electrical engineering from Georgia Institute of Technology in 1965. He also received a master's degree in systems engineering from the University of Southern California in 1971.

While still an undergraduate student at Georgia Tech, Bryson worked for the National Security Agency (NSA) through its student co-op program. After receiving his degree, Bryson joined the Air

Force and was assigned to NSA. Bryson spent the next 24 years involved in a variety of NRO SIGINT projects. After retiring from the Air Force in 1991, Bryson joined the Aerospace Corporation.

At the time of his recognition as a national reconnaissance pioneer, Bryson was a senior vice-president of the national systems group for the Aerospace Corporation, and his interests included photography, bird watching, and building computers. Bryson married the former Vonette Harrison.

**A. Roy Burks**

Mr. Roy Burks was born on 27 April 1931 in Louisville, Kentucky. He attended the University of Wisconsin, where he received a bachelor's degree in mathematics (1953) and a master's degree (1954).

Through his participation in ROTC during college, Burks received a commission in the Army and was assigned to missile programs at Redstone Arsenal in Alabama. Burks joined the CIA in 1956, and he eventually worked on the Corona program as the CIA's technical director for Program B. After retiring from the CIA in 1984, he continued to support NRO operations until 1993 through his work in private industry.

At the time of his recognition as a national reconnaissance pioneer, Burks was consulting for the CIA's Office of Research and Development, and he and his wife, the former Joan Piersol, were traveling extensively. Burks has three children, Bonnie, Robert, and William.

**Frank S. Buzard**

Colonel Frank Buzard, USAF (Ret) was born on 20 April 1921 in St. Joseph, Missouri. He attended Washington University in St. Louis, where he received a bachelor's degree in chemistry and physics in 1943. Buzard attended graduate school at the Air Force Institute of Technology, and in 1958 he received a master's degree in mathematics from the University of Illinois.

Buzard's career in national reconnaissance began when he was assigned to the Air Force Ballistic Missile Division for the Weapon System-117L (WS-117L) program. His other national reconnaissance assignments included the Corona/Discoverer Program Office and the NRO Staff at the Pentagon. He also served as the Program Director for a Corona follow-on activity and finally as the Vice Director, Office of Special Projects, Secretary of the Air Force.

After retiring from the Air Force in 1972, Buzard taught systems management at the University of Southern California until 1991. At the time of his recognition as a national reconnaissance pioneer, Buzard was retired and tutoring third graders in mathematics. Buzard married the former Patricia Ann May. They have three children, Glenn, Gary, and Catherine.

**Connie W. Chambers**

Mr. Connie Chambers was born on 1 March 1935 in Toledo, Ohio. He received a bachelor's degree in mechanical engineering from the University of Arizona in 1957.

Chambers' work in national reconnaissance began in 1962 with the Lockheed Corporation. He worked on the Corona photoreconnaissance program and follow-on programs. Chambers retired from Lockheed Martin in 1997.

At the time of his recognition as a national reconnaissance pioneer, Chambers continued to work as a consultant for the NRO and was spending his free time enjoying golf, skiing, and his several grandchildren. Chambers married the former Millie Ann Flynn; they have three children, Greg, Susan, and Mark.

**John O. Copley**

Colonel John Copley, USAF (Ret) was born on 26 August 1922 in Bangor, Maine. He attended Williams College from 1940 to 1943. In 1955 he graduated from the Air Force Institute of Technology at Wright Patterson Air Force Base, Ohio, with a bachelor's degree in electrical engineering.

Copley began his military career in May 1943. He served in Italy during World War II as a B-24 pilot. Copley served in a variety of engineering and technical assignments with the Air Force, including assignments with the NRO's SIGINT programs. After retiring from the Air Force in

1974, Copley worked at Rockwell International with the Space Systems Division until 1988. From 1990 to 1995, he was a consultant for the Aerospace Corporation in El Segundo, California.

At the time of his recognition as a national reconnaissance pioneer, Copley was retired and pursuing his hobbies of gardening, electronics design and construction, and raising tropical fish. Copley married the former Theresa Bellucci.

**Robert H. Crotser**

Mr. Robert Crotser was born on 15 July 1930 in Anna, Illinois. He attended the University of Southern California and University of California at Santa Barbara.

Crotser joined Lockheed Space Systems Company in 1957, and in 1960 he began work on the Corona photoreconnaissance satellite program. In the late 1960s he became the business manager for proposing an electro-optical program. In 1983 he was promoted to Vice President for Business Operations and then to Vice President & Assistant General Manager in 1986. Croster retired from Lockheed in 1989.

At the time of his recognition as a national reconnaissance pioneer, Crotser was dividing his time between California and Arkansas, playing golf and solving puzzles—jigsaw and crossword. Crotser married the former Norma Nees. They have one child, Lynn.

**John J. Crowley**

Mr. John Crowley was born on 11 October 1906 in Eau Claire, Wisconsin. He attended Eau Claire Normal School (now a branch of the University of Wisconsin) where he earned his teaching degree. Crowley received a master's degree in education from the University of Minnesota.

Crowley's earlier career was as a teacher and administrator in secondary schools in Wisconsin, Minnesota, and California. In 1942 he became a military instructor, teaching radio at the Army/Air Force Technical Training Command at Scott Field, Illinois. Crowley served in the Navy during World War II and after the war continued to work for Department of Defense (DOD) components. In 1964 Crowley joined the CIA, where he served as the CIA Chief of the NRO's Program B. Crowley left the CIA in 1970, but he continued to do consulting work for the CIA until 1983.

Crowley married the former Gladys Gilbertson. He died in 1984.

**Merton E. Davies**

Mr. Merton Davies was born on 13 September 1917 in St. Paul, Minnesota, and grew up in Palo Alto, California. Davies graduated from Stanford University in 1938, receiving a bachelor's degree in mathematics.

In 1947 Davies took a position with the RAND Project in Santa Monica, where he worked on satellite vehicle designs, camera designs, and image interpretation. At RAND, Davies conceived and patented a spin-stabilized panoramic ("spin pan") camera for synoptic earth imaging. Davies continued his efforts in national reconnaissance until 1970. Following his work in national reconnaissance, Davies focused on NASA programs until his retirement from RAND in 1998. Davies died on 17 April 2001 in Santa Monica, CA.

**James C. de Broekert**

Mr. James de Broekert was born in 1930 in Eugene, Oregon. He received a bachelor's degree in electrical engineering from Oregon State in 1952, a master's degree in electrical engineering in 1955, and the degree of professional engineer in 1957.

In 1955 de Broekert became a researcher and lecturer at Stanford University. After Stanford ended its defense contracts in the 1960s, de Broekert cofounded a company that contributed to SIGINT payloads for the NRO. De Broekert founded Advent Systems in 1978.

At the time of his recognition as a national reconnaissance pioneer, de Broekert continued to support the National Reconnaissance Program. His hobbies included photography and flying light planes. De Broekert married the former Marla McGregor.

**Sidney D. Drell**

Dr. Sidney Drell was born on 13 September 1926 in Atlantic City, New Jersey. He received a bachelor's degree from Princeton University in 1946. He received a master's degree (1947) and a doctorate in physics (1949), both from the University of Illinois.

Drell has been a faculty member at Stanford University since 1950. In addition, he has taught at MIT, as well as served as a visiting scientist and professor at a number of American and foreign universities and research centers. Drell has served as a scientific advisor on several government boards, including the President's Foreign Intelligence Advisory Board. In his advisory roles, he was instrumental in building support for several NRO programs. Drell served as a key scientific consultant to the NRO for Program B in the 1960s and 1970s.

Drell married the former Harriet Stainbeck of Minter City, Mississippi.

**Richard L. Garwin**

Dr. Richard Garwin was born on 19 April 1928 in Cleveland, Ohio. He received a bachelor's degree from Case Western Reserve University in 1947, and a doctorate in physics in 1949 from the University of Chicago.

Garwin has served as a scientific advisor on many government boards, including the President's Scientific Advisory Committee (under Presidents Eisenhower, Kennedy, Johnson, and Nixon) and the DoD Science Board. As a key scientific advisor to the NRO's Program B, Garwin established standards for the electromechanical design of exceptionally long-lived spacecraft. Garwin's major contributions to science include research that was instrumental to the development of the hydrogen bomb, and groundbreaking research in experimental physics and computer design.

Garwin continued to consult and advise the government on national defense and disarmament issues, while maintaining his affiliation with IBM even after his retirement in 1993.

At the time of his recognition as a founder of national reconnaissance, he was an adjunct professor of physics at Columbia University, an adjunct professor at the Kennedy School of Government at Harvard University and a Senior Fellow for Science and Technology at the Council of Foreign Relations in New York. Garwin married the former Lois Levy of Cleveland Heights, Ohio.

**Gary S. Geyer**

Colonel Gary Geyer, USAF (Ret) was born 20 March 1943 in Akron, Ohio. Geyer received a bachelor's degree in electrical engineering from Ohio State University in 1966. After graduating, he was commissioned as a second lieutenant in the Air Force.

The focus of Geyer's Air Force career was Air Force Special Projects, his involvement with which culminated with his assignment in 1985 and 1989 as a Systems Program Office Director for two NRO programs. Geyer retired from the Air Force in September 1992. He then worked at Lockheed Martin-Denver from 1992 to 1997 and retired as Vice President and Program Manager for Defense Systems.

At the time of his recognition as a national reconnaissance pioneer, Geyer was teaching space system courses at New Mexico State University, developing a space systems curriculum, and playing golf. Geyer married the former Jan Clark from Shelby, Ohio; they have two children, Gregory and Patrick.

**Thomas O. Haig**

Colonel Thomas Haig, USAF (Ret) was born in 1921 in Ypsilanti, Michigan. He attended the Case School of Applied Science, the University of California at Los Angeles (UCLA) and the Army's Meteorology Instrumentation School. Haig received a bachelor's degree in electrical engineering from the University of Illinois (1955) and a master's degree from George Washington University (1965). He also attended the Industrial College of the Armed Forces at Ft. McNair in Washington, DC.

Haig began his career in the Army Air Corps during World War II. Haig's association with national reconnaissance included his direction of the Air Force program that developed the NRO meteorological satellite system and an assignment as Assistant Director of Research and Development for the NRO. Following retirement from the Air Force in 1968, Haig worked for General Electric (GE) and then took a position at the University of Wisconsin as the Executive Director of the Space Science and Engineering Center where he worked for ten years until his retirement in 1980.

At the time of his recognition as a national reconnaissance pioneer, Haig was focusing on personal projects including converting an old dairy barn into a home, organizing the Madison Print Club, founding Reprise Theatre (a Wisconsin senior citizen acting company), forming the Gerald A. Bartell Community Theater Foundation, and raising money for the renovation of an old movie theater. Haig married the former Barbara Havens; they have 6 children, John, Paul, Robert, Bruce, Charles, and Katherine, as well as 23 foster children.

**Frederick H. Kaufman**

Mr. Frederick Kaufman was born on 25 May 1929 in Washington, DC. He received a bachelor's degree in electrical engineering from Notre Dame in 1951, and a master's degree in electrical engineering from the University of California at Berkeley.

Prior to attending graduate school, Kaufman served for three years in the Navy. After completing his tour in the Navy, he went to work for Ramo-Woolridge, later renamed TRW, where he directed the team that produced SIGINT satellites for the NRO. Kaufman retired from TRW in 1994.

At the time of his recognition as a national reconnaissance pioneer, Kaufman was enjoying playing golf and traveling extensively. Kaufman married the former Ann Reid.

**Robert J. Kohler**

Mr. Robert Kohler was born 29 September 1937, in Rochester, New York. He received a bachelor's degree in photographic science and technology from the Rochester Institute of Technology in 1959.

From 1959 to 1967 Kohler managed the Photo Science and Technology Department for the Itek Corporation in Lexington, Massachusetts. In 1967 Kohler left Itek and joined the CIA to work on space imagery intelligence (IMINT) systems. From 1982 to 1985 Kohler managed the engineering, development, and operation of major technical collection systems in support of the NRO. After Kohler left the CIA in 1985, he held positions at ESL Incorporated, Lockheed Missile and Space Corporation, and TRW. Kohler retired from TRW in 1995

At the time of his recognition as a national reconnaissance pioneer, Kohler was a consultant and served on the board of directors for various corporations. Kohler married the former Grace DeVito.

**Edwin H. Land**

Mr. Edwin Land was born on 7 May 1909 in Bridgeport, Connecticut. He entered Harvard University in 1926, but left to pursue scientific and technical interests on his own. Harvard awarded him an honorary doctorate in 1957.

Land founded the Polaroid Corporation in 1937, where he invented instant photography. While at Polaroid, Land advised several presidents on matters concerning national reconnaissance. In 1968 Land was presented with the National Medal of Science and the Presidential Medal of Freedom for his service to the United States. Land retired from Polaroid in 1982 and went to the Rowland Institute for Science, which he had founded in 1980 in Cambridge, Massachusetts.

Land married Helen Maislen of Hartford, Connecticut. They had two daughters, Jennifer Land Dubois and Valerie Land Smallwood. Land died on 1 March 1991 in Cambridge, Massachusetts.

**Ellis E. Lapin**

Mr. Ellis Lapin was born on 21 February 1917 in Chicago, Illinois. In 1939 he received a bachelor's degree in mechanical engineering from Drexel University. He received graduate degrees in aeronautics (1940) and aeronautical engineering (1941) from the California Institute of Technology.

Lapin began his career with Douglas Aircraft Company as an aerodynamicist. In 1961 he joined the Aerospace Corporation, and eventually became the Program Director for General Systems Engineering and Technical Direction for an NRO IMINT reconnaissance program. He later served in the same capacity for the initial development and operations of the Defense Support Program. Lapin retired in 1983.

At the time of his recognition as a national reconnaissance pioneer, Lapin's activities included woodworking and gardening. Lapin married the former Harriet Taylor and has two daughters, Nancy and Claudia, by a former marriage.

**Lloyd K. Lauderdale**

Dr. Lloyd Lauderdale was born on 21 August 1926 in Beaumont, Texas. He received a bachelor's degree in engineering from the United States Naval Academy in June 1949. Lauderdale received a master's degree in engineering (1957) and a doctorate in engineering (1962) from Johns Hopkins University.

Lauderdale served in the Navy from 1944 to 1955. In 1959 he became a lecturer in electrical engineering at Johns Hopkins University. In 1963 Lauderdale joined the CIA, where he served for six years, attaining the position of Deputy Director of the Directorate of Science and Technology. After leaving the CIA in 1969, Lauderdale became the Director of Corporate Electronics at Ling-Temco-Vought.

Lauderdale died on 19 January 1985 in Lake Proctor, Texas.

**Richard S. Leghorn**

Mr. Richard Leghorn was born on 7 February 1919 in Brookline, Massachusetts. He received a bachelor's degree in physics from MIT in 1939. Upon graduation he joined the Eastman Kodak Company and later became manager of its European Division.

Leghorn served as a combat reconnaissance pilot in the Army Air Corps during World War II, providing reconnaissance for both the invasion of Normandy and the Battle of the Bulge. Leghorn also participated in aerial instrumentation of the first testing of the atomic bomb at Bikini Atoll. In 1952 he led an Air Force plan for peacetime aerial and satellite reconnaissance of the Soviet Union.

In 1957 Leghorn founded the Itek Corporation in Lexington, Massachusetts, which developed the high-resolution photographic system utilized by the Corona photoreconnaissance satellite. During the 1950s and 1960s, Leghorn served in various advisory capacities in the areas of defense and arms control for the White House and the Departments of Defense and State.

**Walter J. Levison**

Mr. Walter Levison was born on 24 July 1918 in New York City, New York. He received a bachelor's degree in physics from City College of New York in 1939.

Levison served as an aerial photographer in the Army Air Corps during World War II. After the war, Levison worked at Boston University Physical Research Laboratories (BUPRL). While at BUPRL, he was called to active duty and served in the Air Force during the Korean War. In 1957 Levison joined several of his BUPRL co-workers at the newly-formed Itek Corporation. While at Itek, Levison designed and served as the project manager for the camera used on the Corona photoreconnaissance satellite. He retired from Itek in the mid-1970s.

After his retirement, Levison founded a solar energy company and established one of the first university-based venture capital programs. He was also the founder of Aegis Venture Funds, a venture capital firm specializing in high-tech businesses. Levison was active in local government, particularly with regard to conservation issues.

Levison married the former Ann Simons; they have two children, Elizabeth and Jane. Levison died on 27 December 1999 in Worcester, Massachusetts.

**Francis J. Madden**

Mr. Frank Madden was born on 17 October 1921 in Boston, Massachusetts. Madden attended Northeastern University and received a bachelor's degree in mechanical engineering in 1951.

Madden served in the Army Air Corps as an aerial photographer during World War II. From 1956 until 1975 he served as chief engineer of the Itek Corporation's camera systems development program. In this position, he directed the design, test, and production of the Corona camera and its improved versions.

After his retirement in 1974, he became a founding member of the Retirees Enhancing Science Education via Experimentation and Demonstration (RESEED). At the time of his recognition as a national reconnaissance pioneer, Madden was still active in RESEED, in addition to working with the Volunteers in Service to America (VISTA). Madden married the former Anne O'Sullivan.

**James T. Mannen**

Colonel James Mannen, USAF (Ret) was born on 20 February 1941 in Washington, DC. In 1964 he received a bachelor's degree in electrical engineering from Virginia Polytechnic Institute, and in 1969 he received a master's degree in electrical engineering from the Air Force Institute of Technology.

In 1965 Mannen was called to active duty in the Air Force and assigned to the 29th Air Division, Air Defense Command. During the remainder of his 30-year Air Force career Mannen was assigned to various positions in the Office of the Secretary of the Air Force for Special Programs (SAFSP) at Vandenberg Air Force Base, California. After retiring from the Air Force in 1995, Mannen was selected to be the Program Director for the National Polar-Orbiting Operational Environmental Satellite System at the National Oceanic and Atmospheric Administration (NOAA). Mannen retired in 1998.

At the time of his recognition as a national reconnaissance pioneer, Mannen was serving as a consultant for the NRO, the Joint Commission to review the NRO, and the civil space community. He was a member of the Intelligence Technical Advisory Group for the Senate Select Committee on Intelligence. Mannen married the former Lynne Howard.

**Paul W. Mayhew**

Dr. Paul Mayhew was born on 26 March 1928 in Barbourville, Kentucky. He graduated from the University of Kentucky in 1951 with a bachelor's degree in electrical engineering. Mayhew received his master's degree (1957) and doctorate (1961) in engineering from Ohio State University. In 1967 Mayhew enrolled in the Executive Program at UCLA.

Mayhew joined TRW as a systems engineer in 1962, where he worked on the design and development of SIGINT reconnaissance satellites. He continued his association with NRO programs throughout his career. Mayhew retired from TRW as Vice President and Assistant General Manager of the Space and Technology Group in 1993.

At the time of his recognition as a national reconnaissance pioneer, Mayhew's interests included experimental aircraft, dirt bikes, woodworking, genealogy, and hunting. Mayhew married the former Imogene Zornes.

**Reid D. Mayo**

Mr. Reid Mayo was born on 18 February 1925 in Albuquerque, New Mexico. Mayo served in the Navy during World War II and saw combat action in the Okinawa campaign and during naval operations that included bombardments of the Japanese homeland.

After leaving the Navy, Mayo moved to Washington, DC, and in 1949 earned a bachelor's degree in engineering from George Washington University. Shortly after his graduation, Mayo joined the Naval Research Laboratory (NRL) as a systems engineer. Mayo later served as a project engineer and technical director of the NRO's Program C.

After his retirement from NRL in 1981, Mayo served as a consultant to the NRL. Mayo married the former Margaret Ann Blok of Grand Rapids, Michigan. Mayo died on 9 February 2001.

**Mark Morton**

Dr. Mark Morton was born on 1 January 1913 in Atlantic City, New Jersey. In 1934 he received a bachelor's degree in mechanical engineering from the Guggenheim College of Aeronautics at New York University. In 1971 he received a doctorate in engineering from Rose Polytechnic Institute in Terre Haute, Indiana.

Morton joined General Electric in 1956, becoming General Manager of the Reentry Systems Department in 1962, a Vice President and head of the Missile and Space Division in 1968, and a Senior Vice President and head of GE's Aerospace Business in 1969. Morton retired from GE in 1978 and then served as a consultant to high-technology industries.

At the time of his recognition as a national reconnaissance pioneer, Morton was pursuing his interests in acrylic painting, metal and wood crafting, and reading science fiction. Morton married the former Ruth Neznamoff.

**Alden V. Munson**

Mr. Alden Munson was born on 9 April 1942 in Kingsburg, California. In 1965 he received a bachelor's degree in mechanical engineering from San Jose State University and later received a master's degree in mechanical engineering from the University of California at Berkeley. Munson completed extensive coursework in computer science at UCLA and has attended executive programs at Harvard and Stanford Universities.

In 1965 Munson joined the Aerospace Corporation, providing system engineering and data systems analysis in support of many SIGINT space programs. In 1973 Munson joined TRW as a program manager for several intelligence systems development projects. Munson served in executive assignments for both TRW and Litton.

At the time of his recognition as a national reconnaissance pioneer, Munson was a consultant in defense and information technology. He also was serving on several corporate boards and was an active member of the Security Affairs Support Association. Munson married the former Jane Stewart; they have one child, Kathryn.

**Charles L. Murphy**

Colonel Charles Murphy, USAF (Ret) was born on 18 July 1918 in Asheville, North Carolina. He had begun his engineering studies at Georgia Tech when he left to serve in World War II. After the war Murphy attended the University of Florida, where he received a bachelor's degree in political science and history, as well as a bachelor's degree in accounting. In addition, Murphy studied engineering at several Air Force schools.

Murphy served in the Air Force as a crewmember on several different types of aircraft and for a variety of missions, including on aircraft during the Berlin Airlift. Murphy began his association with national reconnaissance in 1956 when he was assigned to the CIA Operations Office for work on the U-2 program. Murphy went on to the Corona program, where he became the Launch Director of Space Programs at Vandenberg Air Force Base and finally served as Director of the NRO Control Center at the Pentagon. Murphy retired from the Air Force in 1972 and then joined the CIA. Murphy retired from the CIA in 1978.

At the time of his recognition as a national reconnaissance pioneer, Murphy and his wife, the former Mary Anne White, divided their time between Seal Beach, California and Northern Virginia. The Murphys have a son, Lee, and a daughter, Patricia.

**Frederic C.E. Oder**

Colonel Frederic "Fritz" Oder, USAF (Ret) was born on 23 October 1919 in Los Angeles, California. In 1940 he received a bachelor's degree in geology from the California Institute of Technology (Cal Tech). After graduating, Oder enlisted in the Army Air Corps as a flying cadet. The Corps sent him back to Cal Tech, where he received a master's degree in meteorology in 1941. The Army Air Forces sent him to graduate school at UCLA where he earned a doctorate in meteorology and physics in June 1952.

In 1952 Oder began work with the CIA in the Physics and Electronics Division of the Office of Scientific Intelligence. Later, he became the original director of the Air Force WS-117L Advanced Reconnaissance Satellite Program.

After his retirement from the Air Force in 1960, Oder worked at the Eastman Kodak Company for five years and then at Lockheed Missile and Space Corporation for 20 years. At the time of his recognition as a national reconnaissance pioneer, Oder was residing in Palm Beach, Florida. Oder married the former Dorothy Gene Brumfield, and they have three children, Frederic, Barbara, and Richard.

**Robert M. Powell**

Mr. Robert Powell was born on 28 August 1922 in Cades, Tennessee. He received a bachelor's degree in electrical engineering from the University of Tennessee in 1944, and a master's degree in electrical engineering from the University of Maryland in 1950.

Early in his career Powell held engineering positions at NSA at Fort Meade and the Naval Ordnance Laboratory at White Oak, Maryland. Powell joined Lockheed Missile and Space Corporation in 1956, and in 1958 he began to work in the space reconnaissance field as manager of the Hawaiian Tracking Station in Oahu, Hawaii. Powell continued to work in support of the NRO and other programs at Lockheed. Powell eventually became Lockheed Corporate Vice President, the position he held until his retirement on 31 December 1987.

At the time of his recognition as a national reconnaissance pioneer, Powell was involved in volunteer projects through his church. Powell married the former Margaret Flesher.

**Edward M. Purcell**

Dr. Edward Purcell was born on 30 August 1912 in Taylorville, Illinois. He received a bachelor's degree in electrical engineering in 1933 from Purdue University and a master's degree and doctorate in physics from Harvard University in 1935 and 1938, respectively.

Purcell began his career as a physics instructor and researcher at Harvard in 1938. In 1952 Purcell was the co-recipient of the Nobel Prize in physics for the development of the nuclear magnetic resonance method of measuring certain nuclear properties. While still teaching at Harvard, Purcell served as a member of the President's Science Advisory Committee under Presidents Eisenhower, Kennedy, and Johnson (1957-1965).

Purcell retired from Harvard as emeritus professor in 1980. Purcell married the former Beth Busser. He died on 7 March 1997 in Cambridge, Massachusetts.

**Edward H. Reese**

Mr. Edward Reese was born on 27 December 1937 in Philadelphia, Pennsylvania. In 1959 he received a bachelor's degree in electrical engineering from Villanova University, and in 1963 he received a master's degree from Drexel University.

Reese spent almost his entire career working in the Philadelphia area for General Electric Missile and Space Division (later acquired by Lockheed Martin). He began his national reconnaissance career with the electronic design in support of the Corona photoreconnaissance system.

In 2000 Reese retired as Technical Director for Lockheed Martin Management and Data Systems. At the time of his recognition as a national reconnaissance pioneer, he was serving as a consultant. In addition to consulting, Reese was helping his wife raise funds for the Neighborhood League. Reese married the former Ann Carpenter. They have two children, Carol and Edward.

**Osmond J. Ritland**

Major General Ozzie Ritland, USAF (Ret) was born on 30 October 1909 in Berthoud, Colorado. Ritland studied mechanical engineering at San Diego State College from 1929-1931, and after serving in World War II he graduated from the Industrial College of the Armed Forces.

Ritland completed the Air Corps Advanced Flying School in 1933. As an Air Force pilot, he logged more than 9,400 flying hours. Ritland began his association with national reconnaissance while serving as the Chief of Staff for Development, where he developed the service infrastructure for the U-2 program. He later commanded the WS-117L satellite program and served as Deputy to the Commander for Manned Space Flight, Air Force Systems Command.

After his retirement in 1965, Ritland worked for McDonnell-Douglas Company for five years as the Vice President for Launch.

Ritland married the former Martha Alsup, and they had two daughters, Kathleen and Susan. Ritland died on 23 March 1991 in Rancho Santa Fe, California.

**Robert W. Roy**

Colonel Robert Roy, USAF (Ret) was born on 4 May 1928 in Fort Lauderdale, Florida. He graduated from the United States Naval Academy in 1951 with a bachelor's degree in engineering. He received a master's degree from the Air Force Institute of Technology in 1965.

Roy joined the Air Force in 1951. In 1958 Roy was assigned to the 6595th Test Wing at Vandenberg Air Force Base, where he worked on the Corona/Discoverer launches. After his retirement from the Air Force in 1975, Roy served as a consultant for General Research Corporation and also was an Assistant Professor in the Graduate School at Troy State University.

At the time of his recognition as a national reconnaissance pioneer, Roy was the President of Decision Sciences, a business he cofounded in 1986. Roy married the former Patricia Troll. They have 4 children, Kathryn, Patricia, Robert, and Julia.

**Charles P. Spoelhof**

Mr. Charles Spoelhof was born on 6 August 1930 in Hackensack, New Jersey. He attended Calvin College in Grand Rapids, Michigan, and the University of Michigan in Ann Arbor, where he received bachelor's degrees in engineering physics and engineering mathematics (both in 1953), as well as a master's degree in physics (1954). Spoelhof also attended the University of Rochester and MIT.

Spoelhof began his career at the Eastman Kodak Company in 1954. During his 32 years at Eastman Kodak, Spoelhof contributed to several NRO photoreconnaissance systems and rose to the position of Vice President and Director of Technology Assessment. While at Eastman Kodak, Spoelhof also served on NASA's Hubble Space Telescope Optical Systems Board of Investigation, the Scientific Advisory Committee of the Defense Intelligence Agency, the Fubini Committee on future reconnaissance system, and the Drell Committee on Corona Improvements. Spoelhof retired from Eastman Kodak in 1986.

At the time of his recognition as a national reconnaissance pioneer, Spoelhof was active in lecturing and consulting at Calvin College and was a Fellow of the Rochester Institute of Technology. He also was pursuing his hobbies of amateur astronomy, photography, and travel. Spoelhof married the former Kay Maliepaard.

**Marvin S. Stone**

Dr. Marvin Stone was born on 11 February 1940 in Los Angeles, California. Stone received a bachelor's degree in 1962, a master's degree in 1965, and a doctorate in 1969, all in electrical engineering and all from the University of Southern California.

From 1963-1966 Stone worked for Hughes Aircraft Company, where he was involved in the design and development of missile-seeker electronics. In 1967 Stone joined TRW, where he held a number of positions related to systems engineering of spacecraft electronics payloads, program management, and general management. After retiring from TRW in 1997, he joined the adjunct faculty in the School of Engineering at the University of Southern California, teaching graduate courses in electrical engineering.

At the time of his recognition as a national reconnaissance pioneer, Stone was still teaching at the University of Southern California and enjoying golf, gardening, and traveling. Stone married the former Ann Pyenson in 1963.

**Don F. Tang**

Mr. Don Tang, a first generation American, was born of parents who immigrated to the U.S. from China. He was born on 8 December 1933 in Phoenix, Arizona. He attended the University of Arizona, graduating in 1959 with a bachelor's degree in electrical engineering. He attended the University of Santa Clara for his graduate studies in electrical engineering, and in 1981 he attended Columbia University for its Executive Program.

Tang joined Lockheed Missile and Space Corporation in Sunnyvale, California in 1961. He became the Vice President of Space Systems Division Operations in 1986, the Vice President of Special Programs the following year, and the Vice President and Assistant General Manager of the Space Systems Division in 1988. In 1992 Tang was promoted to President and General Manager of the Space Systems Division, and he served in this position until his retirement in 1995.

At the time of his recognition as a national reconnaissance pioneer, Tang was keeping busy playing golf and babysitting grandchildren. Tang married the former Rose Tang; they have two children, Kevin and Sandra.

**Albert D. Wheelon**

Dr. Albert "Bud" Wheelon was born on 18 January 1929 in Moline, Illinois. He received a bachelor's degree in engineering science from Stanford University in 1949 and a doctorate in theoretical physics from MIT in 1952.

Wheelon's early involvement in space was with TRW, where he focused on missile guidance systems and space projects. From 1962-1966 Wheelon served as the CIA's Deputy Director for Science and Technology (DDS&T). As DDS&T, Wheelon led technical collection and analysis activities for the U-2 and Corona programs. In addition to holding several positions in government and industry, Wheelon served on several DOD and presidential advisory boards.

At the time of his recognition as a national reconnaissance pioneer, Wheelon continued to pursue his interest in science through the publication of books and papers and his association with various scientific associations. Wheelon married the former Cicely Evans.

**Peter G. Wilhelm**

Mr. Peter Wilhelm was born on 26 July 1935 in New York City, New York. He earned a bachelor's degree in electrical engineering from Purdue University in 1957, and he attended graduate school at George Washington University, where he completed the course work for a master's degree in engineering in 1961.

After receiving his bachelor's degree, Wilhelm joined Stewart Warner Electronics as an electrical engineer, where he worked on a NRL, radar test set program. In 1959 Wilhelm accepted a position at NRL where he contributed to the first ELINT satellite program. In 1974 Wilhelm became the head of NRL's Spacecraft Technology Center.

At the time of his recognition as a national reconnaissance pioneer, Wilhelm continued to hold the position of Director, Naval Center for Space Technology at NRL. Wilhelm married the former Linda Greenway.

**Robert W. Yundt**

Colonel Robert Yundt, USAF (Ret) was born on 27 June 1920 in Pittsburgh, Pennsylvania. After serving in World War II, Yundt attended Grove City College.

During World War II, Yundt was stationed in Australia as a fighter pilot. He returned to the U.S. in May 1943. In his junior year at college, the Air Force offered Yundt a position in the Regular Officer Corps, which he accepted.

Yundt's last position in the Air Force was as Director of the Signals Intelligence Project Office in the NRO's Program A. He retired from the Air Force in 1966 and went to work at TRW. Yundt retired from TRW after 20 years. Yundt died in February 2002.

# Appendix 3 The Pioneer Recognition Program

The National Reconnaissance Pioneer Program is sponsored by the National Reconnaissance Office (NRO) and annually honors a select number of individuals who have made significant and lasting contributions to national reconnaissance. The Director of the NRO (DNRO) designates these individuals as Pioneers of National Reconnaissance, and honors them through a public ceremony and by displaying a medallion in Pioneer Hall at the NRO Headquarters. The medallion names the pioneer and commemorates the pioneer's contribution to the discipline of national reconnaissance. In the inaugural pioneer recognition ceremony, held on 27 September 2000, DNRO Keith Hall inducted 46 pioneers in celebration of the NRO's 40th anniversary, and also honored ten Founders of National Reconnaissance.

The Pioneer Hall dedication plaque was signed by Secretary of Defense William S. Cohen, Director of Central Intelligence George J. Tenet, and DNRO Keith R. Hall. The plaque explains the purpose of Pioneer Hall and describes the pioneers as a group:

> "This hall recognizes 'National Reconnaissance Pioneers'—those dedicated individuals who have made significant and lasting contributions to the discipline of national reconnaissance. Their efforts greatly reduced the risk of surprise and uncertainty in the conduct of international affairs, and their contributions proved to be vital to the national security of the United States. The Pioneers of the past contributed significantly to the peaceful conclusion of the Cold War. The Pioneers of the future will continue to maintain global information supremacy for the United States during times of both peace and conflict. Although these Pioneers most often worked in secrecy and without recognition, we now acknowledge and honor them publicly for their commitment, spirit, and innovation. We hereby dedicate this National Reconnaissance Pioneer Hall on 27 September 2000 to commemorate the 40th Anniversary of the founding of the National Reconnaissance Office."

### Selecting the National Reconnaissance Pioneers and the Founders of National Reconnaissance

The DNRO selected the inaugural 46 pioneers and the ten founders based on nominations solicited from across the national reconnaissance community. The first Pioneer Selection Board proposed a list of recommended honorees after reviewing over 200 nominations submitted for the first pioneer selection.

The pioneer selection criteria stipulate that anyone who has made a significant and lasting contribution to national reconnaissance, regardless of organizational affiliation and status, is eligible to be nominated as a pioneer.[1] Eligible individuals include current or retired military, government civilian, contractor, or academic personnel, and the nominees may be living or deceased. The selection criteria also define the types of contributions to national reconnaissance that pioneers typically will have accomplished during their careers. The pioneer must have made a contribution of such significance to change the direction or scope of national intelligence collection and analysis, aerospace and reconnaissance technology, or reconnaissance-based information operations. A pioneer will have played a unique and pivotal role in activities such as those described in the table below. (See accompanying table for criteria.)

In addition to designating 46 individuals as national reconnaissance pioneers, the 2000 Pioneer Selection Board recommended to the DNRO that ten additional individuals who have been advisors to presidents and deserved a different type of recognition, and designated them as Founders of National Reconnaissance. This group of advisors provided counsel to President Dwight D. Eisenhower and his successors. This counsel contributed to the success of the peacetime strategic reconnaissance policy. That policy, formally adopted by President Eisenhower in 1954-1955,

[1] The pioneer program excludes NRO directors, deputy directors, and the legacy program directors, because these individuals already have been honored by the DNRO, and their portraits are on display at the NRO Headquarters.

Figure A3-1. First Pioneer Selection Board with DNRO Keith Hall. The 2000 Selection Board was appointed on the occasion of the NRO's 40th Anniversary to select the initial class of pioneers. Standing, l–r: Major General Nathan J. Lindsay, USAF (Ret) (former Program A Director), DNRO Keith R. Hall, Rear Admiral Thomas C. Betterton, USN (Ret) (former Program C Director); Sitting, l–r: Mr. Robert G. Kaemmerer (former Program B industry executive), the Honorable John McMahon (former Deputy Director of Central Intelligence), Mr. Clarence E. Smith (former CIA executive), Mr. R. Evans Hineman (former Program B Director). Missing from photo: Mr. Jimmie D. Hill (former Deputy Director, NRO). (Photo by Sara Judy, NRO Visual Design Center.)

**Pioneer Selection Criteria**

| *Category* | *Contribution* |
|---|---|
| Systems Development | Greatly improving overhead intelligence collection by conceiving and planning, or successfully developing a new: · sensor · aerial or satellite system · communication system · orbital application |
| Management & Support | Greatly improving overhead intelligence collection by organizing, leading, or successfully managing the teams that designed, fabricated, launched, or operated a new and complex reconnaissance system. |
| Acquisition & Procurement | Greatly benefiting the procurement of overhead reconnaissance systems by conceiving or successfully developing a new acquisition technique or contracting procedure. |
| Information Handling | Conceiving or successfully developing a new technique, procedure, or method that greatly improved the handling of information collected by overhead reconnaissance systems, including: · evaluation · interpretation · information processing · dissemination · cryptanalysis |

reduced the threat of surprise nuclear attack and permitted verification of arms limitation treaties. As scientists, engineers, and innovators, these individuals provided the technical expertise that shaped the emerging discipline of national reconnaissance. They also provided the confidence for founding the National Reconnaissance Office in 1960-1961.

### Pioneer Hall and the 40th Anniversary Plaque

From the conceptual drawings to the final coat of wood stain, Pioneer Hall was a collaborative effort among several NRO staff components. A pioneer committee, comprised of representatives from the NRO staff offices, collaborated with the DNRO to conceptualize and design Pioneer

Figure A3-2. Pioneer Hall, NRO Headquarters. (Photo by Sara Judy, NRO Visual Design Center.)

Hall. The result is a display that honors the pioneers and their accomplishments while highlighting the pioneers' legacy to the current workforce. The NRO 40th anniversary plaque similarly honors the Founders of National Reconnaissance.

**The Elements of Pioneer Hall**

Pioneer Hall exhibits the pioneer medallions and contains the dedication plaque and features a display case containing artifacts and information about national reconnaissance. A reference book in the artifact case provides expanded information about the pioneers and Pioneer Hall.

**The 40th Anniversary Plaque**

The NRO 40th Anniversary plaque, which is located near Pioneer Hall, acknowledges the ten Founders of National Reconnaissance with citations describing their contributions. The plaque depicts five activities of national reconnaissance: imagery intelligence, signals intelligence, airborne collection, space-based collection, and space launch. The Thor Able rocket launched the Grab SIGINT satellite; the recovery bucket returned Corona film; the U-2 aircraft flew reconnaissance missions. The plaque also includes a likeness of President Dwight D. Eisenhower, who made peacetime strategic reconnaissance a national policy.

## Conclusion

The recognition ceremonies for the 46 inaugural National Reconnaissance Pioneers and the ten Founders of National Reconnaissance were rewarding and beneficial for all involved. The pioneers and founders, and their families, expressed pride in the honorees' achievements and appreciation for the NRO's acknowledgement of these individuals'contributions to national reconnaissance:

> "My grandfather's happiest career moments were while working on reconnaissance projects. He was proud of his contributions and would be profoundly touched by this acknowledgement." —Elizabeth Kosich, granddaughter of Osmond Ritland

> "It would have made Ozzie very happy and proud to be honored."
> —Martha Ritland, wife of Osmond Ritland

> "My husband would have been so proud of this honor."
> —Myrnie Browning, wife of John Browning

> "I feel quite honored to have been selected as a Pioneer and recognized in the Hall of Pioneers." —John T. Bennett

> "Thank you for the ceremonies at the NRO's 40th Anniversary celebration, the kind remarks by DNRO Keith Hall and DCI George Tenet, and the Pioneer plaque. You will, I think, be appreciative of my wife's reaction to the remarks and events of that morning—a resurgent feeling of patriotism." —Ellis E. Lapin

> "You have created and recreated a memorable part of our nation's history for the NRO, and the products are enormously appreciated." —William O. Baker

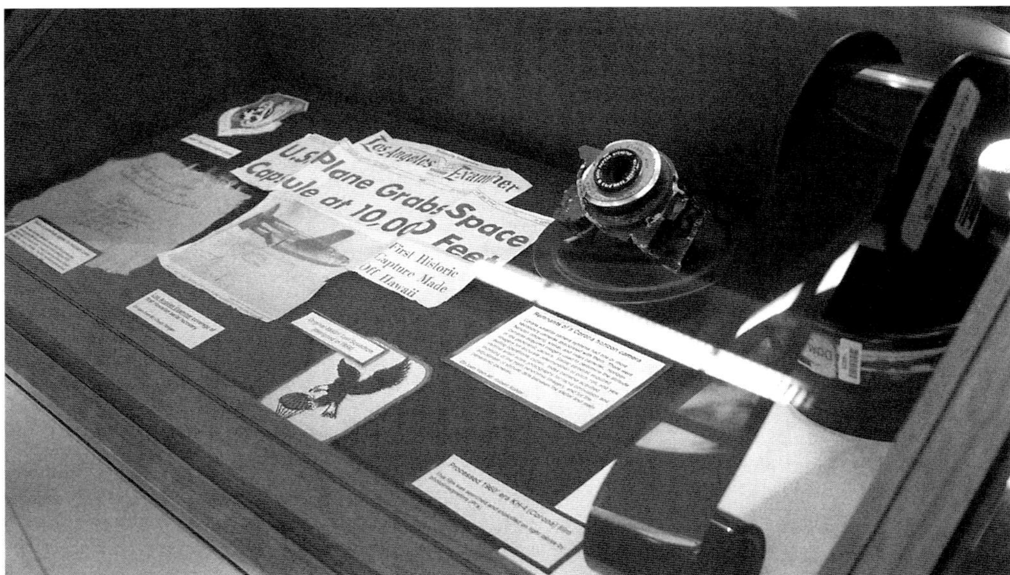


Figure A3-3. The Pioneer Hall 40th Anniversary Artifact Display. Pictured are the 6594th Recovery Control Group patch (upper left), a piece of parachute signed by the aerial recovery crew (lower left), *Los Angeles Examiner* coverage of first Hawaiian-based aerial recovery (left center), the original 6593rd Test Squadron patch (lower right center), remnants of a Corona horizon camera (upper right center), and processed Corona film (right).[2] (Photo by Sara Judy, NRO Visual Design Center.)

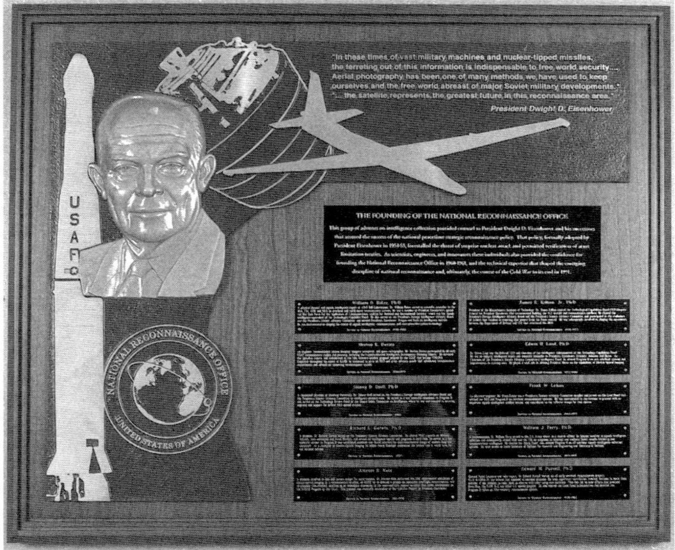


Figure A3-4. The NRO 40th anniversary plaque, NRO Headquarters. (Photo by Sara Judy, NRO Visual Design Center.)

[2] Not pictured but included in the second display case are a congratulatory letter from President Ronald Reagan, National Intelligence Distinguished Service Medal Citation signed by DCI William Casey, *Time* Magazine cover from 6 June 1960 showing a C-119 recovery aircraft, NRO 40th Anniversary citation in the *Congressional Record*, a photograph of a JC-130B recovery crew, and other recovery aircrew patches.

# Index

## ERRATA

Page 314, Fig. 40-5, the caption should read:
"A post-Cold War Atlas launch, circa 2000." (USAF photo).

Page 337, Appendices 1-3, Appendix 1 should read:
"Appendix 1 - **Glossary** and Acronyms"